BATES 25: CELEBRATING 25 YEARS OF BEAM TO EXPERIMENT

Related Titles from AIP Conference Proceedings

508   Hadron Physics: Effective Theories of Low Energy QCD
Edited by A. H. Blin, B. Hiller, M. C. Ruivo, C. A. Sousa, and E. van Beveren, March 2000, 1-56396-927-0

495   Experimental Nuclear Physics in Europe: ENPE 99, Facing the Next Millennium
Edited by Berta Rubio, Manuel Lozano, and William Gelletly, November 1999, 1-56396-907-6

481   Nuclear Structure 98
Edited by C. Baktash, September 1999, 1-56396-858-4

455   ENAM 98: Exotic Nuclei and Atomic Masses
Edited by B. M. Sherrill, D. J. Morrissey, and C. N. Davids, December 1998, 1-56396-804-5

421   Polarized Gas Targets and Polarized Beams: Seventh International Workshop
Edited by Roy J. Holt and Michael A. Miller, January 1998, 1-56396-700-6

412   Intersections Between Particle and Nuclear Physics: 6th Conference
Edited by T. W. Donnelly, December 1997, 1-56396-712-X

To learn more about these titles, or the AIP Conference Proceedings Series, please visit the webpage **http://www.aip.org/catalog/aboutconf.html**

BATES 25: CELEBRATING 25 YEARS OF BEAM TO EXPERIMENT

Cambridge, Massachusetts   3–5 November 1999

EDITORS
T. W. Donnelly
W. Turchinetz
Massachusetts Institute of Technology

Melville, New York
AIP CONFERENCE PROCEEDINGS ■ 520

Editors:

T. W. Donnelly
Center for Theoretical Physics
Massachusetts Institute of Technology 6-300
Cambridge, MA 02139-4307
USA

E-mail: donnelly@mitlns.mit.edu

W. E. Turchinetz
Bates Linear Accelerator Center
P.O. Box 846
Middleton, MA 01949
USA

E-mail: billt@bates.mit.edu

Authorization to photocopy items for internal or personal use, beyond the free copying permitted under the 1978 U.S. Copyright Law (see statement below), is granted by the American Institute of Physics for users registered with the Copyright Clearance Center (CCC) Transactional Reporting Service, provided that the base fee of $17.00 per copy is paid directly to CCC, 222 Rosewood Drive, Danvers, MA 01923. For those organizations that have been granted a photocopy license by CCC, a separate system of payment has been arranged. The fee code for users of the Transactional Reporting Service is: 1-56396-949-1/00/$17.00.

© 2000 American Institute of Physics

Individual readers of this volume and nonprofit libraries, acting for them, are permitted to make fair use of the material in it, such as copying an article for use in teaching or research. Permission is granted to quote from this volume in scientific work with the customary acknowledgment of the source. To reprint a figure, table, or other excerpt requires the consent of one of the original authors and notification to AIP. Republication or systematic or multiple reproduction of any material in this volume is permitted only under license from AIP. Address inquiries to Office of Rights and Permissions, Suite 1NO1, 2 Huntington Quadrangle, Melville, N.Y. 11747-4502; phone: 516-576-2268; fax: 516-576-2450; e-mail: rights@aip.org.

L.C. Catalog Card No. 00-103944
ISBN 1-56396-949-1
ISSN 0094-243X
Printed in the United States of America

Contents

Preface vii
Local Organizing and International Advisory Committees viii
Corporate Sponsors ix

I. NUCLEAR STRUCTURE STUDIES WITH ELECTRONS
Chair: E. W. Vogt

Overview of Nuclear Structure with Electrons 5
    D. F. Geesaman
High Resolution Elastic and Inelastic Scattering 20
    J. H. Heisenberg
Inclusive Scattering 33
    I. Sick
The $A(e,e'p)$ Coincidence Reaction: Past, Present and Future 48
    W. Bertozzi
Coincidence Electron Scattering II: Short Range Correlations in Nuclei 66
    T. Walcher

II. STRUCTURE OF FEW-BODY NUCLEI
Chair: C. F. Williamson

Nuclear Magnetism 87
    G. A. Peterson
Quantum Monte Carlo Calculations of Nuclei 101
    V. R. Pandharipande
Elastic Scattering Studies of Deuterium 115
    M. Garçon
MEC and Relativistic Effects in the Deuteron 130
    J. W. Van Orden
Unpolarized Inclusive Studies of $A=3$ Nuclei 144
    D. H. Beck

III. FEW-BODY NUCLEI AND NUCLEON STRUCTURE
Chair: J. L. Matthews

Recent Results on Spin Dependent Scattering from Few Body
Systems Obtained in Amsterdam 157
    J. F. J. van den Brand
Twenty-Five Years of Progress in the Three-Nucleon Problem 168
    J. L. Friar
Inclusive Scattering from Polarized ^{3}He and Neutron Form Factors 181
    H. Gao
Studies of the Electric Form Factor of the Neutron 196
    H. Schmieden

Essential Differences between the Structure of Nuclei and Nucleons 209
   J. W. Negele

IV. HADRONIC STRUCTURE
Chair: E. C. Booth

Nucleon Electromagnetic Form Factors 225
   K. de Jager
Nucleon Structure Studied Through VCS and the N→Δ Transition 237
   C. N. Papanicolas
Electromagnetic Pion Production: From Yukawa to Goldstone 254
   A. M. Bernstein
Effective Field Theory and χpt 271
   B. R. Holstein

V. PARITY-VIOLATING ELECTRON SCATTERING AND A LOOK FORWARD
Chair: R. D. McKeown

Parity Violation I: Then and Now 291
   P. A. Souder
Parity Violation and Hadron Structure 305
   E. J. Beise
Electrons, New Physics, and the Future of Parity-Violation 313
   M. J. Ramsey-Musolf
A Look to the Future 329
   R. G. Milner

List of Participants 345
Author Index 353

PREFACE

On Nov. 3-5, 1999 a Symposium and party were held to celebrate the twenty-fifth anniversary of taking the first publishable data on high resolution electron scattering and photo-nuclear reactions at the MIT-Bates electron linear accelerator. We note, with gratitude, Bates' narrow escape from fiscal oblivion and mourn the passing of the NIKHEF linac and AMPS ring cut off in the prime of life. Our field is, however, vigorous and prospering with marvelous new data emerging from Bates, Mainz, Bonn and CEBAF and the ghost of NIKHEF. We also celebrate at Y2K the discovery of the electron 102 years ago by Thompson; the Mott cross section 70 years ago; the proposal of a form factor by Guth 65 years ago to explain some data that turned out to be wrong; the Illinois experiment 38 years ago which demonstrated the finite size of nuclei leading to the start of the modern era at Stanford and the work of Hofstadter, Panofsky and their colleagues. Much of what we know about hadronic structure, and most of what we believe, comes from electro-nuclear experiments. In recent years we have developed the technology of spin physics with polarized beams and targets and product polarimeters so that we know in detail how to over-determine the electro-nuclear S matrix. This work has just begun on some of the more promising observables with important new results.

This book contains the lectures given at the Symposium by a group of distinguished speakers and contributors to the developments mentioned above. They were asked to organize their talks around their assigned topics according to the following questions: Where are we, how did we get here, and where are we going? We are pleased that these outstanding contributions are being published so that a wider audience can enjoy both the science and art that made the Symposium such an enlightening and enjoyable event.

We owe thanks to many individuals and organizations for their contributions to the Symposium, parties and these proceedings: to the local organizing committee, especially Ginny Bullard, Sheila Dodson and Jean Flanagan who were essential to the success of the entire endeavor; to Joy Gurrie and Stan Sobcynski who provided artistry; to the international advisory committee of Bates alumnae, alumni and friends who helped plan the scientific program; to LNS, Bates, the Dean of Science Robert Birgeneau and President Charles Vest of MIT who provided support as did the US DOE and NSF; to friends of the laboratory including US Representative John F. Tierney, Captain Raymond Bates (USN ret.) and the numerous corporate sponsors who are listed separately. Thanks to all.

T.W. Donnelly and W. Turchinetz, co-chairmen.

Local Organizing Committee

Ginny Bullard
Sheila Dodson
Bill Donnelly
Jean Flanagan
Richard Milner
Chris Tschalär
Bill Turchinetz

International Advisory Committee

Hartmuth Arenhövel
Douglas Beck
Aron Bernstein
William Bertozzi
Ed Booth
John Cameron
Larry Cardman
Kees de Jager
Peter Demos
Dieter Drechsel
James Friar
Bernard Frois
Haiyan Gao
Justus Koch
Stanley Kowalski
Jean-Marc Laget
Malcolm Macfarlane
June Matthews
Ernest Moniz
John Negele
Vijay Pandharipande
Costas Papanicolas
Robert Redwine
Berthold Schoch
Ingo Sick
Dennis Skopik
Paul Souder
Edward Tomusiak
Thomas Walcher
Dirk Walecka
Claude Williamson

Corporate Sponsors

American Science & Engineering
829 Middlesex Turnpike
Billerica, MA 01821

Bartoszek Engineering
818 W. Downer Place
Aurora, IL 60506-4904

The Clute Company
350 Lincoln Street
Hingham, MA 02043

CML Engineering Sales, Inc.
2020 Ridgecrest Place
Escondido, CA 92029

CPI Communication and Power Industries
607 Hansen Way
P.O. Box 51110
Palo Alto, CA 94303

LeCroy Corporation
700 Chestnut Ridge Road
Chestnut Ridge, NY 10977-6499

DEL Electronics
Del Power Conversion
One Commerce Park
Valhalla, NY 10595

Diversified Technologies
35 Wiggins Avenue
Bedford, MA 01730

Everson Electric Company
2000 City Line Road
Bethlehem, PA 18017

Litton Electron Devices
960 Industrial Road
San Carlos, CA 94070-4194

MDC Vacuum Products Corporation
28842 Cabot Boulevard
Hayward, CA 94545

Stephen A. Stickney & Sons
Middleton Road
P.O. Box 164
Boxford, MA 01921

Total Temperature Control, Inc.
39 W. Water Street
Wakefield, MA 01880

VAT, Inc.
500 West Cummings Park
Woburn, MA 01801-6516

I. NUCLEAR STRUCTURE STUDIES WITH ELECTRONS

Chair: E.W. Vogt

Session I

Overview of Nuclear Structure with Electrons

Donald F. Geesaman*

*Argonne National Laboratory, Argonne, IL 60439

Abstract.
Following a broad summary of our view of nuclear structure in 1974, I will discuss the key elements we have learned in the past 25 years from the research at the M.I.T. Bates Linear Accelerator Center and its sister electron accelerator laboratories. Electron scattering has provided the essential measurements for most of our progress. The future is bright for nuclear structure research as our ability to realistically calculate nuclear structure observables has dramatically advanced and we are increasingly able to incorporate an understanding of quantum chromodynamics into our picture of the nucleus.

To grasp the scientific legacy of the M.I.T. Bates Linear Accelerator Center and its sister electron scattering laboratories in the development of our understanding of nuclear structure, we should look back to our world view of nuclear structure in the early 1970's, to the time when the Bates laboratory was first taking data. For me this is a personal look back to the time when I was a graduate student and first learning what nuclear physics was all about. The classic textbook by De Shalit and Feshbach [1] gives us a vivid picture of those times. After reviewing what we thought we understood, I will point out what were the important elements that were missing in 1974 and give a few examples of the types of experiments which made the difference. The talks which follow will provide much more complete explanations of the data and the physics. What I want to do is give you a sense of where we came from, how we got there, and where we should go into the future.

We begin considering closed shell nuclei which in 1974 were understood in terms of mean-field Hartree-Fock calculations. The nuclear ground state was a Slater determinant of single particle orbitals which interact in a self-consistent mean-field. The major success at the time was the demonstration that such calculations starting with realistic nucleon-nucleon interactions gave an excellent description of the then-measured charge distributions of nuclei. I remember very clearly a seminar by John Negele who asserted that this was now a solved problem. Indeed, while little reliable data on the much-harder-to-measure distributions of neutrons existed, John asserted that we should move on and use the calculated neutron distributions to extract other important nuclear structure information. We will return to the issue of neutron distributions later in this talk where parity violating electron scattering appears to offer a valuable new tool.

Nuclei with several valence nucleons outside a closed shell were the province of shell model calculations with residual interactions obtained from theory (e.g. Kuo-Brown [2]) or experiment (Schiffer-True [3]). Shell model calculations were being extended past the simplest configurations and many experimental phenomena could now be understood, but usually with effective operators that were substantially renormalized from the free nucleon values.

A telling characteristic of the time was that sum rules almost always added up to the full strength expected in simple shell model constructions. Much of the data came from nucleon transfer reactions. For example, everyone believed if one summed the single particle stripping and pickup reaction strengths over states within several MeV of the Fermi surface, one should get 2j+1, where j is the total angular momentum of the single particle orbital. The power of (e,e'p) reactions in examining the single particle strength was well understood [4] but only exploratory experiments had been done.

Hartree-Fock and random phase approximation techniques were the foundation of microscopic descriptions of the collective degrees of freedom and normal modes of the nucleus. The giant dipole resonance which is selectively excited by low energy electromagnetic probes was the only giant resonance which was well in hand, both experimentally and theoretically. Systematics on the giant monopole and quadrupole resonances as well as spin-flip resonances were soon to come. We did know about the importance of the interplay between shell and collective effects [5] in, for example, fission isomers. This last topic has been a central theme in high-spin gamma ray physics over the past two decades.

The most direct measure of bound nucleons in the nuclear medium came from quasifree electron scattering, and here we were all convinced by Moniz et al. [6] that the nucleus looked very much like a Fermi gas of independent nucleons. In the 1976 long range plan for Nuclear Physics, the Friedlander report suggested we just had to do these experiments on a few more nuclei to map out the Fermi momentum vs mass number and we might be done. As an important institutional note, this report, two years after the first beam at Bates, recommended significantly increased funding support for the M.I.T. Bates Laboratory and the doubling of the maximum electron energy.

Finally, we did know in 1974 about the importance of two-body currents in electromagnetic interactions with nuclei. The most striking example at high momentum transfer was in the threshold electrodisintegration of the deuteron, but meson exchange currents were also important at low momentum transfer in the $np \to d\gamma$ reaction. It has taken a long time for the understanding that almost every two-body nucleon-nucleon interaction requires two-body interaction currents to satisfy current conservation to sink in for many of us who dreamed we lived in a one-nucleon current world.

While much of the 1970's viewpoint remains at the center of our understanding of the nucleus today, it is easy to see five vital components of our present understanding of the nucleus that were missing in 1974:

- We had few measurements of radial distributions.

- We had little information on absolute normalizations in reactions.

- The role of short-range Nucleon-Nucleon correlations was considered a problem for theorists, not experimentalists.

- We had only begun to consider microscopic many-body forces.

- We had no consistent framework to deal with the impact of nucleon-substructure in nuclear structure.

In large part, it is these five issues which have substantively changed our view of the nucleus in the past two and a half decades.

The first two points, the lack of radial distributions and absolute normalizations are strongly coupled. I have already discussed that pickup and stripping nucleon transfer reaction sum rules generally added up to 2j+1. What one measures is the product of the squared radial wave function at the strong absorption radius and the normalization of the reaction. Since we always had optical model and reaction mechanism ambiguities we did excellent measurements relative to states near shell closures which we assumed had unit spectroscopic factors and we could build up an apparently self-consistent picture. What absolute matrix elements we measured with electro-weak interactions (beta and gamma decay lifetimes) gave us integral measurements which could only be fit in the shell model with effective charges, typically of 1.5-2.0 for the proton and 0.5-1.0 for the neutron for E2 transitions. This was a clear sign that significant structure effects were left out of our models, most notably, coupling of low-lying states to giant resonances.

Short range correlations were primarily a theorist's ball game. They were essential for handling the strong short range behavior of the nucleon-nucleon interaction in the nuclear medium. But once we had a suitable effective interaction, few seemed to care about the implications, and certainly very few experimentalists. Perhaps part of the problem was that the one place short range correlations were taken seriously was in nuclear matter calculations, and in the early 1970's there was a "crisis in nuclear matter". It had finally become evident that no realistic two-nucleon interaction would be able to simultaneously reproduce N-N phase shifts and the saturation binding energy and saturation density of nuclear matter. This problem along with the difficulties in reproducing the properties of the few-nucleon systems forced the community to consider microscopic three-body forces and the very complicated correlation structure of the nuclear ground state very seriously.

Finally, if we had found a smoking gun failure in our picture of nuclear structure, we had no clear path how to proceed because we did not have a theory of the structure of the nucleon. We could propose ad-hoc changes in the properties of the nucleon in the nuclear medium without any clear idea of how these changes would affect other nucleon properties. Today we believe quantum chromodynamics gives us the framework for a complete description of nucleon, and indeed, nuclear structure. While a realistic calculation of ^{16}O in a QCD basis still remains far in the future, we can now perhaps appreciate the right questions to ask.

From our historical perspective where we know what was missing, it is easy to understand in hindsight why electron scattering was so central to the progress in the past 25 years. Fundamentally, in electron scattering you know what you are measuring, providing the first two of the five links mentioned above that were missing in 1974, because the interaction, quantum electrodynamics, is well understood. Electromagnetic interactions are weak enough for perturbation theory to provide a quantitative tool for extracting the nuclear response. In the one-photon exchange approximation, where a virtual photon with energy ω and three momentum $\vec{q}$ ($Q^2 = |\vec{q}|^2 - \omega^2$) is exchanged the electron scattering cross section is given by

$$\frac{d^3\sigma}{d\Omega dE'} = \frac{4\pi\sigma_M}{M_t}\left\{(\frac{Q^4}{|\vec{q}|^4})W_L(q,\omega) + \left[(\frac{Q^2}{|\vec{q}|^2})/2 + \tan^2\frac{\theta}{2}\right]W_T(q,\omega)\right\} \quad (1)$$

where σ_M is the Mott cross section and M_t is the target mass. The separation between the response, W_L, to longitudinal photons which couple to the charge, and W_T, to transverse photons which couple to convection currents and magnetic moments is very important. In many cases for longitudinal currents the two-body currents are weak and the virtual photon interacts with single nucleon currents, looking deep inside the nucleus at the single particle structure. For inelastic scattering of a fixed multipolarity λ between an initial state $|i\rangle$ and final state $\langle f|$ the response function W_L can simply be considered to be

$$W_L \propto \left[\langle f||a_\alpha^\dagger a_\beta||i\rangle \int \rho_{fi}(r) j_\lambda(qr) r^2 dr\right]^2. \quad (2)$$

We directly measure the Fourier transform of the transition density $\rho_{fi}(r)$ and the magnitude of the particle-hole amplitude $\langle f||a_\alpha^\dagger a_\beta||i\rangle$. To invert the Fourier transform into a coordinate density requires data over a large range of momentum transfers extending out to $1/L$ fm^{-1} where L is less than ~ 0.4 fm. It was the 1970's generation of electron accelerators: Bates, Saclay, and NIKHEF, that had the kinematic range, beam intensity, and experimental equipment to fully exploit this power.

Similarly in proton knockout $(e, e'p)$ reactions in the Plane Wave Impulse Approximation

$$\frac{d^6\sigma}{d\Omega dE' d^3\vec{p'}} = \sigma_{ep}^* S(E_m, \vec{P}_i) \quad (3)$$

the cross section is simply proportional to the electron-nucleon cross section, σ_{ep}^* times the spectral function S which in the independent particle shell model is given by

$$S(E_m, \vec{P}_i) \propto \sum_i Z_i^2 \phi_i^2(\vec{P}_i)\delta(E_m - E_i). \tag{4}$$

where Z_i^2 is the spectroscopic factor and $\phi_i^2(\vec{P}_i)$ is the square of the single particle wave function. $\vec{P}_i$ is the initial proton momentum in the nucleus and E_m is the separation energy of the produced proton-hole state. Here one has to deal with the outgoing proton in the final state through the distorted wave impulse approximation. The dominant (but by no means sole) effect of the distortions is the attenuation of the outgoing protons while passing through the nucleus.

To do real experiments with the extended kinematic range of few hundred MeV electrons, the last essential requirement is experimental detection systems with the resolution and solid angle to make the measurements in a timely fashion. The Bates laboratory set the standard with a superb magnetic spectrometer, ELSSY [7], which led the community to dispersion-matched energy-loss spectrometers and achieved resolutions better than 10^{-4}. The vertical drift chamber detection system [8] was equally innovative. The technique of using a single chamber package to make multiple measurements on a track is now used in all large detector systems to provide the maximum track information with the minimum multiple scattering.

When all these elements came together, one could measure gorgeous experimental spectra, one example of which is shown in figure 1 [9] for electron inelastic scattering from ^{90}Zr. When I first saw such spectra, my eyes popped, even before I noted that they were presented on a logarithmic scale. In Fig. 2, the extracted radial transition densities are shown for the four states of the two-proton $g_{9/2}$ multiplet, $|[\pi g_{9/2}]^2 \otimes J = 2, 4, 6, 8\rangle$. The cross-hatched band is the range of uncertainty in the experimental transition densities and the solid (dotted) lines are calculations with (without) core polarization. One immediately sees that for the 2^+ state there are large core polarization corrections but that the theory has difficulty describing the transition density in the interior of the nucleus. As one goes to higher spin states the transition density looks much more like a single $g_{9/2}$ radial wave function and the calculated effects of core polarization are smaller.

Such inelastic scattering data indicated the simplicity of high-spin excitations and would reveal a significant reduction of the measured particle-hole strength from the mean-field expectation. These observations triggered a reexamination of the nuclear single particle strength near the Fermi surface. In the Hartree-Fock picture of a closed shell nucleus, the wave function was a Slater determinant of single particle orbitals which were occupied up to the Fermi surface and then empty above the Fermi surface. While it was recognized that long range correlations (from surface vibrations) and short-range correlations (from the N-N interaction) would dilute this abrupt transition, the perspective from the success of shell model calculations in the Pb region was that these were not large effects. The electron scattering data gave an overwhelming body of evidence that this was not true. The four types of data that played a key role were:

- Differences in elastic scattering yields for A and A-1 systems such as ^{206}Pb compared to ^{205}Th [10] and ^{208}Pb to ^{207}Pb [11].

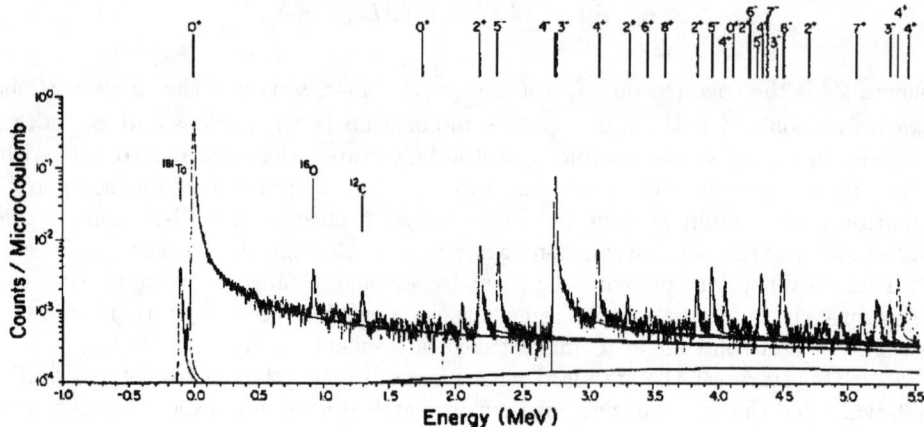

FIGURE 1. ELSSY spectrum of scattered electrons from ^{90}Zr measured at Bates with 150 MeV incident energy electrons [9].

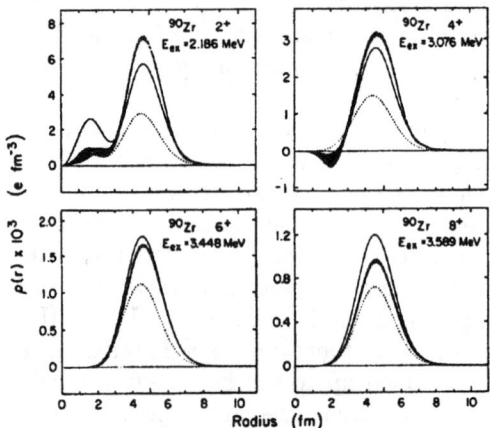

FIGURE 2. Transition charge densities for the 2^+, 4^+, 6^+ and 8^+ levels with dominant configuration $[\pi g_{9/2}]^2$. The cross hatched areas represent the error band on the experimental extraction. The solid (dotted) line is a calculation including (ignoring) core polarization effects [9].

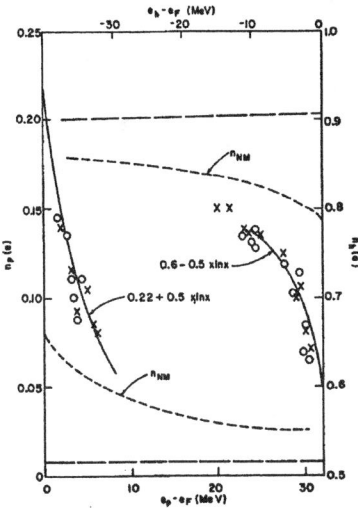

FIGURE 3. Calculated occupation numbers for single particle and hole states near the Fermi surface in Pb [14]. The dashed curve is a nuclear matter calculation including short-range correlations (long dashed) and long-range correlations and the points (crosses for proton states and circles for neutron states) include RPA correlations. The solid line is simply a parameterization of the points.

- Magnetic elastic scattering to high l orbitals [12].

- Inelastic Scattering to "relatively pure" particle-hole states [13].

- Spectroscopic factors for $(e, e'p)$ reactions.

These measurements were brought together in a coherent picture by Pandharipande, Papanicolas and Wambach [14] who showed that the combination of nuclear matter and random-phase approximation calculations shown in Fig. 3 could explain the occupation probabilities of single particle orbitals in the Pb region.

How do modern calculations stack up in describing the absolute normalization and radial transition densities measured in electron scattering? Today the state of the art is ab initio many-body calculations with realistic nucleon-nucleon forces and three-body interactions, free nucleon current operators and two-nucleon exchange currents. Such calculations are becoming a standard non-relativistic model of nuclear structure. The results show excellent detailed agreement with the data in $^6\text{Li}(e, e')$ inelastic scattering transitions (Fig. 4 [15]) and $^7\text{Li}(e, e'p)$ proton knockout data (Fig. 5 [16]). The inelastic longitudinal response functions are dominated by the one-body currents while the transverse response functions show the need for significant contributions from exchange currents to fit the data at the larger momentum transfers. In the knockout reaction the calculation reproduces the $p_{3/2}$ spectroscopic factor of 0.42 that the data require. We have clearly come a long way from effective charges of 2.0 and a proton $p_{3/2}$ occupation probability of 1.0 in the

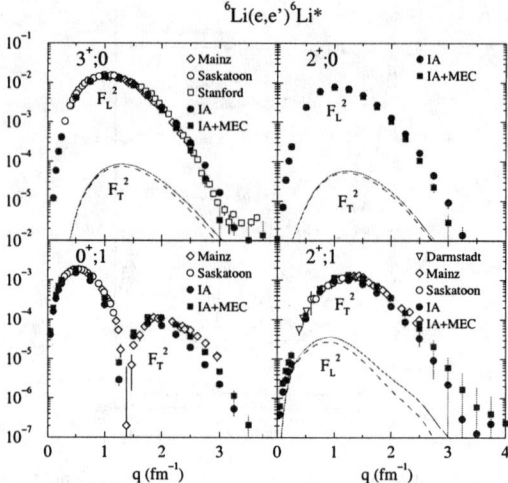

FIGURE 4. Electron inelastic scattering transition from factors in ^{6}Li. The calculations are variational Monte Carlo calculations with one-body (IA, filled circles) and two-body-meson-exchange (MEC, filled squares) transition operators [15].

lithium isotopes.

With the electron data in hand, we can now normalize our hadron inelastic scattering data and go on to extract new information about the isospin structure of nuclear excitations. A prime example is the case of negative parity $1\,\hbar\omega$ excitations. Bill Donnelly and his collaborators [17] pointed out in 1968 that at large momentum transfer the inelastic scattering spectra are dominated by so-called stretched particle-hole states involving the largest angular momentum particles and holes lying just above and below the Fermi surface. These are spin-flip excitations like $|[p_{3/2}^{-1} \otimes d_{5/2}]J = 4^-\rangle$ states in ^{12}C, $|[d_{5/2}^{-1} \otimes f_{7/2}]J = 6^-\rangle$ states in ^{28}Si, and $|[i_{13/2}^{-1} \otimes j_{15/2}]J = 14^-\rangle$ states in ^{208}Pb. Because they were easily observed, one could also study them with probes where such high resolution was not available, as I was doing with pion inelastic scattering at LAMPF or proton scattering at IUCF. The hadronic probes have differing isospin sensitivities and one can combine the electron and hadron data to extract the isoscalar and isovector, or neutron and proton transitions amplitudes for each state.

$$
\begin{aligned}
e: \quad & |M|^2 \propto (\mu_p Z_p + \mu_n Z_n)^2 \\
\pi^+: \quad & |M|^2 \propto (3Z_p + Z_n)^2 \\
\pi^-: \quad & |M|^2 \propto (Z_p + 3Z_n)^2
\end{aligned}
$$

The electron scattering spin-flip matrix element is almost pure isovector—indeed, magnetic-isoscalar transitions were extremely hard to see. However in the pion inelastic scattering case, isoscalar excitations are favored over isovector by a factor

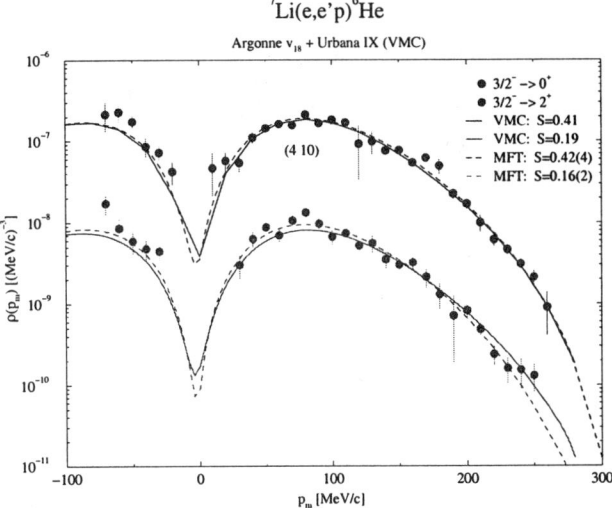

FIGURE 5. Momentum densities measured in proton knockout from ^{7}Li to the ground state 0^+ and first excited 2^+ states in ^{6}He. The calculations are variational Monte Carlo calculations (VMC) and mean field theory calculations (MFT) [16]. S is the spectroscopic factor.

of 4. For these transitions the radial densities should be the same for each isospin combination so pure $\Delta T = 1$ transitions provide perfect normalizations for the pion reaction mechanism. Many groups contributed to this effort. A summary of the separated isovector and isoscalar yields is shown in Fig. 6. Typically 30-50% of the pure particle-hole isovector spin-flip strength is observed primarily due to the occupation of the single particle orbitals discussed above. But only about 15% of the isoscalar strength was observed, an additional quenching of a factor of 2. This can now be reproduced in large scale shell model calculations [19] but no simple explanation in terms of collective degrees of freedom of the nucleus has ever emerged.

Jim Kelly and his collaborators have made the most extensive use of this comparison of electron and hadron scattering to understand the nucleon-nucleon interaction in the nuclear medium. Kelly uses the detailed knowledge of the radial dependence of the transition densities from electron scattering to determine the density dependence of the N-N interaction by fitting proton elastic scattering, inelastic scattering and polarization observables [20]. This has given us powerful insight into the mechanisms of medium modifications and made proton scattering a better quantitative tool for nuclear structure information.

Let me now return to $(e, e'p)$ reactions. While in the one-body current approximation the reaction measures the nuclear spectral function as in eq. 3, Bates studies at high energy loss showed significant strength that seemed to require multi-body

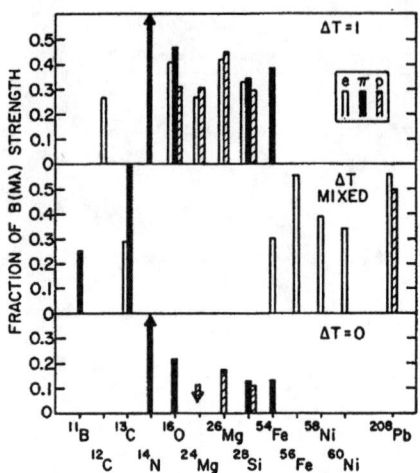

FIGURE 6. The fraction of the single-particle-hole strength contained in inelastic scattering to stretched $1\hbar\omega$ states from electron, proton and pion inelastic scattering is shown for a variety of nuclei. The upper panel corresponds to the summed $\Delta T = 1$ fractional strength and the lower panel to the summed $\Delta T = 0$ fractional strength. The middle panel corresponds to mixed isospin transitions where a $\Delta T = 1$ and $\Delta T = 0$ separation was not made.

mechanisms. The best way to study this is to separate the nuclear response to longitudinal and transverse photons. Since meson-exchange currents affect primarily the transverse response, the longitudinal coupling should give a better picture of the nuclear single particle structure. In the Bates work of Ulmer et al. [21] at $Q^2 = 0.15$ GeV^2, it was observed that for the p-shell proton knockout from ^{12}C the longitudinal and transverse strength were equal, but above two nucleon threshold there was a substantial excess of transverse strength. Everyone knows that L/T separation experiments are tough experiments, and one of the firsts things we did at Jefferson Lab was to repeat this L/T separation at two higher Q^2. In Fig. 7 I show the separated spectral function results from Dipangkar Dutta's thesis [22] on ^{12}C. As the lower panel illustrates, we definitely see an excess of transverse strength compared to longitudinal strength at $Q^2 = 0.6$ GeV^2. At the higher $Q^2 = 1.8$ GeV^2, the transverse strength is reduced. This return to a more purely single particle response is expected as the wavelength of the probe becomes shorter. In Fig. 8 it can be seen the both the longitudinal and transverse response we measured at Jefferson Lab at $Q^2 = 0.6$ GeV^2 are in agreement with the Ulmer et al. results. However in the new data one can see clearly that the longitudinal response extends to at least 80 MeV in missing energy. This long tail of the single particle response is the result of the spreading of the strength due to correlations.

In the 1980's several studies tried to look for evidence of medium modifications of nucleon structure by comparing the longitudinal and transverse strength. In Fig. 9

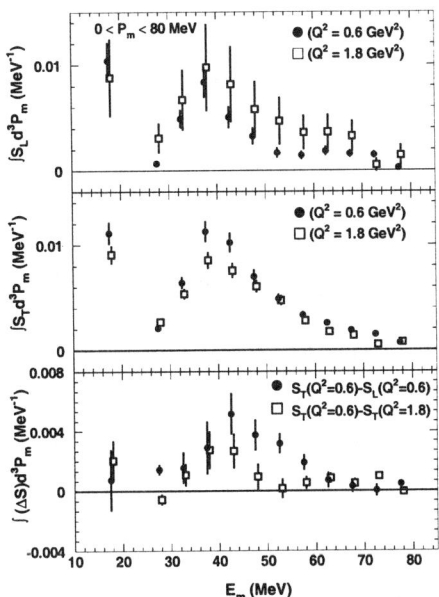

FIGURE 7. The integrals of S_L (top panel) and S_T (middle panel) from $0 < p_m < 80$ MeV are shown at Q^2 of 0.64 (circles) and 1.80 GeV^2 (squares). In the bottom panel the differences: $S_T - S_L$ at 0.64 GeV^2 (circles) and $S_T(Q^2=0.6)-S_T(Q^2=1.8)$ (open squares) are shown [22]. The errors are the sum in quadrature of the statistical and systematic uncertainties. The lowest E_m point is an average over $10 < E_m < 25$ MeV. The response functions at 1.8 GeV^2 are corrected for differences in proton attenuation by factors of 1.075 for $E_m < 25$ MeV and 1.18 for $E_m > 25$ MeV.

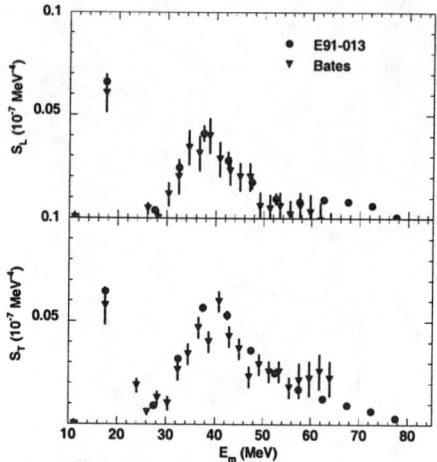

FIGURE 8. S_L (top panel) and S_T (bottom panel) at Q^2 of 0.64 GeV^2 (circles [22]) compared to the results of ref. [21] at Q^2 of 0.15 GeV^2 (triangles). The statistical uncertainties only have been shown. No attempt has been made to correct for different final state proton attenuation effects, but estimates suggest they are similar at the two proton energies.

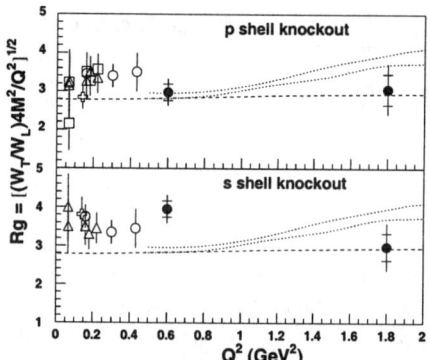

FIGURE 9. $R_G = \sqrt{W_T 4 M_p^2 / W_L Q^2}$ for the ^{12}C$(e, e'p)$ reaction (solid) from the measurements of Dutta et al. [22]. The top panel is for the p shell region and bottom panel is for the s shell region. The inner error bar represents the statistical error and the outer error bar includes the systematic error. Previous data from ^{6}Li (open squares from Nikhef) and ^{12}C (open triangles from NIKHEF, open circles from Saclay and open crosses from Bates [21]) at lower Q^2 are also shown. The dashed line is an older fit to R_G for the free proton while the dotted lines indicate the results of the new JLAB polarization transfer data [23].

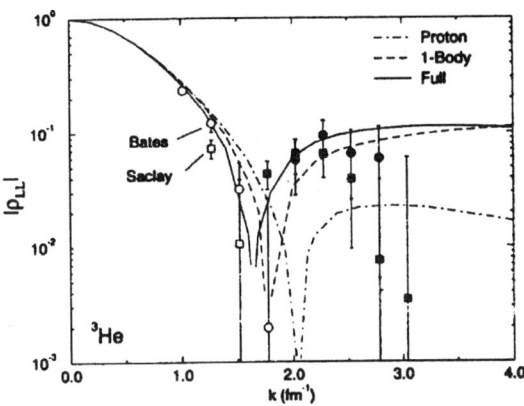

FIGURE 10. Experimental and theoretical longitudinal-longitudinal distribution functions in ^{3}He. The open (filled) symbols represent positive (negative) values of the data. The curves show the proton (dot-dashed), one-body (dashed) and the one-plus two-body contributions (solid) [25].

the ratio of the square root of the transverse to longitudinal response is displayed. If the nucleon electric and magnetic form factors had the same Q^2 dependence, this ratio would simply be the magnetic moment of the proton, 2.80. The Bates and Jefferson Lab results are consistent with this for the p shell knockout, and slightly below, but not inconsistent with, the NIKHEF and Saclay results. For the s shell region, the results are clearly above the free nucleon value at lower Q^2. The recent polarization transfer results [23] from Hall A at Jefferson Lab prove that the nucleon electric and magnetic form factors do not have the same momentum transfer dependence as indicated by the dashed curves in Fig. 9.

As a final example of the effect of correlations in nuclei I want to talk about attempts to directly measure the two-body density matrix or correlation function. With small acceptance spectrometers, this is very difficult to do directly. However it has long been known that by integrating over the longitudinal quasifree electron scattering response, one can use the Coulomb sum rule to extract the two body density. Doug Beck first analyzed the ^{3}He and ^{3}H data from Saclay and Bates to extract the two body density [24] and found significant disagreement with theoretical predictions. The later analysis of Schiavilla, Wiringa and Carlson [25] in Fig 10 showed that neutron contributions, relativistic corrections and meson exchange currents were all important in providing a complete description. The experimentalists will have to work even harder to truly nail down the two-body distributions. Indeed at this time, the largest discrepancies occur for ^{3}H, where the 2-proton correlations are all due to two-body currents.

What I have tried to do in this talk is illustrate with a few key examples that 25 years of electron scattering results from Bates and her sister laboratories have profoundly changed our view of nuclear structure. I have concentrated on the effects of short range correlations which are, perhaps surprisingly, widespread. Have we reached the end? The Program Advisory Committee at Jefferson Lab examined the future of electron scattering in nuclear structure in a recent workshop and was convinced that there are many more exciting revelations to come [26]. We encouraged work in the following areas with nuclear targets

- Testing our standard model of the nuclear many body theory.
- Nuclear single particle structure, particularly at high excitation.
- A decisive measurement of neutron densities in parity violating electron elastic scattering.
- Explicit determination of nucleon correlation functions, possibly from large acceptance $(e, e'pp)$ and $(e, e'pn)$ measurements.
- A decisive measurement of the longitudinal and transverse quasifree response.
- Nuclei as a length scale or a source of nucleon targets for short-lived particles.
- Search for medium modifications of hadrons in nuclei.

and encouraged the users to present their own ideas.

As you listen to subsequent talks, reflect on how our field has changed in the past 25 years due to the work of electron scattering and where we should go in the future. In particular, how can we learn if the structure of the nucleon and quantum chromodynamics does affect nuclear structure, or do we now have a standard model of nuclear structure that will allow us to address all the relevant issues? You will hear that there are lots of exciting questions remaining, and lots more to be done.

This work is supported in part by the Department of Energy, Nuclear Physics Division under contract W-31-109-ENG-38.

REFERENCES

1. A.de Shalit and H. Feshbach, *Theoretical Nuclear Physics*, John Wiley & Sons (New York, 1974).
2. T. T. S. Kuo and G. E. Brown, Nucl. Phys. **85**, 40 (1966).
3. J. P. Schiffer and W. W. True, Rev. Mod. Phys. **48**, 191 (1976).
4. G. Jacob and Th. A. J. Maris, Rev. Mod. Phys. **38**, 121 (1966) ; Rev. Mod. Phys. **45**, 6 (1973)
5. V. M. Strutinsky, Nucl. Phys. A95, 420 (1967).
6. E. Moniz et al., Phys. Rev. Lett. **26**, 445 (1971).
7. W. Bertozzi et al., Nucl. Instr. Meth. **162**, 211 (1979).
8. W. Bertozzi et al., Nucl. Instr. Meth. **141**, 457 (1977).
9. J. Heisenberg et al., Phys. Rev. C **29**, 97 (1984).
10. J. M. Cavedon et al., Phys. Rev. Lett. **49**, 978 (1982).
11. C. N. Papanicolas et al. Bull. Am. PHys. Soc. **26**, 45 (1981).
12. T. W. Donnelly and I. Sick, Rev. Mod. Phys **56**, 461 (1984).

13. J. Lictenstadt et al, Phys. Rev. C **20**, 497 (1979); C. N. Papanicolas et al., Phys. Rev. Lett. **45**. 106 (1980).
14. V. R. Pandharipande, C. N. Papanicolas and J. Wambach, Phys. Rev. Lett. **53** 1133 (1984).
15. R. B. Wiringa and R. Schiavilla, Phys. Rev. Lett. **81**, 4317 (1998).
16. L. Lapikas, J. Wesseling and R. B. Wiringa, Phys. Rev. Lett. **82**, 4404 (1999).
17. T. W. Donnelly et al, Phys. Rev. Lett. **21**, 1196 (1968).
18. D. F. Geesaman et al,. Phys. Rev. Lett. **30**, 952 (1984).
19. J. A. Carr et al. Phys. Rev. C **45**, 1145 (1992).
20. B. S. Flanders et al. Phys. Rev. C **43**, 2103 (1991).
21. P. Ulmer et al., Phys. Rev. Rev.
22. D. Dutta, Ph.D. thesis, Northwestern University (1999) Unpublished ; D. Dutta et al. submitted to Phys. Rev. C.
23. C. Perdrisat, Proceedings of the PANIC 99 Conference, Uppsala, Sweden, June 1999, to be published. M. K. Jones et al., submitted to Phys. Rev. Lett.
24. D. H. Beck, Phys. Rev. Lett. **64**, 268 (1990).
25. R. Schiavilla, R. B. Wiringa and J. Carlson, Phys. Rev. Lett. **70** 3856 (1993).
26. Report of Jefferson Lab Program Advisory Committee XVI, available on the WWW at http://www.jlab.org/exp_prog/PACpage/pac.html?researchers.

High Resolution Elastic and Inelastic Scattering

Jochen H. Heisenberg

Department of Physics, University of New Hampshire, Durham, NH 03824

Abstract.
In this talk I will try to review some of the key experiments that have demonstrated the need to go beyond the mean field approach in the theoretical description. In the second part I show our approach in generating nuclear wave functions that include the correlations.

EXPERIMENTAL RESULTS

I was charged with talking about the high resolution nuclear structure work at Bates and other electron accelerators. The last high resolution experiment at Bates was done about ten years ago, thus this is a talk about past history. I am not a historian and all I can do is give you a personal account of that period. I have selected some experiments not according to importance instead according to economy in making a point about the development of the field. I appollogize to those whose beautiful experiments are not mentioned. Anyway, by now there exist many extensive reviews.

I entered the field when I spent two years as a postdoc with Bob Hofstadter at Stanford. R.Frosch [1] had just measured the elastic scattering of 500 MeV electrons from ^{40}Ca and ^{48}Ca which demonstrated that the charge density for these nuclei showed structure which had to be attributed to the last occupied proton s-shell. During my time in Stanford such structure was also found for ^{208}Pb not only in the ground state but also in the transition to the first 3^- level. The charge densities shown in Figure 1 were taken from the thesis of Cavedon [2] and had resulted from an analysis of the Stanford data augmented by data from Saclay that extended the range of the cross sections down another order of magnitude or in momentum transfer they extended the range by 0.5-1.0 fm^{-1}.

While at first there was amazement that such structures were seen on the charge density, the comparison with mean field theories turned it around to the question: Why do we see so little structure. This held not only for the various ground state charge densities but also for the transition densities, even when compared to the most sophisticated RPA calculations. The diffraction pattern in the excitation of

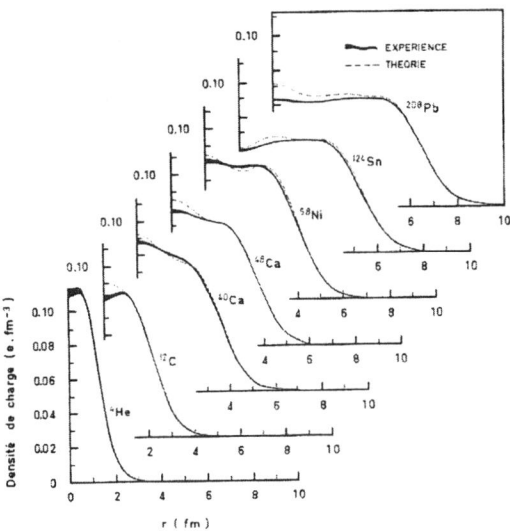

FIGURE 1. Experimental charge densities compared with mean field predicitons.

the first 3⁻ level are shown in Figure 2, again, these are obtained from the extended data set from Saclay as published by Goutte et al. [3]

The last experiment during my stay at Stanford was elastic scattering on the Nd-isotopes ranging from the spherical ^{142}Nd to the strongly deformed ^{150}Nd. The aim was to show how the onset of intrinsic deformations increases the average skin-thickness as seen in the 0⁺-ground state. The analysis was frustrating as for most isotopes the resolution of about 1MeV was insufficient to separate the excited states from the elastic scattering. In the end we managed to see the effect making corrections for the inelastic scattering contributions.

On the other hand, these studies made it clear that in measuring the form factor for the excited rotational states we would be able to determine the deformation of the nuclear surface, namely the $P_2(cos\ \theta)$ deformation from the 2⁺-excitations and the $P_4(cos\ \theta)$ deformation from the 4⁺-excitations.

Shortly after my return from Stanford I learned that a group at MIT was in the process of building an accelerator and a scattering spectrometer with resolutions of $\Delta p/p \approx 10^{-4}$ and currents of up to 100μA. They were also interested in deformed nuclei, and when I got a chance to join this group I certainly did.

At MIT the design was to use the very small phase space of the beam from the accelerator. But rather than cutting out an energy bin of $\Delta p/p \approx 10^{-4}$ to obtain the desired resolution they planned to put a dispersed beam onto the target with the high energy component of the beam up and the low energy component down. This dispersion was adjusted to compensate for the dispersion of the spectrometer

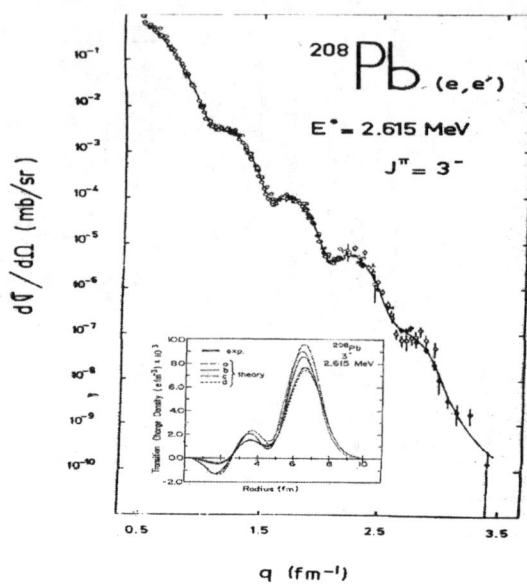

FIGURE 2. Transition charge density and (e,e′) cross sections for the ^{208}Pb 3^- level.

so that the position of the electron track in the focal plane was not a measure of the energy itself but rather a measure of the energy lost in the target. That way it was possible to obtain resolutions of $\Delta p/p \approx 10^{-4}$ with an energy spread in the beam of 1%. This was simultaneously to solve the problem with beam intensity and with obtaining the required resolution.

Together with the precision optics one needed a detector capable of determining the position of the track in the focal surface to much better than 1mm in order to not deteriorate the resolution of the instrument. At that time the MIT group developed the "vertical drift chamber", an ingenius design for a proportional wire chamber that was copied many times in the years to come. It shifted the analysis for the position of the scattered electron in the curved focal surface to an off-line analysis which due to the increasing power of computers even with the rather high event rate seemed quite reasonable.

The result even exceeded the design. Figure 3 shows a spectrum of scattered electrons from a target of ^{142}Ce. The elastic peak shows a resolution of $\Delta p/p = 5 \times 10^{-5}$.

The earlier spectra did not yet utilize the full capability; nevertheless, the data collected were a tremendous improvement over the capabilities available at Stanford. Figure 4 shows a spectrum of scattered electrons from ^{238}U [4], a benchmark case in which the rotational states are clearly separated. In a measurement at Stanford the whole area shown was embedded in a single peak. It is obvious that this increased resolution opened a new world in nuclear structure investigations.

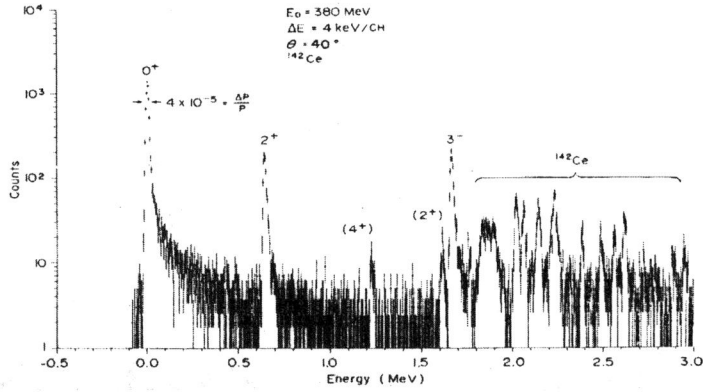

FIGURE 3. Energy loss spectrum for (e,e′) from ^{142}Ce.

These measurements coincided with the development of mean-field calculations for deformed nuclei. The density profile from such calculations done by Negele and Rinker [5] are shown in Figure 5. This density can be expanded as:

$$\rho(r,\theta) = \sum_\ell \rho_\ell(r) P_\ell(cos\ \theta)$$

and assuming rigid rotation one can derive that the transition density to the rotational state of angular momentum ℓ is given by $\rho_\ell(r)$. This allows to compute the (e,e') form factor shown in Figure 6. The good agreement seems to indicate that the mean field wave functions yield a rather reasonable description of the densities as far as the nuclear radius and the skin thicknes are concerned.

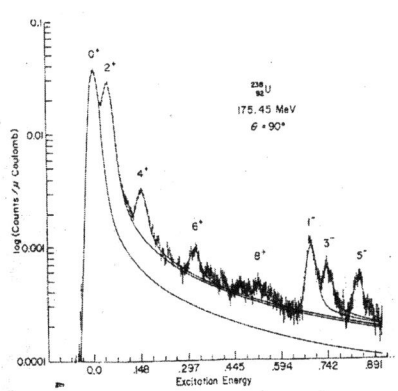

FIGURE 4. Energy loss spectrum for (e,e′) from ^{238}U.

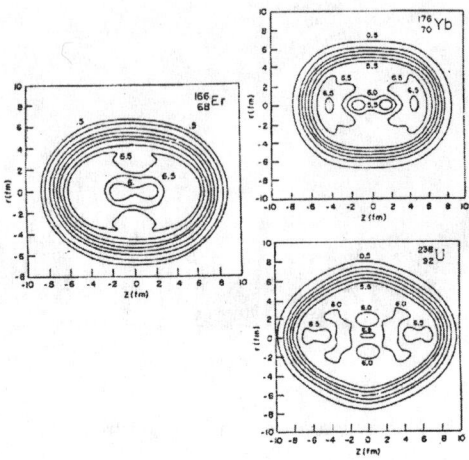

FIGURE 5. Density contours for three deformed nuclei.

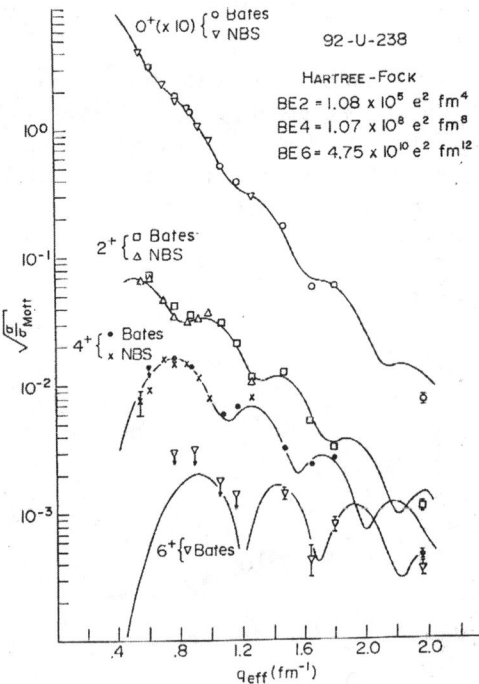

FIGURE 6. Cross sections predicted for ^{238}U and compared to experimental ones.

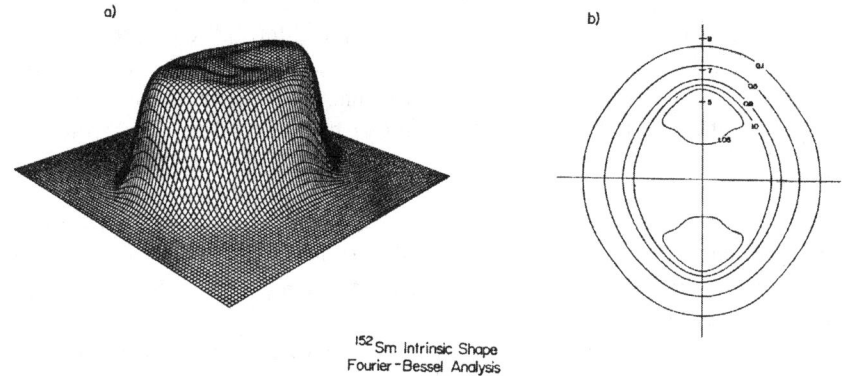

FIGURE 7. Experimental density for the deformed nucleus ^{152}Sm.

There was one case where the measurements have been used to determine the deformed intrinsic nuclear density in a "model independent" way. This is the ^{152}Sm density [6] shown in Figure 7. It shows that the density is not uniform in the interior, but it has the lobes with higher density in the elongated part.

When one looks at spectra in backward direction some levels show up as large spikes. This is demonstrated in the spectrum of ^{90}Zr in Figure 8. They originate

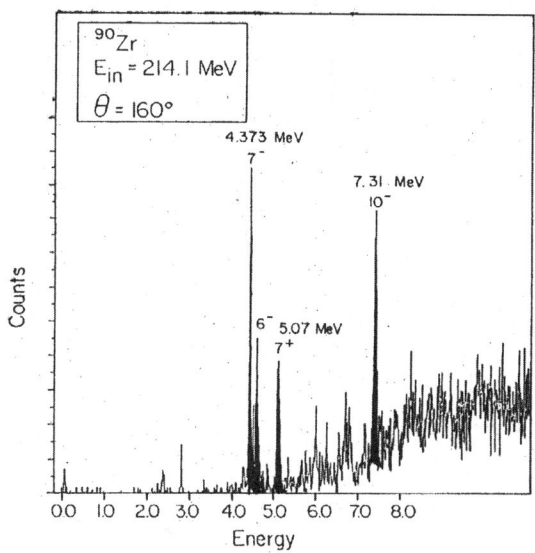

FIGURE 8. Backward scattering spectrum from ^{90}Zr.

from the excitations of the "stretched high spin states".

Actually, the first one was measured by Sick and McCarthy on ^{12}C [7], however, the record in highest multipolarity is the 14^- excitation in ^{208}Pb [8] with the dominant neutron configuration $j_{15/2}i_{13/2}^{-1}$. Figure 9 shows a spectrum for ^{208}Pb together with the form factors predicted with the mean field wave functions from Negele's mean field calculations using the DME interaction [9]. Even though the shapes of the form factors are rather well described by the mean-field wave functions, the strength is only about half of the strength expected for such a configuration. Most of similar measurements showed also that about half of the transverse strength was missing from the mean field predictions while the longitudinal strength was much closer to the prediction.

Another very pretty use of the high resolution was in the measurement of the

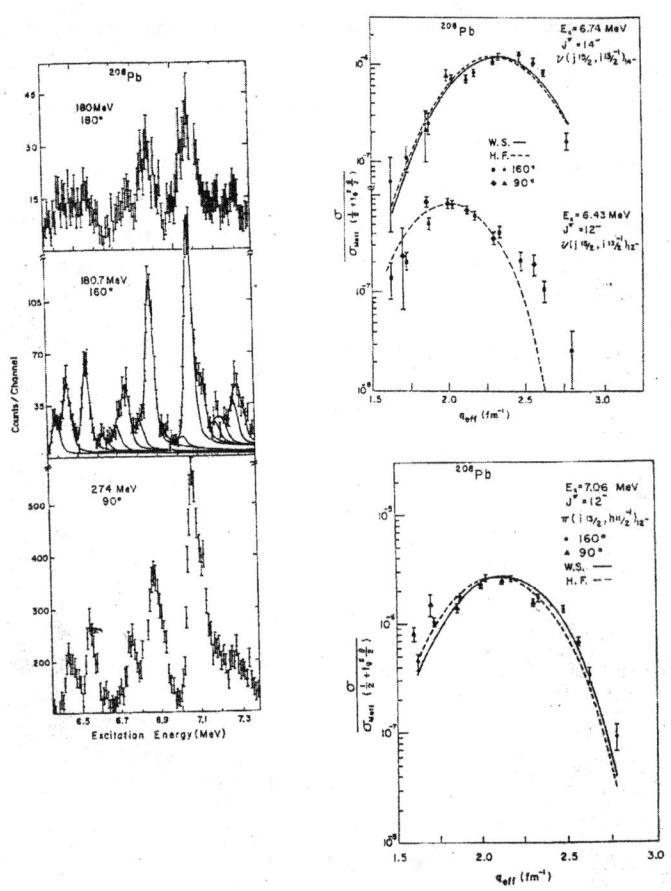

FIGURE 9. High spin states in ^{208}Pb.

stable oxygen isotopes. Figure 10 shows a spectrum of scattered electrons from a target containing all three isotopes: ^{18}O, ^{17}O, and ^{16}O [10]. Due to their different mass the recoil energy loss is different, and the elastic scattering peaks are nicely separated in this spectrum. Since the isotopic abundance was accurately known, this measurement allowed a very precise determination of the ratio of the elastic cross sections for the three isotopes.

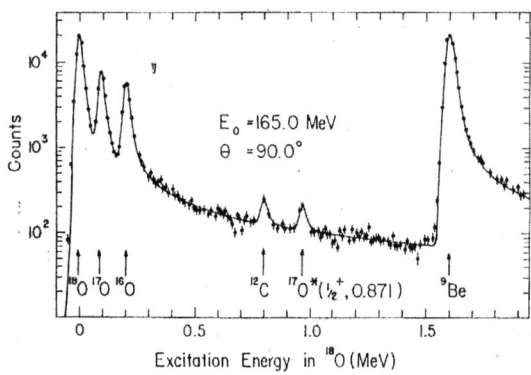

FIGURE 10. Scattered electrons from a mixture of isotopes ^{18}O, ^{17}O, and ^{16}O.

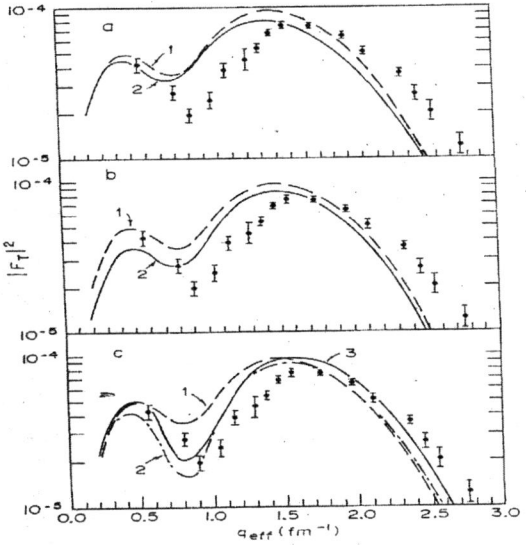

FIGURE 11. Magnetic form factor for ^{17}O.

In fact, the nuclei around A=16 provided a rich area for nuclear structure investigations. The magnetic form factor of ^{17}O [11] due to the $d_{5/2}$-proton orbit is shown in Figure 11. The curves were obtained in a mean field approach, where a Saxon-Woods potential had been fitted to reproduce to experimental binding energies as well as the experimental nuclear charge radius. These curves miss the data badly. This does not mean that the data are inconsistent with the single particle picture, it only means that such a single particle wave function can not be obtained in this mean field approach.

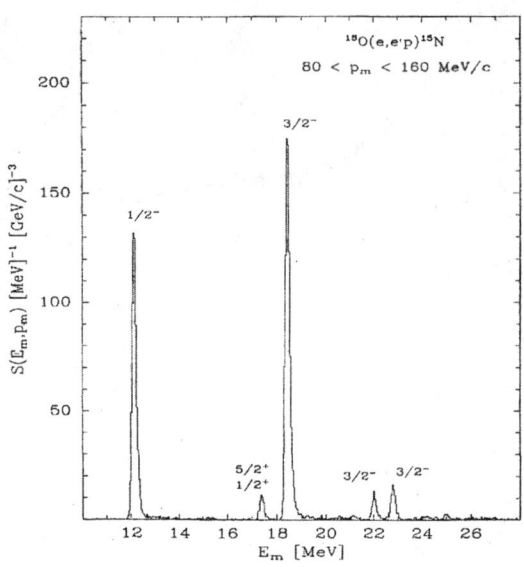

FIGURE 12. Missing mass in proton knock out from ^{16}O.

Another place where the mean field wave functions failed to describe the magnetic scattering form factors was in the measurement of ^{15}N done at NIKHEF. There are two curves shown with the data [12]: The predicted form factor for a harmonic oscillator wave function, and a fit to the data using several harmonic oscillator wave functions. This shows again, that the typical mean field wave functions are unable to account for the measured magnetic form factors in this region.

At NIKHEF the emphasis was in obtaining high resolution in (e,e'p) experiments. Figure 12 shows one of the impressive results from the proton knock out from ^{16}O [13],citeref:leusch2 allowing to separate the final nuclear state. The observed levels not only show the dominant $p_{1/2}$ and $p_{3/2}$ proton knock out, but it also shows a small but measurable strength for $d_{5/2}$ and $2s_{1/2}$-proton knock out. The separation for the two states can be performed on the basis of their different missing momentum distribution as shown in Figure 13.

All these results seemed to demonstrate that correlations play an important role

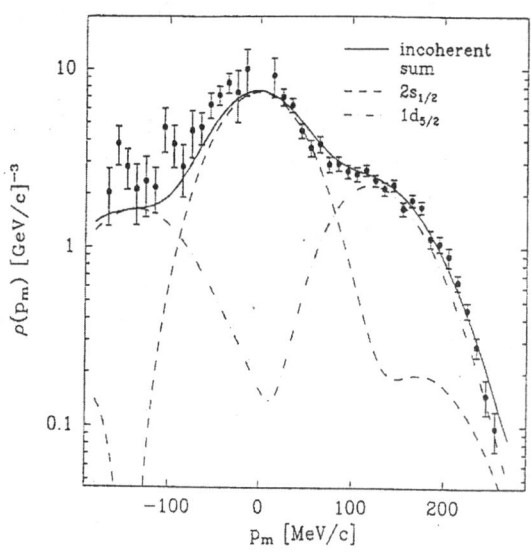

FIGURE 13. Decomposition of d- and s-proton knock out.

in the nuclear dynamics resulting in occupation of orbits above the Fermi-level and of deoccupation of orbits below the Fermi level. It also leads to a picture where particles represent "quasi-particles" with rather complex structure.

This complex structure was explored further in experiments on ^{89}Y [15]. Here the two proton single particle transitions of $2p_{3/2}$ to $2p_{1/2}$ and $1f_{5/2}$ to $2p_{1/2}$ were mapped out separating the longitudinal and transverse parts of the form factors. These experimental results were augmented by some model calculations that allowed a more detailed interpretation [16]. It showed that there are two processes going on: First, the single particle properties are quenched. This is observed in the transverse parts as well as in the longitudinal charge transition density as evidenced in the interior part of it. On the other hand, the single particle couples to the collective phonons or the core-deformations. These, however, only increase the strength at the nuclear surface, bringing it back up close to the observed strength. One conclusion was that the concept of "core-polarization" is much more descriptive of the actual process than the concept of "effective charge".

THEORETICAL ADVANCES

These findings were summarized in the paper by Pandharipande et al. [17] giving a qualitative description as well as making estimates of the size of the effects. However, a confirmation has to come from a quantitative and exact calculation of the many body problem. Such calculations are now available for nuclei with A≤8 from the Argonne group [18] and for closed shell nuclei such as ^{16}O, ^{14}C, ^{14}O, and ^{12}C from our calculations using the Coupled Cluster Expansion(CCE) [19], [20], [21], [22].

Our calculations were done with the Argonne V18 potential and also include the Urbana IX three-body potential. They were carried out in a configuration space of $50\hbar\omega$ limiting the orbital angular momentum to $\ell \leq 12$. The center of mass corrections for the charge density are included using a many-body expansion approach [20]. The meson exchange contributions are included as well in calculating the charge form factor. The resulting DWBA form factor for ^{16}O is shown in Figure 14 together with the data from Sick and McCarthy [23].

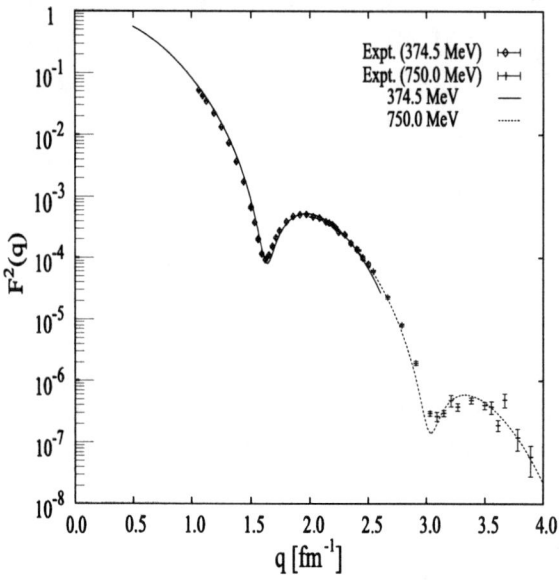

FIGURE 14. Calculated elastic cross section for ^{16}O.

Such calculations do give reasonable correlations. The calculated occupations above and below the Fermi level are entirely consistent with the observations and the calculation of the Coulomb sum rule also is in full agreement with the data as shown in Figure 15 [22].

The agreement with the data appear to be excellent. However, elastic charge scattering or the Coulomb sum rule sample only average properties of the nucleus. Thus we are in the process of calculating the hole-nuclei around ^{16}O such as ^{15}N. The magnetic form factor in ^{15}N samples specifically the $p_{1/2}$ proton orbit which was only poorly described in the mean field approach. Our results are rather encouraging. In our calculation of the magnetic form factor we include all terms up to second order in the correlation amplitudes. We also do the center of mass corrections according to the many-body expansion as outlined in our second paper

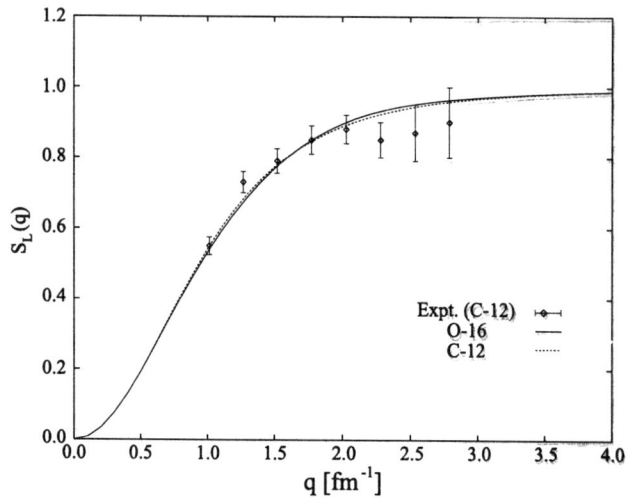

FIGURE 15. Calculated Coulomb sum rule for ^{12}C.

[20]. We have not yet included the meson exchange contributions. Nevertheless, the predicted form factor shown in Figure 16 is in reasonable agreement with the data.

At the same time we predict the overlap for $p_{1/2}$ proton removal to be 0.77 in close agreement with the expectations from the expermental results.

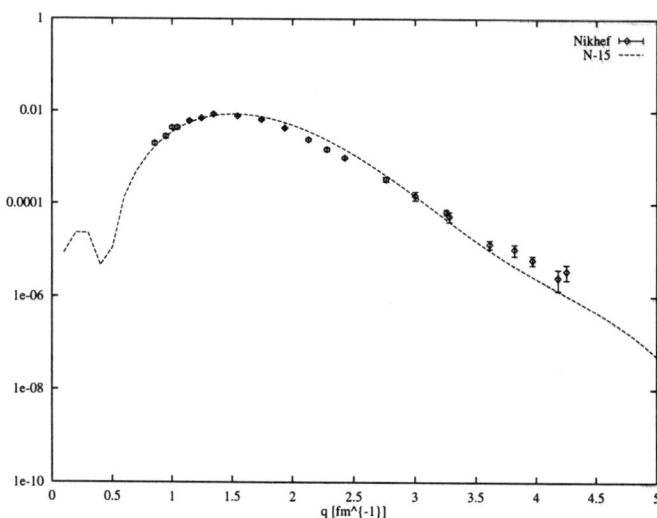

FIGURE 16. Calculated magnetic form factor for ^{15}N.

In summary, I tried to show that from the many high resolution experiments of the seventies and early eighties we have a large body of precise data. Since that time the efforts have shifted to the theoretical side where the calculations due to the ever increasing power of computers have become more and more precise, trying to do justice to the accuracy of the experiments. And these calculations are catching up and provide us now with precise wave functions as warranted by the experiments.

REFERENCES

1. R. E. Frosch et al., Phys. Rev. **174**, 1380 (1968)
2. J. M. Cavedon, Ph. D. Thesis, Centre d'Orsay, (1980)
3. D. Goutte et al., Phys. Rev. Lett. **45**, 1618 (1980)
4. C. W. Creswell, "Electron Scattering Studies of ^{166}Er, ^{176}Yb, and ^{238}U." Ph. D. Thesis, MIT (1977)
5. J. W. Negele and G. Rinker, Phys. Rev. **C 15**, 1499 (1977)
6. X. H. Phan et al., Phys. Rev. **C 38**¡ 1173 (1988)
7. I. Sick et al., Phys. Rev. Lett. **23**, 1117 (1969)
8. L. Lichtenstadt et al., Phys. Rev. **C 20**, 497 (1979)
9. J. W. Negele and D. Vautherin, Phys. Rev. **C 5**, 1472 (1972)
10. B. E. Norum, "Inelastic Electron Scattering form Oxygen 18" Ph. D. Thesis, MIT (1979)
11. M. V. Hynes, "Electron Scattering from the Ground State Magnetization Density of ^{17}O" Ph. D. Thesis, MIT (1978)
12. H. P. Blok and J. H. Heisenberg, "High resolution inelastic electron scattering and nuclear structure", in "Modern Topics in Electron Scattering", B.Frois and I.Sick, eds., World Scientific, Singapore
13. M. B. Leuschner, "Nuclear Structure Studies of the ^{16}O Nucleus observed with the Quasielastic (e,e'p) Reaction" Ph. D. Thesis, UNH (1992)
14. M. Leuschner, et al., Phys. Rev. **C 49**, 955 (1994)
15. O. Schwentker et el., Phys. Rev. Lett. bf 50, 15, (1983)
16. J. Heisenberg, "Electron Scattering and the Shell Model", in Comments on Nuclear and Particle Physics Vol. **13**, 267 (1984)
17. V. R. Pandharipande, C. N. Papanicolas, and J. Wambach, Phys. Rev. Lett. **53**, 1133 (1984)
18. B. S. Pudliner et al., Phys. Rev. **C 56**, 1720 (1997)
19. B. Mihaila, "Ground State of ^{16}O", Ph. D. Thesis, UNH (1998)
20. J. H. Heisenberg and B. Mihaila, Phys. Rev. **C 59**, 1440 (1999)
21. B. Mihaila and J. H. Heisenberg Phys. Rev. **C 60**, 054303 (1999)
22. B. Mihaila and J. H. Heisenberg Phys. Rev. Lett. *to be published* (2000)
23. I. Sick and J. S. McCarthy, Nucl. Phys. **A 150**, 631 (1970)

Inclusive scattering

Ingo Sick

*Dept. für Physik und Astronomie, Universität Basel,
CH-4056 Basel, Switzerland*

Abstract. We discuss the understanding of quasi-elastic inclusive electron-nucleus scattering. Particular emphasis is placed on the scaling property of the data, the status of the Coulomb sum rule and the puzzle of the excess transverse strength.

I INTRODUCTION

Quasi-elastic electron-nucleus scattering is a particularly "simple" process. The cross section is dominated by the contribution from incoherent electron-nucleon elastic scattering processes. Much of the behaviour of the inclusive cross section is dominated by the kinematics of quasifree electron-nucleon scattering. Accordingly, the interpretation of the data is relatively straightforward, a feature that greatly contributes to the appeal of quasielastic scattering.

The physics of interest to inclusive scattering is related to basically three areas:
• The properties of the nuclear spectral function $S(k, E)$ are responsible for much of the shape of the quasi-elastic response, and selected parts of $S(k, E)$ can be measured in part of the kinematical region accessible.
• The cross section contains information of the bound-nucleon form factor which potentially differs from the one for the free nucleon.
• The cross section receives contributions from processes other than quasi-free scattering, such as final state interactions (FSI) or meson exchange currents (MEC); due to the simplicity of the quasi-free kinematics this information can often be extracted.

Much of the interest in quasi-free scattering comes from the fact that analogous processes occur in other areas of physics: electron-atom scattering at keV-energies, neutron scattering from liquids and solids at eV-energies and deep-inelastic electron-nucleon scattering at GeV-energies. The analogies and differences between these different areas are fascinating, and provide much insight.

A large number of experiments on quasi-elastic scattering from nuclei with mass number A=2 - 238 have been carried out, at Bates, JLAB, Saclay, SLAC etc [1].

---
[1] for references see [1]

These data provide a reasonably complete coverage of the kinematical domain (see *e.g.* fig. 1). The data set is not so complete when it comes to data separated into the

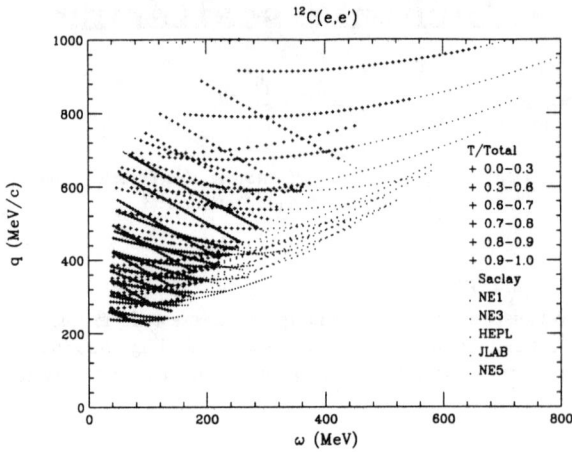

FIGURE 1. Kinematical coverage of inclusive scattering data on Carbon. The symbols are coded according to the source of the data and, for $-200 < y < 600$ MeV/c, for the T/(L+T) ratio.

longitudinal (L) and transverse (T) contributions. Here serious difficulties arise at the larger momentum transfers q, where a measurement of the small L-component becomes difficult given the domination of the T-term.

II SCALING

In general, the inclusive cross section is a function of two independent variables, the momentum transfer q and the energy transfer ω of the electron. (The L- or T-nature of the scattering might be considered as a third independent — discrete — variable). When observing only the scattered electron, and when the momentum transfer is big as compared to typical nucleon momenta in the initially bound nucleus, the scattering takes place on an individual target nucleon which is ejected from the bound system.

Scaling means that the cross section is a function of only *one* independent variable $y(q,\omega)$, which in turn depends on q and ω. This scaling property is basically a consequence of momentum- and energy-conservation in the quasi-free scattering process.

Inclusive scattering by a "weakly" interacting probe such as the electron can often be interpreted in terms of the plane wave impulse approximation, PWIA. Quantitative derivations of scaling in PWIA have been given in several places (see *e.g.* the review paper by Day *et al.* [2]). Here, I limit myself to a *qualitative* consideration, which however contains all of the basic physics. Energy- and momentum conservation for a quasi-free scattering process off an initially bound

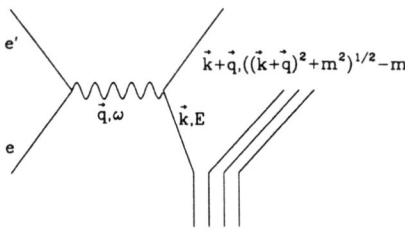

FIGURE 2. Quasi-elastic scattering from an off-shell nucleon in IA.

nucleon (fig.2) leads to [3] $\omega = [(\vec{k} + \vec{q})^2 + m^2]^{1/2} - m + E + recoil$. Splitting $\vec{k}$ into its components $k_\parallel$ and $k_\perp$ parallel and perpendicular to $\vec{q}$ and assuming $q, \omega \to \infty$, such that the $k_\perp^2$ and recoil- and removal-energy terms can be neglected, yields $(\omega + m)^2 = k_\parallel^2 + 2k_\parallel q + q^2 - m^2$. Schematically this equation means that $k_\parallel = y(q, \omega)$; q and ω no longer are independent variables. Combinations of q and ω given by the known function y correspond to the same $k_\parallel$. The cross section $\sigma(q, \omega)$ divided by the electron-nucleon cross section $\sigma_{eN}(q)$ gives a function $F(y)$ that only depends on y. This function F has an easy interpretation: it represents the probability to find in the nucleus a nucleon of momentum component y parallel to $\vec{q}$.

The quantitative derivation [2] of y-scaling is more involved, but brings little additional qualitative insights on the origin of scaling. The main quantitative information it provides concerns the region of $S(k, E)$ which does indeed contribute to $F(y)$; fig. 3 shows an example.

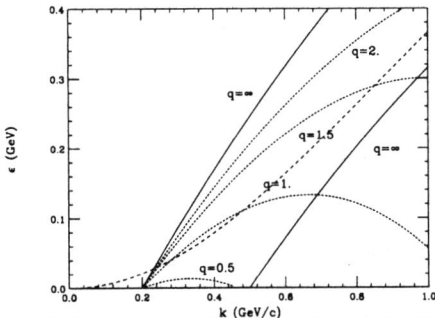

FIGURE 3. Region of k, E (below corresponding curves) contributing to (e,e') at given q and y [2].

As an illustration, fig. 4 shows some of the inclusive scattering data for ^{3}He. The data cover a large dynamical range in cross section, and the quasi-elastic peak shifts over a large range of ω with increasing q. The same data, plotted as a function of the scaling variable y (fig. 5) shows an impressive scaling behaviour for $y < 0$, *i.e.*

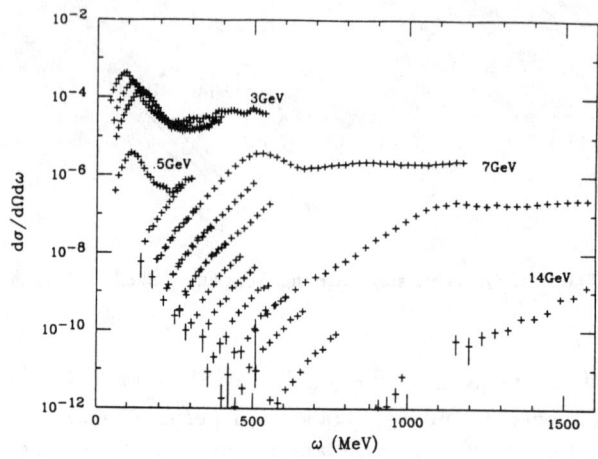

FIGURE 4. Cross sections for ^{3}He(e,e') as function of incident electron energy and energy loss.

for the low-ω side of the quasi-elastic peak. Cross sections that differ by several orders of magnitude define the *same* function $F(y)$. For $y > 0$ the values strongly diverge.

This scaling property allows one to exploit the data in several directions: The presence or absence of scaling tells us something about the reaction mechanism (we assumed PWIA without FSI to derive scaling), a residual q-dependence of $F(y)$ can tell us something about the q-dependence of the in-medium nucleon form factor, and the function the data scale to provides information of the nuclear spectral function. I address below in more detail some of these points.

Reaction mechanism. The data show a marked deviation from scaling behaviour for $y > 0$, thus indicating that there processes other than quasi-free scattering, such as MEC, pion production, Δ-excitation and DIS, contribute. Fig. 6, which shows the scaling function derived from calculated cross sections for MEC [4], illustrates this point for the $y < 0$ region; pronounced non-scaling occurs despite the fact that the calculated cross sections cover only about half the q-range of fig. 4. We thus may conclude in particular that, at places where the data approximately scale, the contributions of reaction mechanisms other than quasi-free scattering are smaller than the residual non-scaling of the data. Thus, scaling gives us direct information on the reaction mechanism, an input we must understand before we can use the data for quantitative interpretation.

Constituent form factor. Scaling is obtained after dividing the experimental cross section by the electron-nucleon cross section. An incorrect q-dependence of the latter will lead to non-scaling, at least as long as the range of q covered is large enough to lead to a variation of the elementary cross section over a large range. In order to exploit this idea quantitatively, one can compute the scaling

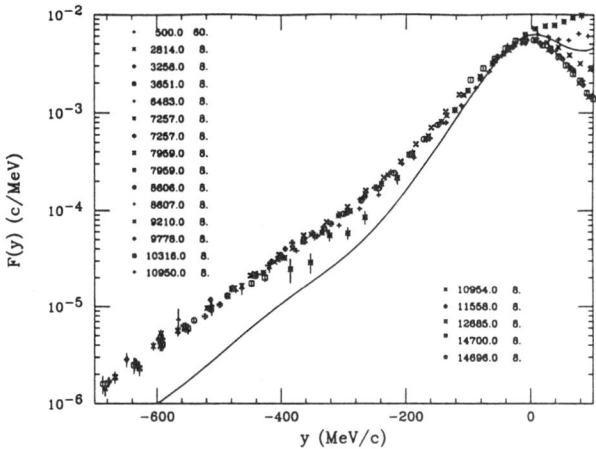

FIGURE 5. Scaling function for ^{3}He(e,e').

function using a modified nucleon form factor, using some suitable parametrization depending on a parameter describing the assumed change of the bound-nucleon form factor (size) and then fit $F(y)$. The minimum of χ^2 of this fit gives the best value for the parameter. As an example, we show in fig. 7 the χ^2 for iron obtained when varying the size of the nucleon (a change made by varying what, in a dipole parametrization of $G_{p,n}(q)$, would be the usual radius parameter). The minimum of χ^2 is found near zero change, and a consideration of the systematic errors of this approach allows the statement that the nucleon radius (which essentially here refers to the *magnetic* radius as the data are dominated by the transverse cross section) differs by less than 3% from the one of the free nucleon. Given the number of models that do predict a sizeable influence of the nuclear medium on the nucleon form factors, the information provided by scaling is rather constraining.

Nucleon FSI. The final state interaction of the knocked out nucleon in general is of minor importance in inclusive scattering; the electron carries information only about the FSI that takes place within a distance of order $1/q$ from the scattering vertex. Subsequent interactions of the recoil nucleon on its way out of the nucleus (which are important for (e,e'p)) are not influencing the scattered electron.

At low energy loss FSI does, however, play a role. While the distribution of the spectral function $S(k, E)$ in E leads to a convergence of $F(y,q)$ from below with increasing q, the FSI leads to a convergence from *above*. Fig. 8 demonstrates this for data for iron which, in a recent experiment at JLAB [5], have been extended to very large q and y. The figure shows that with the momentum transfers now achieved convergence has been reached for values of $-y$ up to 600MeV/c.

The effects of FSI can be quantitatively studied at large q; there the high-momentum recoiling nucleon can be described using Glauber theory [6]. The FSI

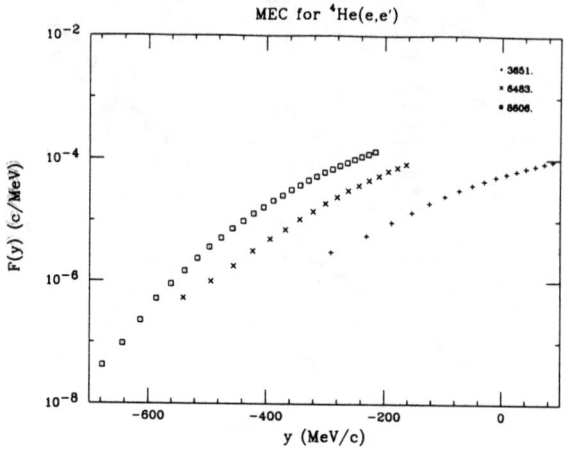

FIGURE 6. Scaling function for MEC cross sections calculated for ^{4}He(e,e').

in Correlated Glauber Theory CGT can be shown to lead essentially to a folding of the PWIA response, with a folding function that at large q (large recoil momentum) becomes independent of q due to the approximate momentum-independence of the N-N total cross section. This then indicates that at large q one can expect to find y-scaling, but the scaling function is not equal to the momentum distribution $n(k_\parallel)$ as expected in PWIA [16].

A quantitative understanding of the scaling function including the effects of FSI is possible in particular for infinite nuclear matter, where (together with A=2,3) a reliable spectral function is available, and where a treatment of FSI using CGT is feasible [6]. It turns out to be very important to include not only the short-range

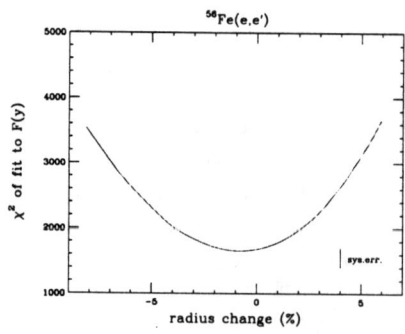

FIGURE 7. χ^2 of fits to $F(y)$ for iron as a function of the assumed change of the bound-nucleon size.

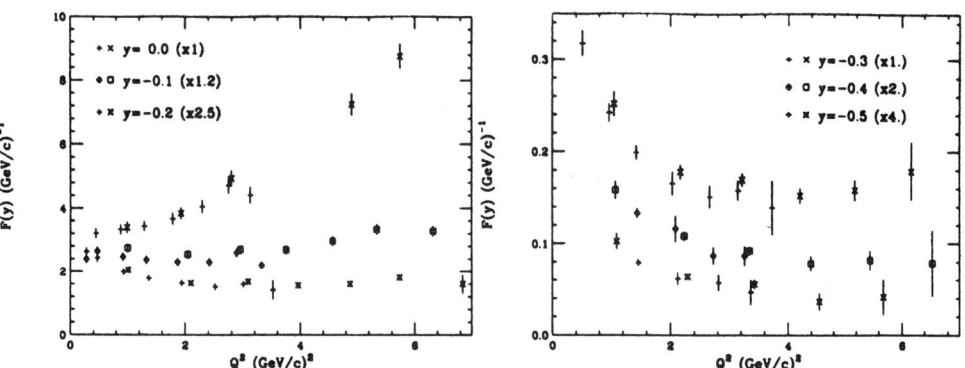

FIGURE 8. Convergence of $F(y,q)$ with increasing q, for different values of y [5].

N-N correlations in the initial state (they are accounted for in the spectral function calculated via CBF); the short-range correlations also must be accounted for in the FSI, as the nucleon hit by the electron initially travels through a "correlation hole" in the $A-1$ nucleus. This leads to an appreciable reduction of FSI, as the inclusively scattered electron only sees the FSI that occurs close to the point of the e-N interaction. Fig. 9 shows that the scaling function at large q and its convergence properties can be quantitatively understood.

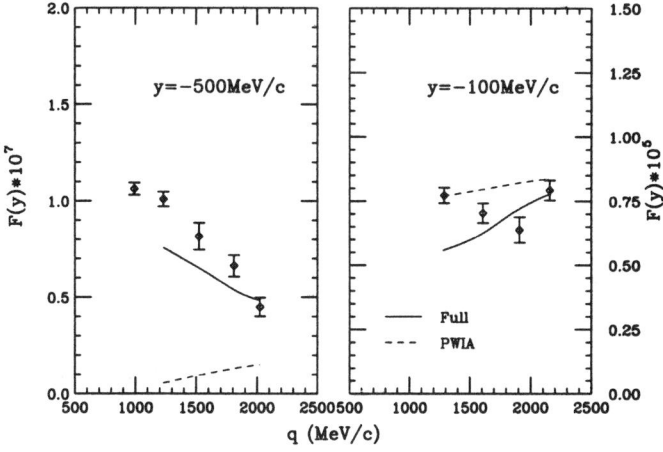

FIGURE 9. Convergence of experimental and CGT $F(y,q)$ for nuclear matter [6].

Spectral function at large k. The properties of $S(k,E)$ at large k are relevant for the behaviour of the inclusive cross section at large q (several GeV/c) and comparatively low ω (several hundred MeV). This is immediately obvious when considering the limit of $\omega=0$. Transferring in PWIA a large momentum $\vec{q}$ to a nucleon, with a nucleon in the final state that has little energy and small momentum $\vec{k} + \vec{q}$, is only possible if $\vec{k} \sim -\vec{q}$. The region of the low-ω tail of the quasielastic response is best studied considering the ratio of the nuclear and deuteron response (the latter being well known experimentally and accurately calculable for any N–N potential).

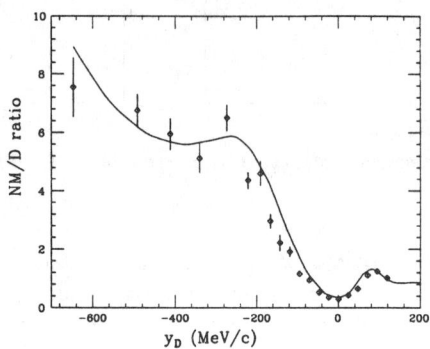

FIGURE 10. Cross section ratio of nuclear matter and the deuteron, as a function of the scaling variable y. The result of the CBF calculation is shown as a solid line [7].

Figure 10 shows the ratio for one of the kinematics at large q, small ω where data are available. Data and theory are plotted as a function of the variable y_D, which is basically the component of the nucleon momentum k parallel to $\vec{q}$. The dip at $y = 0$ corresponds to the maximum of the quasielastic peak dominated by nucleons of low momentum k, the region of $y < -300 MeV/c$, corresponds to the high-momentum tail of $n(k)$ of interest here.

Figure 10 shows that the particular nuclear matter spectral function used here [8] agrees well with the data. Tests carried out by renormalizing the spectral function at $k > k_F$ have shown that the cross section ratio at $y < -300 MeV/c$ is essentially proportional to $S(k, E)$ at $k > k_F$. We conclude that the experimental ratio nuclear matter-to-deuteron for $k > k_F$ is 5.5 ± 0.8, as compared to the value of 6.0 as given by the CBF theory.

Together with similar data taken at other kinematics, the electron-scattering data fix the nuclear matter-to-deuteron ratio with $10 \div 15\%$ accuracy. This uncertainty in the ratio is much smaller than the spread between modern calculations for $n(k)$, which amounts to about a factor of two [9]. This shows that quasielastic electron scattering does provide information on the short-range properties of nuclear matter that is most valuable in constraining theory.

III SUPERSCALING

A recent extension of scaling analyses [10] has featured the topic of "uperscaling". The basic idea — as in other areas of physics where scaling-type ideas are used (hydrodynamics, for instance) — is to use *dimensionless variables*, i.e. to remove length scales from the problem. Scaling of the first kind occurred because the length scale defined by the electron, $1/q$, is much shorter than the one defined by the nucleus. Scaling of the second kind can occur if different nuclei differ mainly in length scale (given by the Fermi momentum), which can be removed by dividing out this scale using y/k_F as a variable. The analysis of superscaling actually has been done using not y/k_F, but ψ' — a variable originating from the relativistic Fermi gas model and involving a somewhat different (better) treatment of the removal-energy aspect — but the difference between the two variables for the superscaling-aspect of interest here is irrelevant.

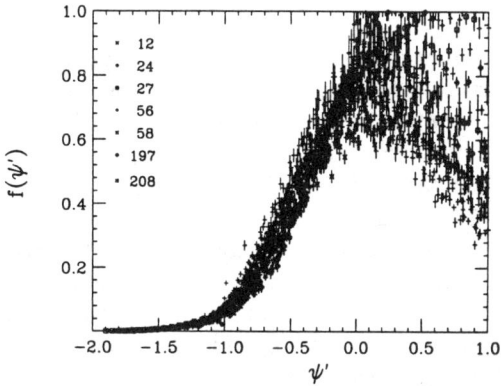

FIGURE 11. Scaling function $f(\psi')$ as function of ψ' for all nuclei $A \geq 12$ and all kinematics. The values of A corresponding to different symbols is shown in the insert.

Superscaling allows one to compare the scaling function of different nuclei. While ordinary scaling (of the first kind) studies the degree to which data for one nucleus at different q and ω define a universal function, scaling of the second kind asks how well the scaling functions $f(\psi')$ of different nuclei agree. For these studies we concentrate on nuclei with $A > 4$, as the lightest nuclei are known to have spectral functions that are very far from the "universal" one which is at the basis of the superscaling idea. The data for the nuclei $A = 12 \ldots 208$ have analyzed using for k_F 220, 230, 235 and 245 MeV/c for C, Al, Fe, Au, with intermediate values for the intermediate nuclei; for E_{shift}, which has a minor effect, we use 15, 15, 20, 25 MeV for the same nuclei.

Figure 11 shows the scaling function $f(\psi')$ for all kinematics suitable for the present study and all A available. We clearly observe a scaling behavior for values of $\psi' < 0$: while the cross sections at a given ψ' vary over more than three orders of magnitude, the values of $f(\psi')$ are essentially universal. For $\psi' > 0$, on the other

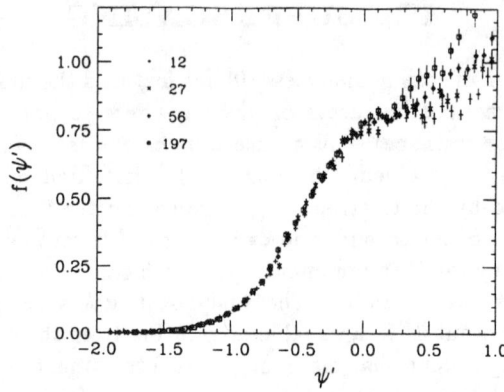

FIGURE 12. Scaling function for C, Al, Fe, Au and fixed kinematics [11]. The correspondence of symbol and mass number of the nucleus is shown in the insert.

hand, the scaling property is badly violated, as expected, since there processes other than quasi-elastic scattering – meson exchange currents, Δ-excitation, deep inelastic scattering – contribute to the cross section. The scaling applies only to processes having the behavior of electron-nucleon quasi-free scattering.

In order to separate some of the effects leading to less-than-perfect scaling at negative ψ', in fig. 12 we show the function $f(\psi')$ for the series of nuclei $A = 12\ldots197$, but for fixed kinematics (3.6 GeV, 16°, and hence nearly constant q). The quality of the scaling in the region $\psi' < 0$ is quite amazing. This shows that the removal of the A-dependence, i.e. scaling of the second kind, actually is *better* realized in nature than ordinary scaling. The deviations from scaling observed in fig. 11 are *not* from an A-dependence.

A part of the A-dependent increase of $f(\psi')$ at positive ψ' results from the increase of k_F with A, yielding an increase of the width of the quasi-elastic and Δ peaks, and a correspondingly increased overlap with non–quasi-free scattering processes (Δ-excitation, π-production, ...). At the same time, the increasing average density of the heavier nuclei also leads to an increase in contributions of two-body MEC processes which are strongly density-dependent (i.e., do not scale with k_F in the same way the one-body knockout processes do [12]).

Figure 13 shows the data for $A = 4, 12, 27, 56, 197$ on a logarithmic scale for the kinematics of fig. 12 and demonstrates that the scaling property extends to large negative values of ψ', corresponding to large momenta of the initial nucleon. This feature clearly cannot be predicted within the RFG model, since there the response is restricted to $|\psi'| < 1$. However, there are indications of this behavior from theoretical studies of the nuclear matter spectral function as a function of density. For different nuclear matter densities and large k, the spectral functions are similar in shape [7] and the tail of the momentum distribution $n(k)$ at $k > k_F$ (corresponding to $\psi' < -1$) is a near-universal function of k/k_F [13]. For finite nuclei and large momenta we can employ the Local Density Approximation (LDA),

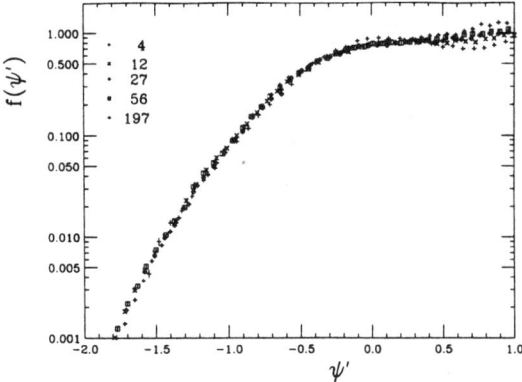

FIGURE 13. Scaling function for nuclei $A = 4 - 197$ and fixed kinematics on logarithmic scale.

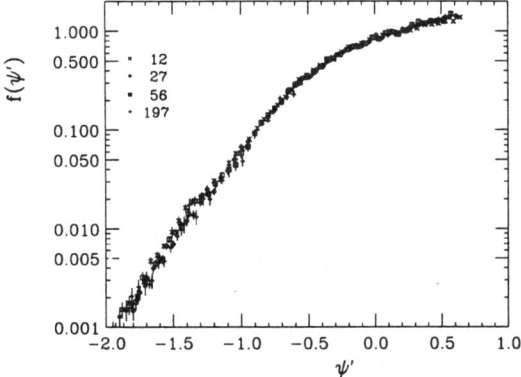

FIGURE 14. Scaling function for nuclei $A = 4 - 197$ at higher momentum transfers (3.6 GeV, 25°).

as at large k we are dealing with short-range properties of the nuclear wave function [7]. Within LDA, the nuclear momentum distribution (spectral function) is then a weighted average over the corresponding nuclear matter distributions. This means that the large-momentum tail of the nuclear spectral function also scales with k_F, a dependence that is removed when using ψ'. Previous work [6] has shown that in the extreme tail of the quasi-elastic peak FSI play an increasingly important role, and lead to a slow convergence of $F(y,q)$ with q. Figure 13 indicates that the effects of FSI on scaling of the second kind are less pronounced. This is presumably due to the fact that it is the FSI of the nucleon immediately after the scattering that counts, and these FSI for different A are near-universal, modulo surface effects.

In order to emphasize the quality of this superscaling in the tail, in fig. 13 we have also included the data on ^{4}He, taken under the same kinematical conditions [11] (k_F=200 MeV/c, E_{shift}=20 MeV). While at ψ'=0 $f(\psi')$ for ^{4}He is about 10% higher than for heavier nuclei as a consequence of the sharper peak of the spectral function at $k \sim 0$, the scaling function for ^{4}He agrees perfectly with the one for heavier nuclei for $\psi' < -0.2$.

We thus find that this A-independence of the superscaling function is much better realized than the q-independence of the normal scaling function. In other words, scaling of the second kind seems to be much less sensitive to scaling violations resulting from processes such as MEC, nucleon FSI and the spread of $S(k,E)$ in E. While this reduced sensitivity is plausible, it needs to be understood better theoretically.

IV COULOMB SUM RULE

The Coulomb sum rule makes a statement about the longitudinal quasielastic strength integrated over the quasi-elastic peak. Assuming that the contributions of MEC, Δ excitation, *etc.* are small in the longitudinal piece, and assuming that the momentum transfer is $q >> 2k_F$ to eliminate Pauli suppression, the Coulomb sum rule reads:

$$\int_0^\infty \sigma_L^A(q,\omega)d\omega \; / \; \tilde{\sigma}_L^p(q) = Z$$

where $\tilde{\sigma}_L^p$ is the (appropriate combination of) longitudinal e-N cross section corrected for relativistic effects. This sum rule has received a lot of attention as some results [14] seemed to indicate a violation of $\sim$40%! Several mechanisms — often rather artificial ones — have been evoked to explain this result.

Difficulties arise from the fact the L<<T, so an extraction of the L-piece is subject to the blow-up of the systematic errors. Additional difficulties arise from the upper limit of the integral, which is not reachable.

J. Jourdan [15] has performed an improved analysis of the data. The results are shown in the table. When analyzing the Saclay data alone, but using the correct e-p cross section, when including the well-known relativistic corrections, when accounting for the fact that the upper cutoff of the integral is not ∞ and when doing the Coulomb corrections using the exact calculations of the Ohio group (*not* LEMA) J. Jourdan finds that there is *no* violation of the Coulomb sum rule; within experimental errors the Coulomb sum, divided by Z, derived from the Saclay data alone is compatible with 1.

A more precise Coulomb sum can be determined by including the SLAC data [11], which have been taken at more forward angle, and in a more reliable data acquisition mode (event mode, not histogram mode). This allows for a more reliable extraction of the longitudinal contribution. The result, shown in the last line of the table, shows that the larger angular range of the world data leads to a two times smaller error bar on the Coulomb sum. The sum rule is fulfilled for the case (the iron nucleus) where the biggest deviation had been claimed. Similar fulfillment is found for all other nuclei where the data base allowed a reliable extraction.

So far for the good news; the bad news is that, in order to exploit the sum rule for a study of short-range N-N correlations, on would have to do a three times more accurate determination. Given the main problems remaining at present — rescattering of the electrons inside the spectrometer, reliability of Coulomb corrections in both the radiative corrections and the L-T separation — this does not seem to be practical at present.

V REACTION MECHANISM

The resolution of the "problem" with the longitudinal strength leaves us with the true, 30 year old, difficulty to understand the transverse one. While the contribution of the Δ to the transverse response in the region of the Δ peak is presumably understood, the filling in of the "dip" between quasi-elastic and Δ-peak, and the excess of the transverse strength in the quasi-elastic peak region are not understood.

With the discovery of superscaling, we can make further progress in understanding this transverse strength [1]. In PWIA, *both* the L- and the T-response should scale to the *same* function (scaling of the zero'th kind). Having superscaled longitudinal responses allows us to take an average over all nuclei, the resulting quantity having reasonably small error bars. This universal longitudinal response $f_L(\psi')$ can be subtracted from the transverse one, to yield the transverse *excess* that has been giving us problems for such a long time. Fig. 15 shows the *difference* between the transverse and longitudinal scaling functions, $\Delta f(\psi')$.

Fig. 15 shows that part of the excess transverse strength does display a peak at the location of the maximum of the quasielastic response, $\psi'=0$. The strength at larger ψ', corresponding to larger electron energy loss and of presumably different origin, rises rapidly with increasing q. Much of the strength of Δf at $\psi' < 0$ is below the threshold for pion production on a nucleus with $A \geq 12$ (and even more so for quasifree production). This is shown by the arrows in fig. 15 which indicate, for the various q's, the position of the π-production threshold both on the nucleus and on the free nucleon — we consider the latter to be the more relevant one, since coherent production on the entire nucleus is expected to be very small. The presence of large excess transverse strength *below* π-threshold means that some other mechanism must be identified as its source. The perhaps most likely origin are MEC processes which, due to the strong Δ-N coupling via π-exchange, feed strength back into the

TABLE 1. Results for the Coulomb sum at $q=$ 570 MeV/c for iron, divided by Z.

Result of Meziani (^{56}Fe)	0.60±0.20
Dipole G_{ep} replaced by Simon	0.64±0.20
Relativistic correction added	0.69±0.20
Tail contribution added	0.75±0.20
Coulomb correction added	0.81±0.20
....	....
SLAC–data added	0.97±0.12

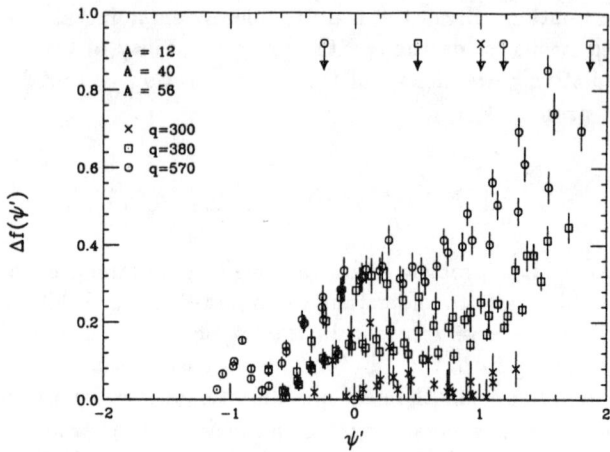

FIGURE 15. Difference between the transverse and longitudinal scaling functions, $\Delta f(\psi')$ [1]. The arrows indicate the values of ψ' for π-production on the free nucleon and the nucleus at the three values of q.

one-nucleon knockout channel. This remains to be investigated quantitatively in the future.

VI CONCLUSIONS

Quasi-elastic electron-nucleus scattering has provided us with a wealth of information, both on the reaction mechanism and the properties of the nuclear ground state. The (super)scaling of the data has allowed us to study many questions in considerable detail. The major open question concerns the understanding of the excess transverse strength in the region of the top of the quasi-elastic peak and the "dip", which has eluded a quantitative description for more than 3 decades.

Acknowledgements. Much of the work described in this paper was done in collaboration with T.W. Donnelly.

REFERENCES

1. T. W. Donnelly and I. Sick. *Phys. Rev. C* 60:065502, 1999.
2. D. Day, J.S. McCarthy, T.W. Donnelly, and I. Sick. *Ann.Rev.Nucl.Part.Sci.*, 40:357, 1990.
3. I. Sick, D. Day, and J.S. McCarthy. *Phys. Rev. Lett.*, 45:871, 1980.
4. T.W. Donnelly, J.W. Van Orden, T. de Forest, and W.C. Hermans. *Priv. comm. and Phys. Lett.*, B76:393, 1978.
5. J. Arrington, C.S. Armstrong, T. Averett, O. Baker, L. deBever, C. Bochna, W. Boeglin, B. Bray, R. Carlini, G. Collins, C. Cothran, D. Crabb, D. Day, J. Dunne,

D. Dutta, R. Ent, B. Fillipone, A. Honegger, E. Hughes, J. Jensen, J. Jourdan, C. Keppel, D. Koltenuk, R. Lindgren, A. Lung, D. Mack, J. McCarthy, R. McKeown, D. Meekins, J. Mitchell, H. Mkrtchyan, G. Niculescu, T. Petitjean, O. Rondon, I. Sick, C. Smith, B. Tesburg, W. Vulcan, S. Wood, C. Yan, J. Zhao, and B. Zihlmann. *Phys. Rev. Lett.*, 82:2056, 1999.
6. O. Benhar, A. Fabrocini, S. Fantoni, G.A. Miller, V.R. Pandharipande, and I. Sick. *Phys. Rev.*, C44:2328, 1991.
7. O. Benhar, A. Fabrocini, S. Fantoni, and I. Sick. *Nucl.Phys.*, A579:493, 1994.
8. O. Benhar, A. Fabrocini, and S. Fantoni. *Nucl. Phys.*, A505:267, 1989.
9. I. Sick. *Nuclear Astrophysics*, GSI, edts. M. Buballa *et al.*:37, 1998.
10. T. W. Donnelly and I. Sick. *Phys. Rev. Lett.*, 82:3212, 1999.
11. D. Day, J.S. McCarthy, Z.E. Meziani, R. Minehart, R. Sealock, S.T. Thornton, J. Jourdan, I. Sick, B.W. Filippone, R.D. McKeown, R.G. Milner, D.H. Potterveld, and Z. Szalata. *Phys. Rev.*, C48:1849, 1993.
12. J.W. Van Orden and T.W. Donnelly. *Ann. Phys.*, 131:451, 1981.
13. M. Baldo, I. Bombaci, G. Giansiracusa, U. Lombardo, C. Mahaux, and R. Sartor. *Phys. Rev. C*, 41:1748, 1990.
14. Z. Meziani, P. Barreau, M. Bernheim, J. Morgenstern, S. Turck-Chieze, R. Altemus, J. McCarthy, L.J. Orphanos, R.R. Whitney, G.P. Capitani, E. de Sanctis, S. Frullani, and F. Garibaldi. *Phys. Rev. Lett.*, 52:2130, 1984.
15. J. Jourdan. *Nucl. Phys. A*, 603:117, 1996.
16. O. Benhar. *Phys. Rev. Lett.*, 83:3130, 1999.

The $A(e,e'p)$ Coincidence Reaction: Past, Present and Future

William Bertozzi

Department of Physics and Laboratory for Nuclear Science
Massachusetts Institute of Technology
Cambridge, MA 02139, USA

Abstract. A brief review is presented of the ideas that motivated the first studies of the $(e,e'p)$ reaction type, but with proton probes. The reasons for the evolution to the electron probe are presented along with some interesting results that were observed. These results provided many of the reasons for building Bates. The present status of the field is discussed where interference and polarization observables play a prominent role and define much of the future course of research.

I INTRODUCTION

Where and how did it start? What were the ideas? These are often the questions we hear when we review a field as prominent as that defined by the variety of $(e,e'x)$ reactions we use these days to study nuclei and the nucleon. As usual in physics, the reasons form a complexity of many when we view them in the context of a literature. But a few stand out as clearly effective views in fundamental terms, and I shall endeavor to bring these forth. Some of these as yet remain unrealized.

It all began when people started to question the ideas of the shell model and its phenomenal success. Although this questioning was happening all along the course of shell model development, people were caught up in the successes, such as that of intermediate coupling in the 1p-shell, the deformed shell model and the tremendous growth of knowledge through the 1950's, and did not pay too much attention to these problems of concept. As usual, it was more interesting to carry out yet one more study in an area that was tremendously fashionable and successful. All we really knew about the shell model experimentally concerned a few of the low lying states of nuclei and only the dynamics of the last few particles introduced to make a specific nucleus. Stripping reactions and other strong reactions dealt only with the outermost nucleons on the nuclear surface. When we studied complex reactions we invoked the "Cloudy Crystal Ball" models that did a great deal of averaging to avoid detailed questions.

One question of conceptual concern was: "Do the nucleons really move in shell model orbits when they are in the dense center of the nucleus?" After all, nucleons have a matter distribution extending to about one fermi in radius and their centers on the average are spaced by about 2.2 fermi. Most of the time they are shoulder to shoulder like passengers in a crowded subway! There were quantum mechanical arguments based in part on the Pauli principle that could be invoked to help explain the transparency of nucleon motion deep in a nucleus, but the direct experimental evidence was indeed missing. It became important to measure the distribution of momentum of a deep nucleon and see it behave as that of a semi-pure orbital.

By the late 1960's it was also realized that the nucleon-nucleon correlations due to the strong short range part of the two-body interaction were playing an important role in mixing configurations. The simple shell model description was only about 60% of the nuclear wave function. Studying the nature of these correlations also became an important goal. This goal remains unrealized to this day. The study of these correlations forms one of the areas of study in which the electromagnetic probe is uniquely suited.

II   THE FIRST EXPERIMENTS

Interestingly enough the first experiments were not of the $(e, e'p)$ variety but rather $(p, p'p)$ experiments. This was in part a result of the fact that there were many easily accessible proton beams and they were comparatively easy to use. In the first experiments a simple idea was tried by people like M. Riou [1], T. Berggren and B. Tyren [2] and interpreted by G. Jacob and T. A. J. Maris [3]. Another early player was B. Gottschalk from Northeastern University [4]. These experiments used protons varying from about 160 MeV to 500 MeV, depending on the laboratory, and scattered them from nuclei in a geometry like that in figure 1. The central idea is that in free proton scattering the angle between the two protons is 90 degrees. If the opening angle between the protons was bigger or smaller than 90 degrees it was due to an initial momentum of the struck proton. These early experimenters used ±42 degrees relative to the incident beam for their reference directions rather than±45 degrees, to account for relativistic effects and binding effects (refraction at the nuclear surface). The results were spectacular! In figure 2 we show a reproduction from Gottschalk [4] reviewing the $12C(p, p'p)$ results. Notice that in the top figure the conditions are that the two proton angles are equal and at 42.5 degrees. In this case an initial momentum results in a lack of equal energy sharing. The p-shell and the s-shell both show up over a wide range of energy sharing (the parameter on the curves). The momentum distributions are very clear on the remaining two curves. For the p-shell there is a minimum at zero initial momentum. For the s-shell there is a maximum at zero initial momentum. This is what one expects from a Fourier decomposition of the p-shell and s-shell spatial wave functions, respectively. The distribution of binding energies is shown in figure 3 for a beam energy of 440 MeV. Nowadays we call this the missing energy distribution. You see the clear

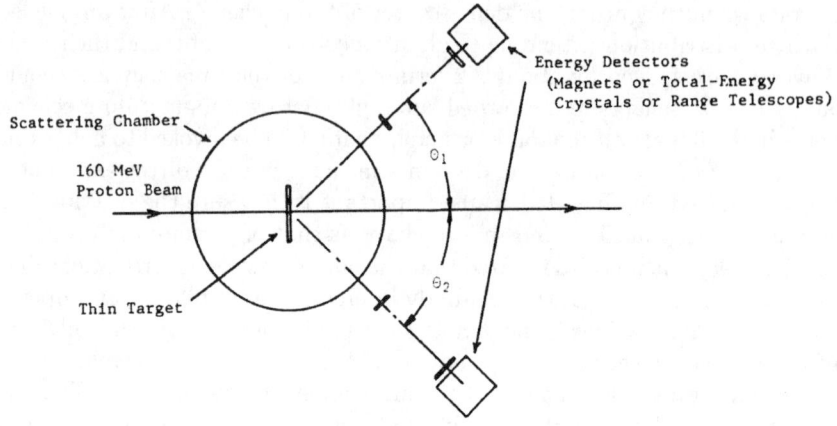

FIGURE 1. Example of a $(p, p'p)$ kinematic setup. [4]

identification of the two groupings of strength, the p-shell and the broad s-shell. Notice that for the first time one had identified an eigen-energy (identified as Eb on the figures) with a momentum distribution characteristic of a specific orbital angular momentum. The surface p-shell is observed in a narrow range of binding and the deeply bound s-shell is broadly distributed as expected for a deep excitation with a very short lifetime.

III THE $(E, E'P)$ REACTION STUDIES

One can think of many difficulties with using protons as the probe of internal structure of the dynamics of a nucleus. These all originate from the fact that protons interact with the nucleus so strongly that one can not think of them as a probe in the sense of first order perturbation theory. Basically the "noise" is as big or bigger than the "signal". For this reason we migrated to using electrons as probes where the interaction with a nucleon in a nucleus is 1/137 times smaller and the perturbative ideas are appropriate. Of course the outgoing proton remains a strong interaction problem, but there is only one trajectory to modify in the $(e, e'p)$ process, not three as in the $(p, p'p)$ process. Also, by requiring energetic protons we can do the experiments at an energy where the nucleon-nucleon interaction is a minimum, 200 to 500 MeV, and correct for distortions via a good DWBA calculation. Of course, with electrons there is the need to correct for radiative processes which are non-negligible. However, these are in principle processes that can be evaluated quantitatively.

There was one very big obstacle to doing the electron experiments. Most of

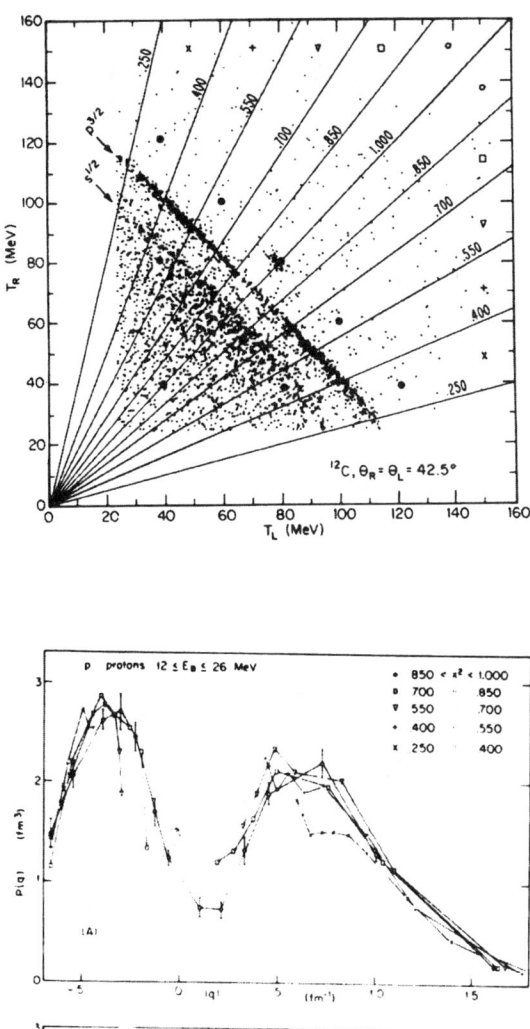

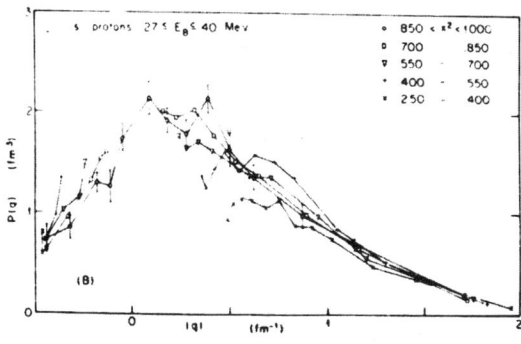

FIGURE 2. Scatter plot of raw data from a 12C$(p, p'p)$ experiment at 160 MeV. Distributions of initial momenta ($|q|$). [4]

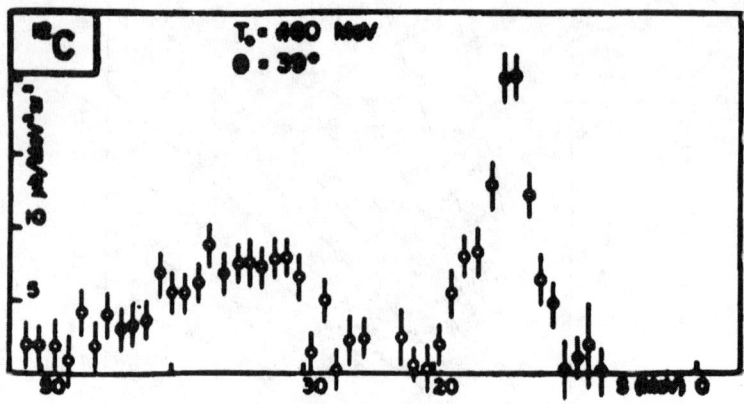

FIGURE 3. Distribution of binding energies from a 12C$(p,p'p)$ experiment using 460 MeV protons. [4]

the higher energy machines were electron linear accelerators with duty ratios less than 10^{-3} and the accidental counting rates were simply unacceptable. Here was the first big reason for Bates. Some of you may remember we designed for a 5% duty ratio accelerator. This was a big technological step at that time. The factory testing of the radio frequency system was at 5,000 pulses per second and 5% duty. It was only the practical problem of operational expenses coupled with component life time that limited us to 1% duty ratio, still a viable duty ratio for $(e, e'p)$ coincidence experiments.

As usual in physics, when a good idea comes along, someone finds a way of implementing this idea. Our colleagues at Frascati were the first to do an $(e, e'p)$ experiment and demonstrate the practicality of the work. The names we remember here are Ugo Amaldi, Gloria Campos Venuti, Georgio Cortelessa and several others. They used the internal beam of the synchrotron at Frascati and a monoenergetic proton channel of .16 to one steradian. Those of you using internal beams at NIKHEF and planning for BLAST would do well to read these early papers. I will show some results from Frascati as reported by Georgio Cortelessa at the MIT 1967 Summer Study: "Medium Energy Nuclear Physics With Electron Linear Accelerators" [5]. They show in figure 4 the layout of the experiment indicating the proton range telescope, the electron beam and the electron spectrometer. In figure 5 they show the plastic scintillators that form the proton "onion" detector, circles concentric about the momentum transfer. In figure 6 they show the results for 12C$(e, e'p)$ for the momentum distributions of the p-shell and s-shell. The qualitative results are the same as with the $(p, p'p)$ experiments. Notice however how the p-shell results are much closer to zero at p=0. This is due to the much stronger signal from $(e, e'p)$ because the proton interactions are much smaller. They went on to examine many nuclei up to Ca and produced the results in figure 7, a com-

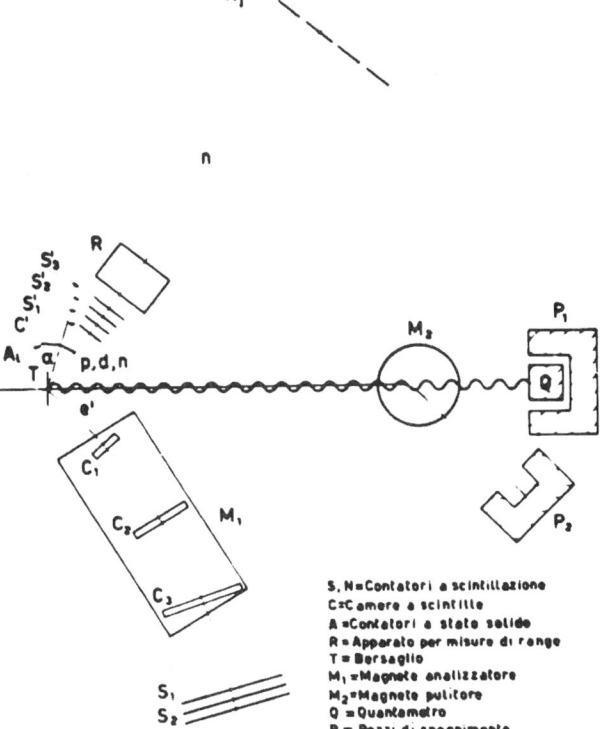

FIGURE 4. The experimental arrangement for the Frascati $(e, e'p)$ experiments showing the proton channel and the scattered electron channel. [5]

pendium of orbital binding energies for the 1s, 1p1/2, 1p3/2, 2s, 1d3/2 and 1d5/2 shells in nuclei. These results were confirmed and much more accurately established by precise experiments at Saclay when that accelerator became operational in the 1970's. Much of that work is described in a review article by S. Frulani and J. Mougey [6]. I show you one result of the Saclay work in figure 8 [7]. It is the measurement of the momentum distribution of the 2s shell in 40Ca in the region of the missing energy (binding energy) corresponding to that eigen-energy. Notice the peaking at $p = 0$. Notice the small second maximum in the distribution. The 1s shell has no second maximum of a comparable size. In fact, if we think of the harmonic oscillator as a reasonable approximation to the shape of the mean field nuclear potential (not bad for these nuclei), we remember that the 1s has no second maximum while the 2s does have a second maximum. In this last figure we also present the momentum distributions for other regions of missing energy to show the changing character and the dominance of other orbital angular momenta.

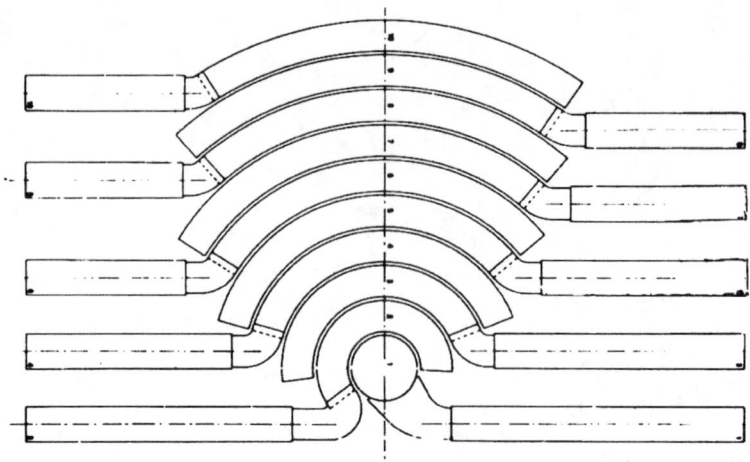

FIGURE 5. The 'onion' proton hodoscope made of plastic scintillators and used in the $(e,e'p)$ experiments at Frascati. [5]

In these results we have the first indications of approximate single particle energy eigenstates and corresponding angular momentum orbitals for deeply bound shell model particles.

Our friends from NIKHEF [8] also contributed to this field with great originality. In figure 9 we show their results for 208Pb$(e,e'p)$. The different states of the residual system are easily resolved and comparisons of the accurate data with calculations [9–12] are shown using orbitals ranging from the 3s up to the h11/2 and g7/2. We see that the orbital structure of the shell model exists deep inside the heavy nucleus of lead. Of course I remind you that only about 60% of the predicted 3s strength is observed. Where the missing strength resides in energy and the nature of its momentum structure is related to the fundamental question of short range nucleon-nucleon correlations.

At MIT-Bates we took a different approach to the $(e,e'p)$ reaction study. We were struck by two facts. The first is that in 3He(e,e') inclusive scattering, the transverse and longitudinal structure functions are equal in the quasi-elastic peak. This is in contrast to (e,e') with 4He and heavier nuclei where the transverse structure function is 40% larger than the longitudinal. In the so-called dip region between the quasi-elastic peak and the delta, the cross sections for (e,e') are much bigger than we could calculate even when including meson exchange currents. How would these features show up in the $(e,e'p)$ reaction? Would we get a hint about the currents responsible for these problems by studying the $(e,e'p)$ reaction? These questions were the focus of our investigations.

We proposed to study the $(e,e'p)$ reaction in the He isotopes and carbon. Our focus was appropriate also because we did not have the experimental resolution to compete with Saclay and NIKHEF. We used OHIPS and MEPS in the new south

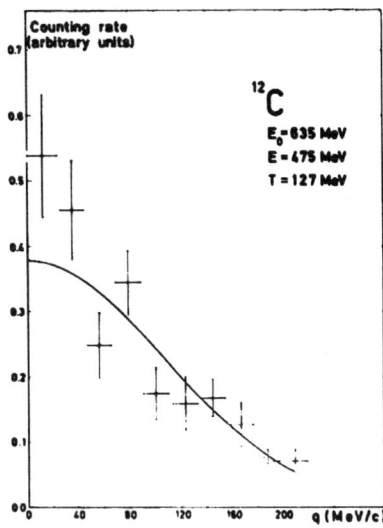

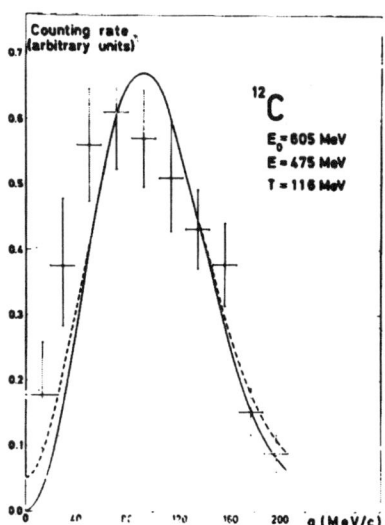

FIGURE 6. Initial momentum (q) distributions determined by the experiments at Frascati [5]. The lower figure corresponds to the p-shell peak at the lowest binding energy. The upper figure corresponds to the broad bump at a binding energy of about 35 MeV, as in figure 4 and identified as the s-shell.

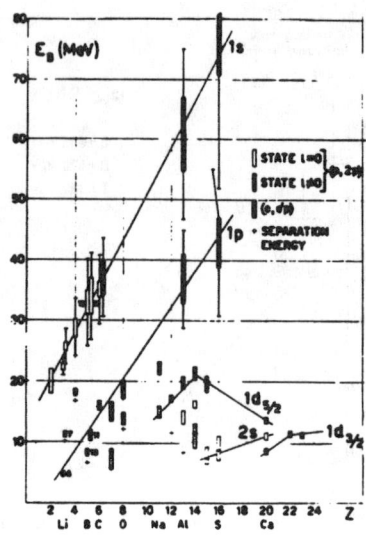

FIGURE 7. Locations of the binding energies of the different orbitals as determined by the $(p,p'p)$ and $(e,e'p)$ experiments. [5]

hall and we succeeded in proving the feasibility of the work. Advised by the PAC to select one nucleus, we chose carbon and found some astounding results. In figure 10 I show you the results from Paul Ulmer's Ph.D. thesis [13] at 400 MeV/c of three-momentum transfer. The transverse response function for the 1s shell in $(e,e'p)$ is about 40% larger than the longitudinal and this is reflected in the difference of the spectral functions. However, for the p-shell there is no difference. At this symposium you heard D. Geesaman report on results from Jefferson Lab verifying our results at one GeV/c of momentum transfer. We have no explanation of this result in meson exchange currents or in nucleon-nucleon correlations as calculated by Ryckebush [14].

This mystery deepens as we examine our recent results from Mainz for $(e,e'p)$ with the helium isotopes [15]. For both isotopes the transverse and the longitudinal spectral responses for the two-body and many-body reactions are both equal up to initial momenta of 160 MeV/c. This is similar to the p-shell in carbon but not like the s-shell. Yet the 4He(e,e') reaction, as we explained above, has a 40% greater transverse response than longitudinal response, like carbon and in contrast to the 3He(e,e') reaction. Some mysterious currents are at work that we do not understand, and they represent a large part of the (e,e') process.

Finally, I turn to our recent work at Jefferson Lab on the $(e,e'p)$ reaction with oxygen at 1 GeV/c of momentum transfer. We chose q and w to be on the quasi-free peak. In figures 11 and 12 [16] you see respectively the cross sections for the p-1/2 and p-3/2 shells and the Alt asymmetry (the transverse/longitudinal interference

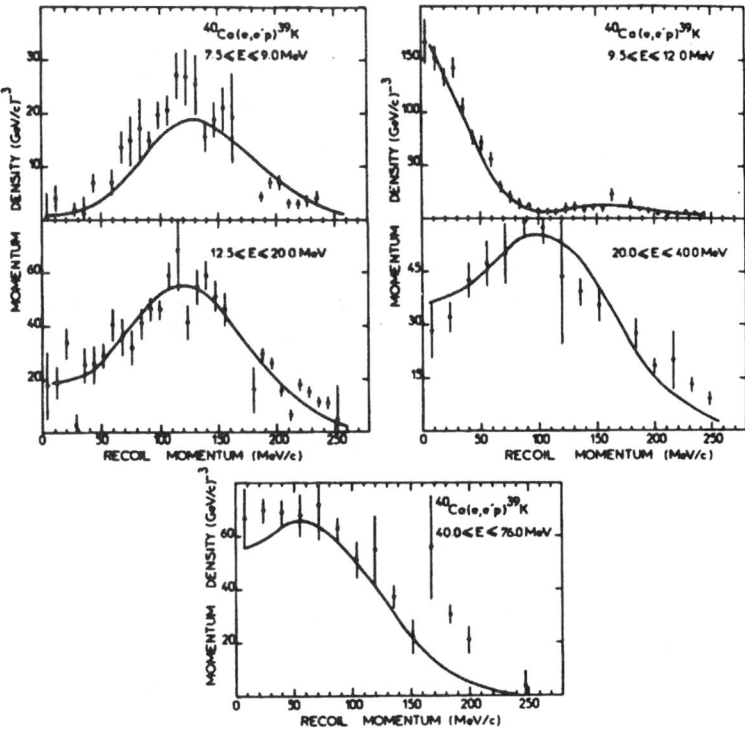

FIGURE 8. Distributions of recoil momenta for different regions of missing energy (binding energy) [7]. Note that the term 'recoil momentum' in this figure is some times called the 'initial momentum' and also the 'missing momentum' by some authors. Reprinted from Ref. [7], Copyright 1976, with permission from Elsevier Science.

asymmetry) as a function of missing momentum. The cross sections range over five orders of magnitude. They are compared with the ab-initio calculations in the Dirac formalism of Udias and collaborators [17] using reasonable mean field optical potentials from proton scattering. The agreements are much better than we see using non-Dirac approaches as typified by Kelly [18] using relativistic corrections in a Schroedinger formalism. Notice the striking agreement with the Udias calculations for the Alt parameter. The results show us for the first time the need to dynamically include the small component spinors both in the initial state and final state with proper deformation.

I shall summarize the results for the s-shell in our oxygen work at Jefferson Lab [19] by saying that our results from carbon at Bates were verified by a similar behavior in the transverse/longitudinal comparisons. In addition, as we look at deeper missing energies and at higher missing momentum, we observe very large cross sections that are not explained by correlations or standard meson exchange currents. This latter feature is illustrated by a comparison in figure 13 [20] of data

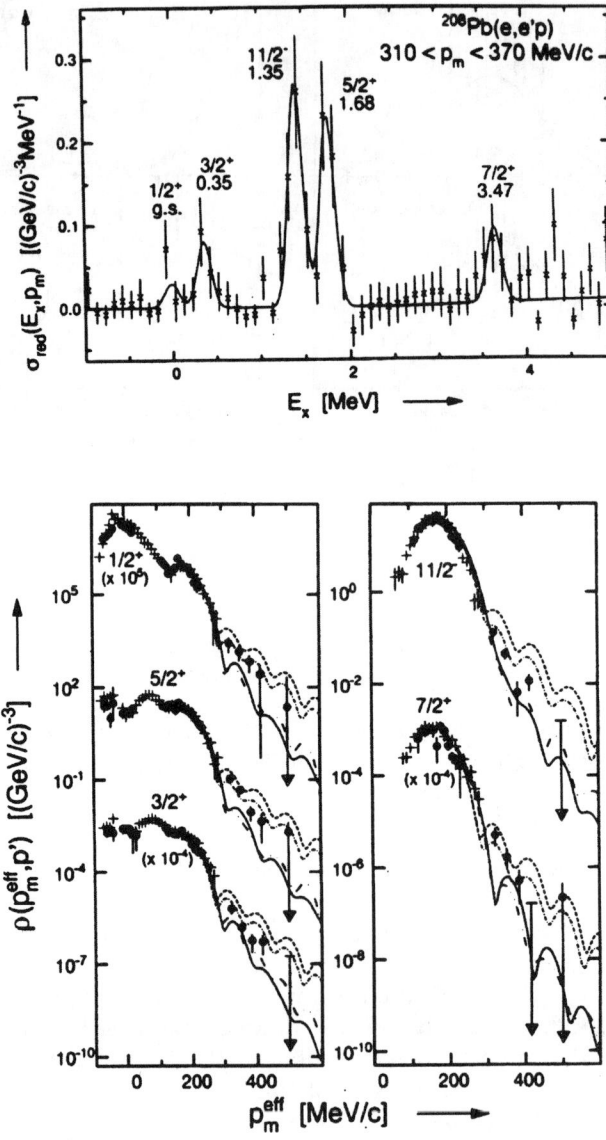

FIGURE 9. Distributions of excitation energy, Ex, (also called missing energy) and distributions of effective missing momentum, Pm, (also called recoil momentum or initial momentum) from the Pb$(e,e'p)$ reaction studies at NIKHEF [8]. The calculations are from: +++ [9]; -..-.. [10]; ---- [11]; -.-.-. [12].

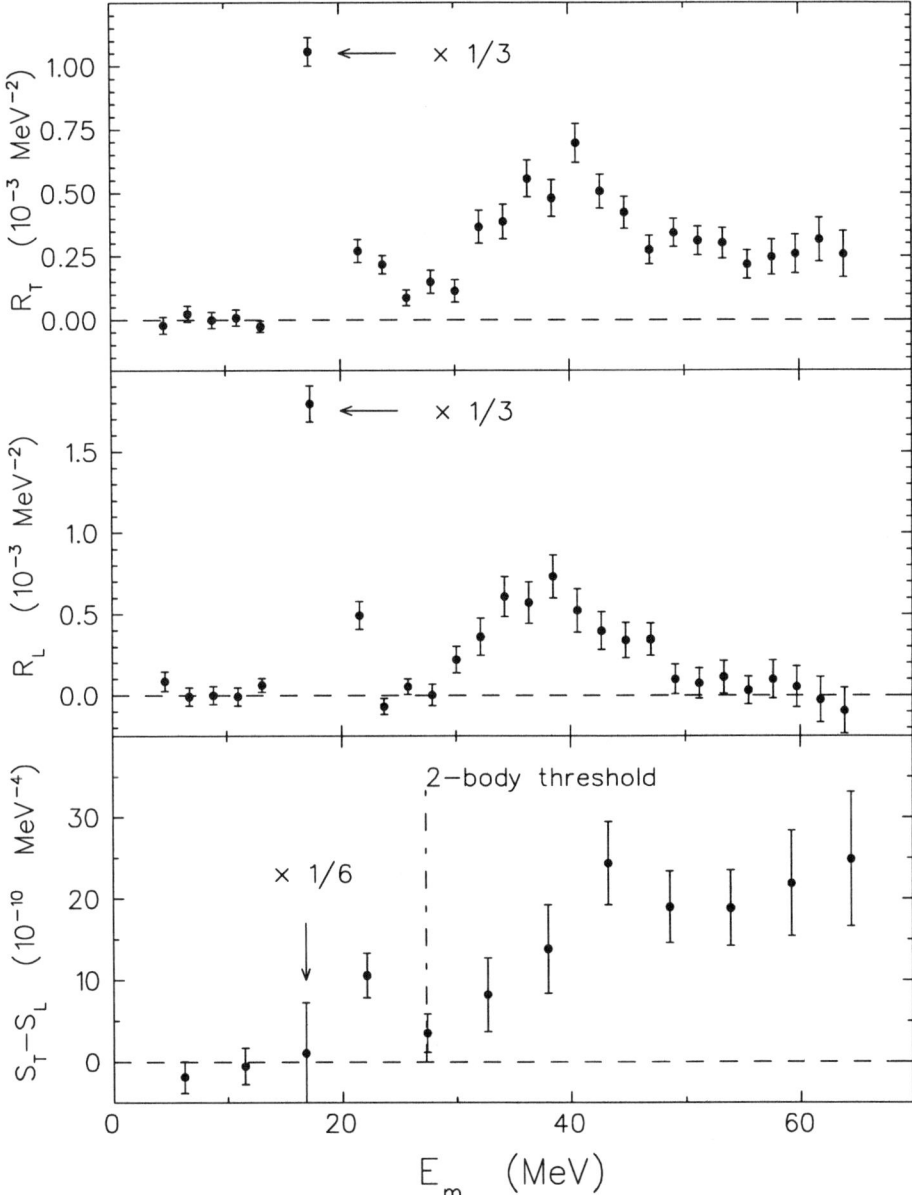

FIGURE 10. The response functions Rt and Rl are shown along with the difference of the corresponding spectral functions, St-Sl. The data are from [13] reporting results from the 12C$(e, e'p)$ reaction studies at MIT-Bates.

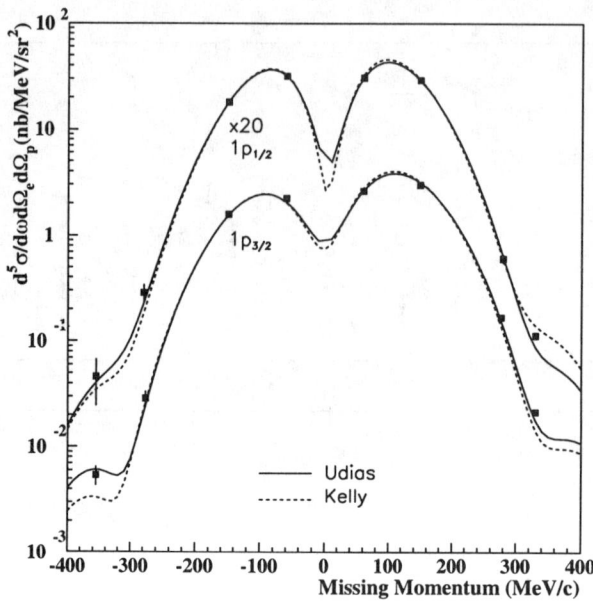

FIGURE 11. Cross sections for the $16O(e, e'p)$ reaction studied at Jefferson Lab [16].

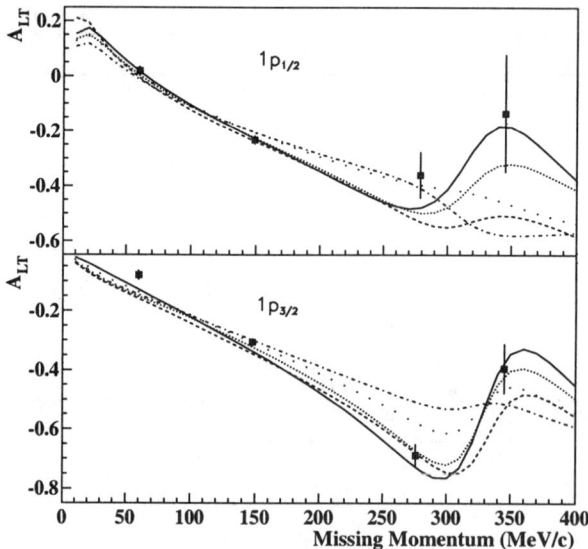

FIGURE 12. The Asymmetries, Alt, measured in the $16O(e, e'p)$ reaction at Jefferson Lab. The solid curve is from Udias [17] and the other curves from Kelly [18] and based on different assumptions about the relativistic corrections applied to the the Schroedinger formalism.

from carbon at Bates at momentum transfer of 1 GeV and at $w = 475$ MeV, the

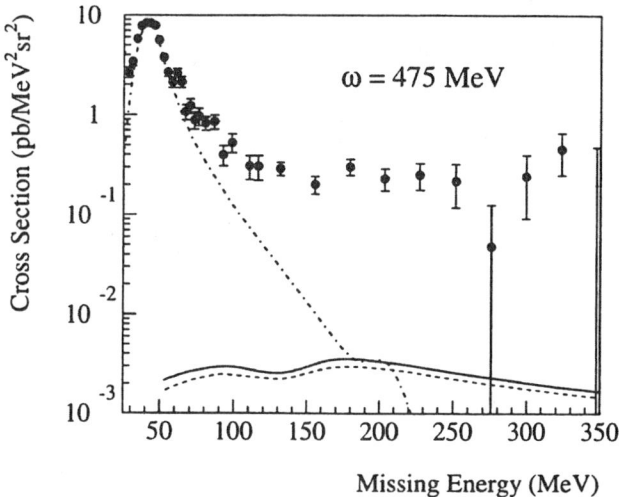

FIGURE 13. Results from 12C$(e, e'p)$ at 1 GeV/c momentum transfer [20]. The curves are from Ryckebush [14] assuming: a standard mean field wave function, -.-.-.; the other curves show the contribution from Jastrow-like correlations with and without meson exchange currents.

quasi-free peak, with calculations of Ryckebush [13]. Notice the disparity of more than two orders of magnitude. The same is true with our recent data on oxygen at the Jefferson Lab.

IV TODAY AND THE FUTURE AT BATES

During the past decade we have pioneered two new areas at Bates. The first is with the measurement of the proton polarization in $(e, e'p)$ studies. This makes use of carbon scattering in the focal plane of the proton spectrometer (FPP) as a polarization analyzer. The second is the development of the out-of-plane-spectrometer system (OOPS). In addition to the advent of a polarized beam, each of these represent a new departure from the standard electron scattering of the previous generations wherein the cross sections were separated into at most transverse and longitudinal components. With OOPS one can measure five physical observables with the polarized beam: the standard transverse and longitudinal components, the interference components Rlt and Rtt, and the helicity dependent interference part Rlt'. With the focal plane polarimeter (FPP) we add the three components associated with the three directions of polarization. If the proton spectrometer were able to measure out of plane with the FPP and also involve the electron beam polarization, we could measure a total of 18 physical observables, 16 more than

than the standard situation a decade ago. This is a great deal more information about the currents in nuclei and in the nucleon in the $(e,e'p)$ process.

Let me show you some examples. In figure 14 we show you the normal component of the polarization in the $p(e,e'p)pi0$ reaction [21]. With this one result we es-

$H(\vec{e},e'p)\pi^0$ final state polarization

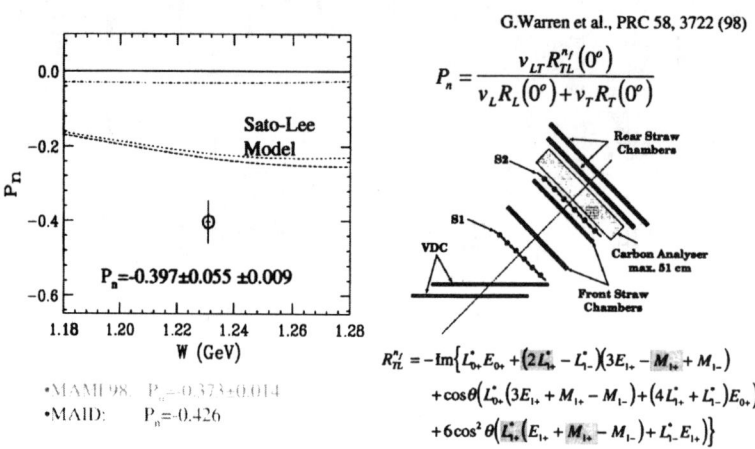

FIGURE 14. This figure diagrams the focal plane polarimeter used at MIT-Bates to study the delta excitation and the results of that experiment for the normal component of proton polarization, Pn [21].

tablished that the truncation assumptions previously used to derive the Coulombic contributions to this reaction on the delta resonance are incorrect. If no background existed at the resonance this polarization would be zero. Notice the poor comparison one particular model achieves with this data [21]. Focal plane polarimetry has recently become standard at Jefferson Lab and at Mainz.

With this same physics, delta excitation leading to neutral pion decay, we have also used OOPS to derive the longitudinal/transverse interference asymmetry, Alt. For this observable one measures the proton yield to the left and right of q and the asymmetry is the difference over the sum of the cross sections. In figure 15 we show you our measurement compared to theoretical models [22]. The MAID 2000 model was adjusted to earlier data at smaller opening angles yet you see the difficulty it has in predicting our most recent result. This disagreement is not understood as yet.

Turning to experiments on the deuteron, $D(e,e'p)$, I focus on the transverse-transverse asymmetry. This requires the OOPS system to be out of plane. You can see in figure 16 the comparisons of data [23] to different model assumptions by Arenhovel [24] and by Tjon [25] dealing with the sensitivity of these results to

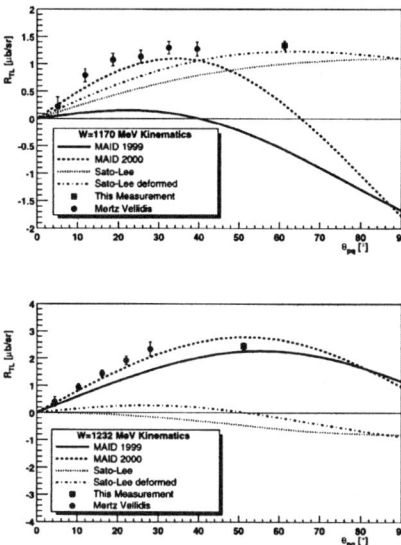

FIGURE 15. The Rlt response function measured at two different excitation energies in the delta region. The point referred to in the text is the one at the highest angle between p and q. For more details and references to the calculations please see [22].

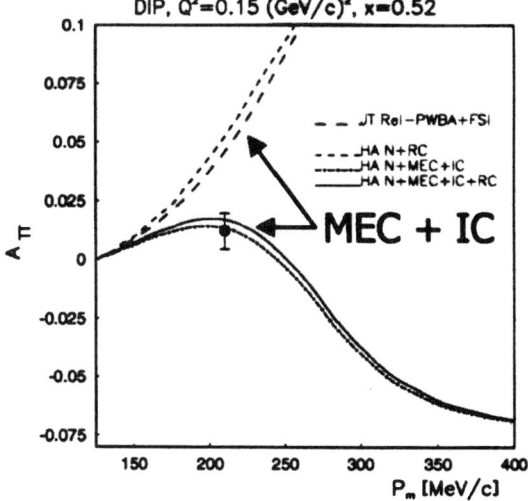

FIGURE 16. The asymmetry Att as measured in the D(e,e'p) reaction with the OOPS system at MIT-Bates [23]. The curves labeled AH are from [24] and illustrate the influence of two-body currents. The curve labeled JT is from [25] and does not include two-body currents.

meson exchange currents and to isobar currents. Our first result is already definitive and we plan for much more data at different values of (q, w). The sensitivity to the delta-nucleon interaction is a new feature that these experiments can provide.

V SUMMARY

We have discussed briefly the evolution of the field of electron scattering. This field was once considered peripheral to the strong interaction many body problem. The field is now rich in physical observables and it is providing, perhaps, the defining data in studies of strongly interacting systems of nucleons. The advent of good resolution and 100% duty accelerators has been of central importance. The use of polarization observables and interference observables has added immeasurably to the richness of the data and the physics. I make the following observations:

- The electromagnetic probe is one of the preferred ways to study the strongly interacting system. It is specific, detailed and relatively unambiguous.

- Many new responses are being measured and these add considerably to the challenge presented to fundamental nuclear theory.

- Polarization and out of plane responses are powerful new tools.

- We have observed unexplained currents in nuclear systems that may prove to require new theoretical ideas.

- Bates, Jefferson Lab and Mainz form a complimentary complex of facilities entering a very productive era of new and exciting measurement and discovery.

REFERENCES

1. M. Riou, Rev. Mod. Phys. 37 (1965)
2. T. Berggren and B. Tyren, Ann. Rev. Nucl. Sc. 16,(1966)153
3. G. Jacob and T. A. J. Maris, Rev. Mod. Phys. 38,(1966)121
4. B. Gottschalk, Medium Energy Nuclear Physics With Electron Linear Accelerators; MIT 1967 Summer Study, USAEC TID-24667, Edited by W. Bertozzi and S. Kowalski, pgs.443 and 455
5. Georgio Cortelessa, Medium Energy Nuclear Physics With Electron Linear Accelerators; MIT 1967 Summer Study, USAEC TID-24667, Edited by W. Bertozzi and S. Kowalski, pgs, 474 and 478
6. S. Frulani and J. Mougey, Adv. Nucl. Phys., 14(1984)1
7. J. Mougey et al., Nucl. Phys. A262(1976)461
8. I. Boreldjk et. al., Phys. Rev. Lett. 73(1990)167
9. E. N. M. Quint, Ph.D. Thesis, University of Amsterdam, 1988
10. V. R. Pandharipande and S. C. Piper, Nucl. Phys. A507(1990)167
11. Z. R. Ma ans J. Wambach, Phys. Lett. B256(1991)1

12. C. Mahaux and R. Sartor, Adv. Nucl. Phys. 20(1991)1
13. P. Ulmer et al., Phys. Rev. Lett. 59, (1987)2259
14. J. Ryckebush, University if Ghent, Belgium, private communication
15. R. Florizone, Ph.D. Thesis, MIT, December, 1998: and R. Florizone et al., Phys. Rev. Lett., 83,No.12(1999)2308
16. J. Gao, Ph.D. Thesis, MIT, February, 1999
17. J. M. Udias, University of Madrid, Spain, private communication
18. J. Kelly, University of Maryland, College Park, MD, private communication
19. N. Liyanage, Ph.D. Thesis, MIT, January, 1999
20. J. H. Morrison et al., Phys. Rev. C59,No.1(1999)221
21. G. Warren et al., Phys. Rev. C58(1998)3722 9
22. C. Kunz, Ph. D. Thesis, MIT, January, 2000
23. J. Chen, Ph.D. Thesis, MIT, January 1999
24. H. Arenhovel, University of Mainz, Mainz, Germany, private communication and; F. Ritz, H. Goller, Th. Wilbois and H. Arenhovel, Phys. Rev. C55, (1997)2214
25. T. A. Tjon, Private communication and; E. Hummel and T. A. Tjon, Phys. Rev. C42, (1990)423; C49, (1994)21

Coincidence Electron Scattering II
Short Range Correlations in Nuclei

Thomas Walcher

Mainz Microtron MAMI
Institut für Kernphysik
Johannes-Gutenberg Universität Mainz, Germany

Abstract. Short range correlations in nuclei are interesting because they are necessary to explain the stability of nuclei against collapse. From this follows that they are related to the internal structure of nucleons which is still poorly understood at low energies. Three experiments relevant to this context are presented. In an exploratory experiment of the $^{12}C(e,e'pp)$ reaction with a 4π-BGO ball and a high resolution magnetic spectrometer of the three spectrometer setup at the Mainz Microtron MAMI the optimal kinematical region for a study of short range correlations is found. In an selective experiment of the $^{16}O(e,e'pp)^{14}C$ reaction using all three spectrometers models of the reaction mechanism and effects of a hard and soft core of the NN potential are studied. In the $^{12}C(e,e'p\pi)^{11}C$ reaction again using all three spectrometers evidence for narrow bound Δ states in ^{12}C is derived which may be related to the very short mean free path of the Δ-decay πs in nuclei.

I INTRODUCTION

The necessity to assume a short range repulsive interaction between nucleons in nuclei was recognized in the early days of nuclear physics from the saturation of nuclear binding energies. A nucleus without short range repulsive nucleon-nucleon interaction or "hard core" would collapse to the size of the range of the attractive interaction [1]. This hard core is, of course, known from the study of the scattering of free nucleons. For nucleons bound in nuclei it contributes to short range correlations which manifest themselves in high relative momenta. These high momenta are off shell and extend to the region of the Δ resonance, the role of which is still not very well understood. The intricate interplay of the nucleon-nucleon attractive potential, the exchange potential and the repulsive core in the many body system nucleus require an in situ study of short range correlations.

However, the interest of such a study goes beyond the question of the stability and binding of nuclei. It is widely believed and also experimentally supported that the color charge of strong interactions is felt out to a nucleon radius of about 0.8 fm [2]. This radius should be compared to the average distance of two nucleons in nuclei of about $1.8\ fm$. Even if the main attraction at a range of $R_{2\pi} = \lambda_{compton} = \hbar c/2m_\pi \approx 0.7\ fm$ is dominated by the π exchange due to chiral dynamics, the repulsive core should be related to the quark-gluon structure of the nucleon. Ideas how this could happen have been worked out by Faessler et al. [3]. On the other hand the empirical models, describing the nucleon-nucleon scattering best, assume a ρ, ω-meson exchange to produce the hard core. It is, therefore, of great interest to study the nucleon-nucleon short range correlations, the meson-nucleon correlations, and also the baryon-nucleon correlations in nuclei, where the baryon stands for a hyperon or a nucleon resonance. Such studies cover a different kinematical range than the free particle scattering and allow also to vary the baryon, i.e. also quark, quantum numbers. This may lead to new insights into the hadron structure at low energies. But they are certainly interesting as studies of the nucleus as a multi body system determined by the two scales of the attraction and the repulsive core.

The situation sketched so far has been realized since many decades, but it was always difficult to access a direct signature of short range correlations. An indirect method has been the observation of the depletion of the occupancy of shells in nuclei [4]. There exists, however, an old proposal by Gottfried to use two nucleon knockout reactions [5]. The best suited reaction of this class appears to be the $(e, e'pp)$ reaction. In first nonrelativistic approximations, meson exchange currents do not contribute in the pp channel. Furthermore, the important $pp \to \Delta^+ p \to pp$ contribution is suppressed if the two protons are in a relative ^1S_o state [6,7]. Since the two protons represent an isovector state, also tensor correlations are absent if the relative NN angular momentum l is even (odd), when the angular momentum L of the pair with respect to the residual nucleus is even (odd).

A rather different kind of short range correlations than the repulsive core comes into focus if one considers the Δ resonance in nuclei. The mean free path of a pion emitted with the momentum corresponding the decay with the resonance energy is $\langle \lambda \rangle = \lambda_{absorption} = 0.4\ fm$. It is, therefore, interesting to also consider the excitation of Δs in nuclei again in the described context. The $(e, e'\Delta) \to (e, e'p\pi^-)$ reaction provides a well controlled kinematics with good energy resolution. The desire for good energy resolution may surprise, given the half width full maximum of $\Gamma_\Delta = 110\ MeV$ but it allows to clearly sort out the final state of the residual nucleus. This in turn permits to select quantum numbers and energy and momentum transfer for the propagation of the Δ in the nucleus.

This paper presents three experiments motivated by the still puzzling compositeness of nuclei reminded to above:

1. the reaction $^{12}C(e, e'pp)X$ using a magnetic high resolution spectrometer for the electron e' and a close to 4π-BGO ball for the two protons pp,

2. the reaction $^{16}O(e, e'pp)^{14}C$ using three high resolution, large solid angle spectrometers for e', and pp,

3. the reaction $^{12}C(e, e'p\pi^-)^{11}C$ using the same three high resolution spectrometers as for 2.

2. AN EXPLORATORY EXPERIMENT OF $^{12}C(E, E'PP)X$

2.1 Experiment

This experiment [8–10] was the first performed with one of the high resolution spectrometers of the three spectrometer setup [11] at the Mainz Microtron MAMI. It used the Spectrometer A in coincidence with a BGO-crystal ball originally built for the Los Alamos Meson Physics Facility LAMPF [12]. These spectrometers represent the result of a long evolution toward large solid angle, broad momentum acceptance with good momentum resolution. This evolution has been decisively driven by developments at MIT [13] and Bates [14]. With a solid angle of $\Delta\Omega = 28\ msr$, a momentum acceptance of $\Delta p/p = 20\ \%$ and resolutions of $\delta p/p \leq 10^{-4}$ and $\delta\theta = \delta\phi \leq 1.5\ mrad$ for 700 MeV electrons for spectrometer A, a limit of this kind of instruments may have been reached.

The protons were detected by the close to 4π-BGO ball having an inner radius of 6 cm in whose center the ^{12}C target was placed. The energy resolution of $\delta E/E \approx 5\ \%$ for protons with a maximal energy $E_p^{max} \approx 185\ MeV$ was sufficient for the aim of the experiment to get a global overview. The very good time resolution of the BGO crystals of $\delta t_{BGO} \sim 1\ ns$ matched the excellent timing resolution of the spectrometers of $\delta t_A \sim 0.5\ ns$.

The decisive new quality, however, making such an experiment for the first time feasible at an electron accelerator was the excellent beam quality of the Mainz Microtron MAMI. Due to the low accelerating gradient of the normal conducting linear accelerator sections, this accelerator delivers a beam with a duty factor 1 and an emittance of $\epsilon_x \leq 12\ \pi\ nm\ (1\sigma)$ and $\epsilon_y \leq 0.7\ \pi\ nm\ (1\sigma)$. The many recirculations in the microtron clean the beam effectively so that the halo is about 10^{-5} of the intensity outside a radius of 1 mm. The residual background stemming from the Møller electrons was suppressed by a mini orange consisting out off six wedge-shaped NdFeB permanent magnets by a factor of 20 to 140 kHz per BGO crystal. In this way a current of up to 300 nA could be used resulting in a luminosity of $\approx 10^{33}$ nucleons $cm^{-2}s^{-1}$.

2.2 Results and Discussion

The measurements have been performed at an electron energy $E_e = 705\ MeV$ and angle $\theta_{e'} = 34.4^0$ in the "dip" region between the quasi-free nucleon knockout and the quasi-free Δ production at an average energy transfer $\langle\omega\rangle = 225\ MeV$, momentum transfer $\langle q\rangle = 412\ MeV/c$ and a virtual photon polarization $\langle\epsilon\rangle = 0.8$. In analogy to the one nucleon knockout reaction, one introduces the missing energy $E_m = \omega - T_1 - T_2 - T_X$ with T_1 and T_2 the kinetic energies of the two protons and T_X the recoil energy of the residual system. The missing momentum is then $\vec{P}_m = \vec{q} - \vec{p}_1 - \vec{p}_2$ with $\vec{q}$ the three-momentum transfer and $\vec{p}_1$ and $\vec{p}_2$ the momenta of

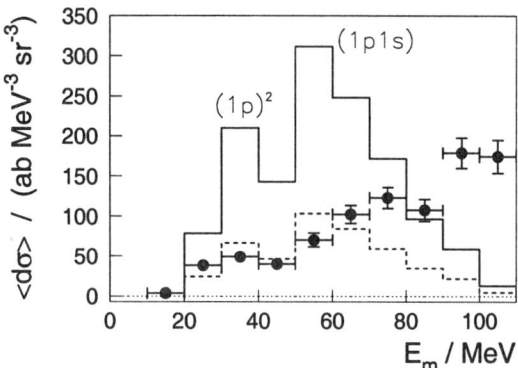

FIGURE 1. Nine-fold differential cross section for the $^{12}C(e,e'pp)X$ reaction as function of missing energy E_m. The units of the cross section are attobarns, $1\,ab = 10^{-46}\,m^2$. The horizontal bars show the integration interval for E_m, the vertical bars the statistical error. The solid line shows the calculation for a 0.6 fm hard core and the dashed line the calculation for a soft core with the Variational Monte Carlo method.

the two emitted protons. Figure 1 shows the ninefold differential laboratory-cross section, averaged over the acceptances of the electron spectrometer and the BGO ball.

The histograms in figure 1 show calculations using a plane wave approximation for the protons in the final state [15,16]. The virtual photon couples to one of the protons knocking out the second one due to the nucleon-nucleon correlations. Additionally the knockout through the excitation of a Δ resonance is included. As seen from figure 1 the assumption of a nucleon-nucleon correlation with a "hard core" of 0.6 fm radius [17] is clearly ruled out. On the other hand a "soft-core" correlation obtained with the Argonne V_{14} potential, using the cluster Variational Monte Carlo method (VMC) [18] is in reasonable agreement with the data up to $E_m \approx 70\,MeV$. The configurations with 2 holes in the $1p_{3/2}$ shell, $(1p)^2$, and 1 hole in the $1p_{3/2}$ and one in the $1s_{1/2}$ shells, $(1p1s)^1$ are weakly indicated. Above 70 MeV the $(1s_{1/2})^2$ configuration could contribute. However, the calculations show that this contribution is small. It is reasonable to assign the large cross sections at this high missing energy to higher order processes as e.g. three-nucleon emission.

Intuitively the highest sensitivity to short range correlations should be found in the relative momentum of the two nucleons $\vec{p}_{rel} = \vec{p}_1^{\,lab} - \vec{p}_2^{\,lab}$ since high $|\vec{p}_{rel}|$ should be equivalent to small $\vec{r}_{rel}$. Figure 2 shows the average cross section as a function of this variable. Again the 0.6 fm "hard core" is excluded. As "soft core" correlations two models are shown: the already introduced VMC method and a G-matrix calculation (GM) [19]. Both hardly differ and reproduce the data reasonably well. The significance of this variable becomes even more questionable if one switches off the correlations meaning that the two proton knockout happens

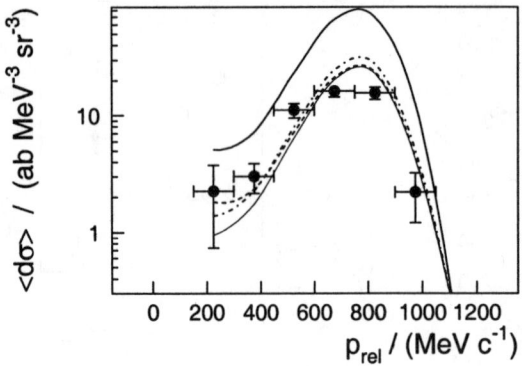

FIGURE 2. The averaged differential cross section as a function of the relative proton-proton momentum in the $^{12}C(e,e'pp)$ reaction for missing energies $E_M < 70\ MeV$. The curves show the results of the following calculations: 0.6 fm hard core (solid line); soft core G-matrix (dashed dotted line); soft core Variational Monte Carlo method (dashed line); no correlations, Δ contribution only (thin solid line)

via the virtual Δ excitation and the $\Delta^+ p \to pp$ decay only.

However, this dominating Δ contribution can be suppressed by choosing the right angular correlation between the two protons. The M1 excitation of the $\Delta^+ (I = 3/2, J = 3/2^+)$ produces preferentially a $\Delta^+ p$ pair in a relative S state. Therefore, the decay $\Delta^+ p \to pp$ will be suppressed due to angular momentum and parity conservation at the correlation angle $\gamma = 180^0$.

This expectation is indeed seen in figure 3. The Δ contribution alone cannot describe the data. But also two other soft core models, the Variational Monte Carlo (VMC) and the Fermi Hyper Netted Chain (FHNC) [20] correlation functions, fail to reproduce the data at large correlation angles. Only the GM model is in good accord with the experimental results. It is interesting to note that the GM correlation function is not as soft as the VMC and FHNC ones. The GM correlation function is zero at very small relative nucleon distances, whereas the other two still have a finite probability to completely overlap at zero relative distance.

In summary this exploratory experiment has shown two points:

1. It is possible to distinguish different model ideas about short range correlations in the $(e,e'pp)$ reaction.

2. The kinematics, however, has to be restricted to large correlation angles in order to suppress the Δ contribution and gain sensitivity to the different correlation functions.

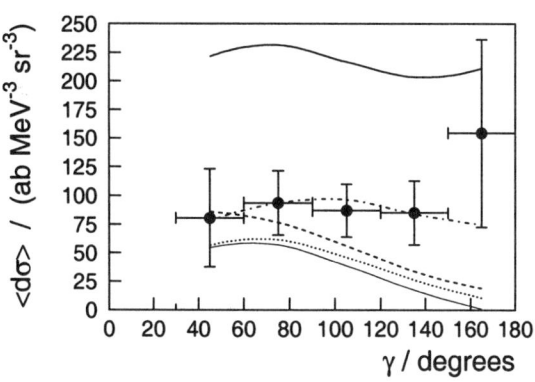

FIGURE 3. The average differential cross section as a function of the correlation angle γ between the two protons in the $^{12}C(e, e'pp)X$ reaction for missing energies $E_M < 70$ MeV. The emission direction of one of the protons was in direction of the electron momentum transfer. The lines of the curves for the calculations are as in figure 2 with additionally the "soft core" of the Fermi Hyper Netted Chain model (FHNC) (dotted line).

3. A SELECTIVE STUDY OF $^{16}O(E, E'PP)^{14}C$

This insight was used to investigate the $^{16}O(e, e'pp)^{14}C$ with the three spectrometer setup of the A1 collaboration in super parallel kinematics [21,10]. This setup is not only well suited for this kinematics but allows also to resolve final states in ^{14}C and this way further narrow down the reaction mechanisms.

3.1 Experiment

The study of the $^{16}O(e, e'pp)^{14}C$ reaction took full advantage of the three high resolution spectrometers. The excellent beam quality (little halo, duty factor 1) and imaging properties of the spectrometers allowed to use currents of up to 80 μA. This beam was swept periodically across a 54.2 mg/cm^2 waterfall target allowing luminosities of up to $3 \cdot 10^{37}$ nucleons $cm^{-2}s^{-1}$. The large solid angles and momentum acceptances (spectrometer A: 28 msr, 20%; B: 5.6 msr, 15%; C: 28 msr, 25%) made the observation of triple coincidences between the three magnetic spectrometers possible. For the missing energy spectrum in figure 4 the spectrometers were set at laboratory scattering angles $\theta_{e'}^B = 8^0, \theta_p^A = 38.8^0$, and $\theta_p^C = -141.2^0$. This means that $\theta_p^A - \theta_p^C = 180^0$ and $\theta_p^A = \theta_q$ define the superparallel kinematics found favorable to suppress the Δ contribution. The central four-momentum transfer was $\langle Q \rangle^2 = \langle (\omega, \vec{q}) \rangle^2 = (215\ MeV/c, 316 MeV/c)^2 = 0.054\ (GeV/c)^2$ and the average virtual photon polarization $\langle \epsilon \rangle = 0.91$. The four kinematical settings with central missing momenta at $P_m = 0, 125, 225$, and 340 MeV/c were chosen so that the rel-

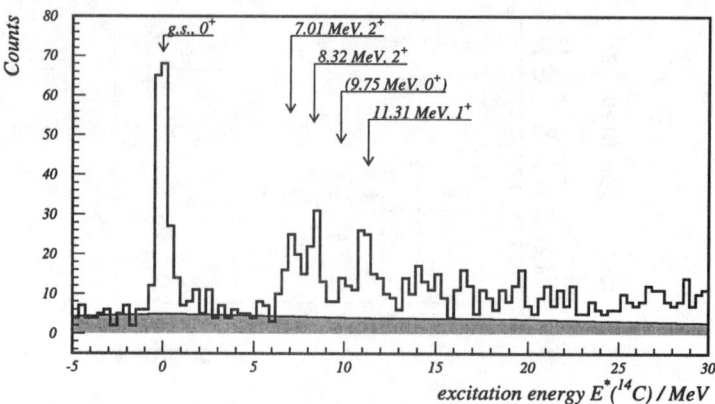

FIGURE 4. An excitation energy spectrum of the $^{16}O(e,e'pp)^{14}C$ reaction at $E_e = 855\ MeV$, $\omega = 215\ MeV$, $q = 316\ MeV/c$, and a missing momentum interval $-100\ MeV/c < P_m < 100\ MeV/c$. The gray band shows the background of accidental coincidence events.

ative proton-proton momenta changed in the range from $p_{rel} = 200$ to $600\ MeV/c$. The coincidence time resolution between any pair of the three spectrometers was about $1\ ns$ so that the true-to-random coincidence ratio was, even at the highest luminosities, always better than $2:1$.

3.2 Results and Discussion

Figure 4 shows an excitation energy spectrum constructed from the missing energy according to $E^* = E_m - T_{^{14}C} - \varepsilon_1 - \varepsilon_2$, where $T_{^{14}C} = \frac{P_m^2}{2M_{^{14}C}}$ is the kinetic recoil energy of the ^{14}C nucleus and $\varepsilon_1, \varepsilon_2$ are the separation energies of the two protons from the $1p_{1/2}$ shell. A clean excitation spectrum is visible which offers the determination of missing momentum distributions for definite final states. The selection of final states opens the possibility to distinguish different quantum numbers of the pp pair and consequently a deeper insight into the nucleon-nucleon correlations. It is important to realize at this point that the correlations are a property of the initial state, whereas the different excited final states are part of the reaction model. The resolution of the final states is, therefore, mandatory in order to control the reaction model and separate it from effects of the short range correlations.

The assignment of the peaks seen in figure 4 is derived from nucleon transfer reactions [22] following ref. [10]. The states with dominating two hole structure are the ground state (main configuration $(1p_{1/2})^{-2}$, the two 2^+ states at $E^* = 7\ MeV$ and $8.3\ MeV((1p_{1/2},1p_{3/2})^{-1})$ and a 1^+ state at $11.3\ MeV$ $((1p_{1/2},1p_{3/2})^{-1})$ [23]. The experimentally observed splitting of the 2^+ state is probably due to core polarization. For the comparison with theoretical calculations both states have been summed. The $(1p_{3/2})^{-2}$ configuration can produce $0^+, 1^+, 2^+$ states. These states could contribute to the peaks observed at 9.75, 11.31 and around 15 MeV in figure 4 (see also ref. [23]). All these states are only weakly excited in superparallel kinematics. It is surprising that in the NIKHEF experiment [7] states above 5 MeV are so strongly seen. This could be due to the smaller correlation angle in that experiment and the emphasis on higher proton partial waves following from this.

The knocked out protons can be in relative $^1S_o, ^3S_1, ^1P_1, ^3P_{0,1,2}$, etc. states depending on the quantum numbers of the final state and the angular momentum L of the pair with respect to the final nucleus. The theoretical calculations [23,24] predict that the ground state of ^{14}C is predominantly due to a $^1S_0$ pair with $L = 0$. The first 2^+ state around 7.7 MeV should also decay with a $^1S_0$ pair but with $L = 2$. The 1^+ state at 11.3 MeV shall be mainly reached with $^3P_{0,1,2}, L = 1$ pairs. These considerations are valid for the lowest partial waves. If one looks at small P_m favoring the low angular momenta L, one can select the corresponding pair configurations. For $L = 0$ one expects $^1S_0$ pairs at $\langle P_m \rangle = \sqrt{L(L+1)} \cdot \hbar/R^p_{12C} = 0$, for $L = 1$ ^3P pairs at $\langle P_m \rangle \approx 150$ MeV/c, and for $L = 2$ $^1S_0$ pairs at $\langle P_m \rangle \approx 250$ MeV/c.

The short range correlations will be most clearly visible for $^1S_0$ pairs since here the radial overlap of the two protons is largest and the Δ^+-proton correlation does not contribute. In contrast the ^3P pairs will be dominated by the Δ^+-proton correlation and thus offer a welcome test of the correct modeling of the Δ^+ contribution.

Figures 5, 6, and 7 show the measured missing momentum distribution for the three states considered. They are compared to two model calculations:

A. Pavia model [24]
This model is based on the description of the reaction dynamics of ref. [25]. The short range correlations are included from the wave functions derived from a G-matrix calculation using the Bonn A nucleon–nucleon potential [23]. The contribution of the Δ was included in a dynamical manner following ref. [26]. The distortions of the outgoing proton waves were calculated using the optical model parameters of ref. [27].

B. Gent Model [28]
The reaction model is described in ref. [28]. The short range correlations were implemented via the G-matrix correlation functions of ref. [19]. The nuclear structure input was, as for model A, taken from ref. [23]. The Δ contribution considers nuclear medium effects. The final state distortions are calculated as a partial wave expansion in terms of 2 particle-2 hole states.

The overall comparison of experiment and theory shows that model A seems to describe the short range correlations dominating at $P_m \approx 0$ MeV/c for the ground state of ^{14}C in figure 5 and at $P_m \approx 250$ MeV/c for the two 2^+ states around 7.7 MeV in figure 7 reasonably well. Whereas model B seems to have the better description of the Δ contribution as visible from the rough agreement at

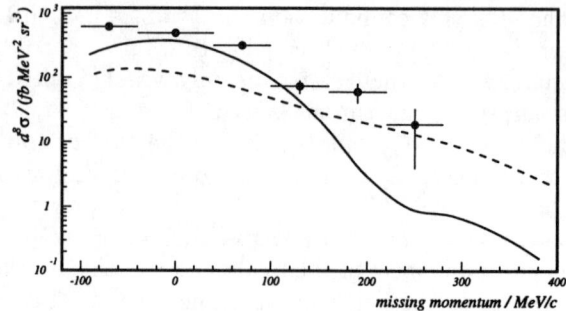

FIGURE 5. The differential cross section of the $^{16}O(e,e'pp)^{14}C$ reaction as a function of the missing momentum for the 0^+ ground state of ^{14}C. The curves indicate the results of the Pavia model A (solid) and the Gent model B (dashed).

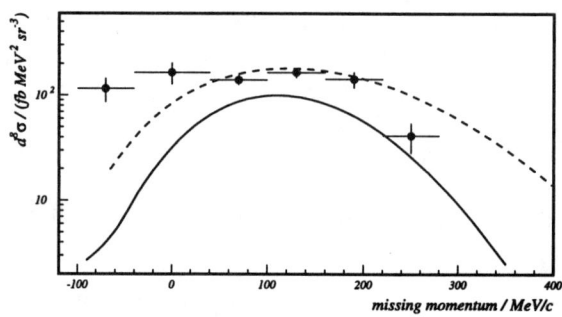

FIGURE 6. The same as figure 5 for the 1^+ state at 11.3 MeV in ^{14}C.

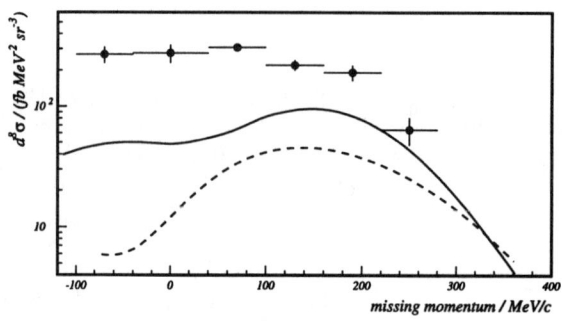

FIGURE 7. The same as figure 5 for the two 2^+ states at 7.7 MeV in ^{14}C.

$P_m \approx 150\ MeV/c$ for the 1^+ state at 11 MeV. This observation is also consistent with the reasonable agreement of model B with the experiment for the ground state in figure 5 for $P_m > 150\ MeV/c$ where the Δ^+ contributes relatively stronger [24]. However the dramatic underestimation of the experimental cross section for small P_m for the two 2^+ states of both models is a puzzle.

4. SEARCH FOR NARROW BOUND $\Delta(1232)$ STATES IN $^{12}C(E, E'P\pi^-)^{11}C_{GROUND\ STATE}$

4.1 Motivation

As mentioned in the introduction, a very different kind of short range correlations appears if one considers the π propagation in the region of the $\Delta(1232)$ resonance in nuclei. This correlation is not directly related to the small distance structure of the nucleon, but rather to the question of the nature a resonance, i.e. a decaying state of a composite system, in nuclei. An analogy may be an excited atom trapped in a hohlraum resonator interacting with an electromagnetic field. For such a system the spontaneous emission can be suppressed and the decay width reduced by an order of magnitude [29]. The analogy to a baryon in the mean field of nucleus seems to be strange in the first moment. In order to develop the idea on can start with the well known Λ hypernuclei. The Λ is stable against strong decays and replaces a neutron in the nucleus feeling a somewhat weaker mean field potential than the nucleon and practically no spin-orbit interaction [30]. This intriguing result has led to detailed investigations (see e.g. ref. [31] for a summary). The effective Λ-nucleus potential could be explained in the frame work of a self consistent picture of the Λ baryon system starting from the elementary Λ-N interaction.

However, a big challenge of this classical idea of hypernuclei was the claim that narrow Σ hypernuclear states exist [32] since in the nucleus the strong $\Sigma N \to \Lambda N$ transition is possible and should broaden these states to 30 MeV [33]. Though some theoretical calculations seemed to explain 5 MeV wide Σ states [34] most theories could not model the experimental evidence [33]. Unfortunately, all calculations did consider the "quasi-free production" only. This means, it was assumed that the nucleon on which the strangeness exchange reaction happened takes all the momentum of the incoming K^- or π^+ in the strangeness exchange reaction. However, the experiments in which the evidence for "bound Σ-states" was found had all a good mass resolution such that "substitutional states", i.e. states in which a nucleon is replaced by a hyperon of the same wave function or momentum distribution, could be separated from the "quasi-free" production [32]. The subsequent experiments at CERN were done close to the magic kaon momentum $p_{K^-} = 270\ MeV/c$ at which the momentum transfer is zero, i.e. a "recoilless" production [35]. In this situation the hyperon is produced with almost the same momentum distribution, i.e. wave function, as the nucleon. In other words, the population of "substitutional states" is favored.

But, as pointed out by B. Povh [36] the significance of recoilless production goes beyond kinematics. In complete analogy to the Mössbauer effect, one can identify the "quasi-free" situation with the normal resonance absorption shifted by the recoil energy and the "recoilless Σ production" by the "recoilless photoabsorption". These two situations represent actually two different quantum mechanical states. Therefore, any consideration of narrow Σ states has to start from a Σ state bound in the average potential of a nucleus. This means that the theoretical proofs that narrow Σ hypernuclear states are impossible [31] can be questioned.

On the other hand the experimental attempts to confirm the narrow Σ hypernuclear states failed too (for a summary see [31], [37,38]). Only in ^4He a state was found and interpreted as a special effect due to the spin-isospin part of the ΣN interaction [39]. An explanation of these negative results could be that all these experiments have been done at large momentum transfer so that the substitutional states were suppressed compared to the quasi-free ones. Additionally, a good resolution with low π background may have been essential for the positive experiments [32,35].

Considering this situation, it was tempting to think about the propagation of a "recoilless Δ", i.e. a Δ bound in the mean field potential [40]. An experiment was proposed [40] and realized in which the π^+ was absorbed on ^{12}C and the reaction $\pi^+ + ^{12}C \rightarrow ^{12}C_{\Delta^+} + p^{proton}_{forward} \rightarrow X + p^{proton}_{sideward} + p^{proton}_{sideward} + p^{proton}_{forward}$ was investigated [41]. This experiment was, however, inadequate to represent a serious test of the idea.

The idea of narrow Δ states in the nucleus is, of course, even more against the established picture than narrow Σ states. As for the $\Sigma N \rightarrow \Lambda N$ one expects the $\Delta N \rightarrow NN$ transition. But this partial width is smaller than the intrinsic strong decay width of the $\Delta \rightarrow N\pi$, with $\Gamma_{\Delta free} = 110\ MeV$, and seems to make any attempt to look for narrow Δ states crazy.

On the other hand, the three spectrometer setup with the excellent MAMI beam was designed to study the $^{12}C(e, e'^{12}C_{\Delta^0}) \rightarrow ^{12}C(e, e'p\pi^-)^{11}C$ reaction with good resolution. The major problem of this study lays in the fact that $|\vec{q}| > \omega$ and this means that the transition form factor or Deby-Waller factor giving the probability to produce a "recoilless Δ state" is small. Nevertheless an experiment was started to conduct a search.

4.2 Experiment

The basic experimental idea is that the two initial "quasi-free" and "bound Δ" states will contribute differently in the phase space of the decaying Δ. The first reaction mechanism favors an emission of the decay p and π in forward direction, whereas the second is almost equally distributed over 180^0 [42]. Therefore, the second mechanism can be favored over the first one by putting the two magnetic spectrometer A and C under large emission angles for the p and π, respectively. Such measurements have been performed in the years 1996 and 1997 and very

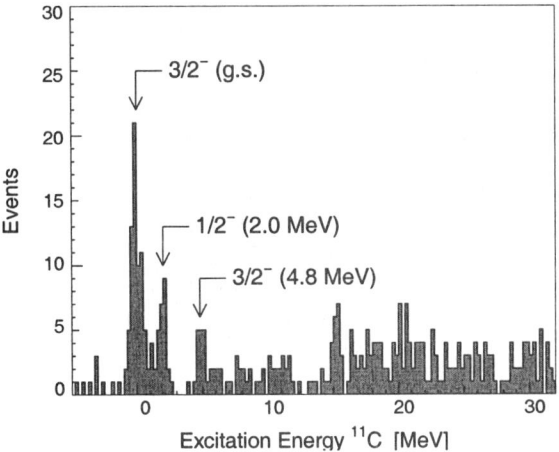

FIGURE 8. An excitation energy spectrum of ^{11}C in the $^{12}C(e, e'p\pi^-)^{11}C$ reaction at $E_e = 855\ MeV$; $E_{e'} = 520\ MeV$, $\theta_{e'} = 18^0$ (spectrometer B); $E_p = 348\ MeV$, $\theta_p = -55.7^0$ (A); $E_\pi = 186.2\ MeV$, $\theta_\pi = 56.5^0$ (C); $q = 385\ MeV/c$. The bin width is $1.5\ MeV$.

recently in August 1999. The performance of the three spectrometers for triple coincidence measurements has been already described.

The electron spectrometer was set under a scattering angle $\theta_{e'}^B \approx 18^0$ resulting in a momentum transfer of $|\vec{q}| \approx 0.35\ GeV/c$. A carbon target with a thickness of $45\ mg/cm^2$ and a beam current of up to $60\ \mu A$ was used giving a luminosity of 10^{37} nucleons $cm^{-2}s^{-1}$. More details of the experiment can be found in ref. [42].

4.3 Analysis and Results

The events were reconstructed by using the standard transfer matrix of the spectrometers. Then cuts on the target vertex and the nominal acceptance of the spectrometers were applied. After the identification of the particles the triple e-p-π coincidences were cut out. From these events excitation spectra, as shown in figure 8, could be produced. It is important to realize that, beside the very small background, all events in this spectrum are due to Δ excitation in ^{12}C. The ground state of ^{11}C is most strongly populated in the Δ decay. In order to construct the excitation spectrum of $^{12}C_\Delta$ one more cut on the ground state of ^{11}C was applied. This cut accepts only events which had no energy transfer and little angular scattering for the emitted p and π. This means final state interactions are suppressed.

Figure 9 depicts the $^{12}C_\Delta$ spectrum. The structure visible in this spectrum shows two peaks of about 5 MeV width and a broad bump. First the statistical significance had to be checked. Assuming a constant background of 3 events/$1 MeV$ bin for the peak at $282\ MeV$ and integrating the peak over 8 bins results in a significance of

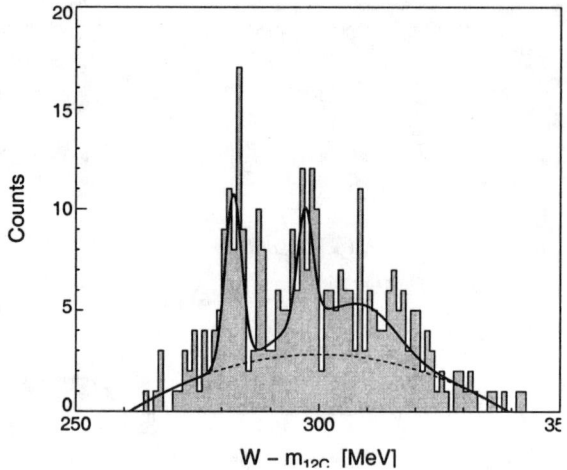

FIGURE 9. The missing mass spectrum, i.e. excitation energy of the $^{12}C_\Delta$ nucleus at $q = 360\ MeV/c$. The bin width is 1.0 MeV. The line represents a fit with a quadratic background, two narrow Gaussian peaks and two broad peaks as suggested by the schematic model discussed in the text.

5 standard deviations. With a background of 4 events/1MeV bin for the peak at 296 MeV 4 standard deviations are derived (for details see [42]).

In order to see whether the peaks could have anything to do with narrow Δs, a simple single particle model on the basis of the known Δ-nucleus interaction [43] was used [42]. It turns out that the energies of the peaks can be reproduced within 1 MeV if a potential depth of 43 MeV is taken. However, in distinction to ref. [43] the spin-orbit interaction was set to zero. The solid line in figure 8 shows a fit to the spectrum using this single particle model.

Though the appearance of two 5 and 4 standard deviation peaks in one spectrum is very unlikely, independent spectra are needed to confirm this finding before one would believe such a provocative result. In ref. [42] such an independent spectrum, consistent with the one of figure 8, is presented. On the other hand, the fast preliminary analysis of the measurement in August 1999, having about the same statistics as figure 8, does not seem to show the structure again. If one realizes that each spectrum represents a beam time of about 300 hours, it becomes clear that only experimental setups with larger efficiency will finally decide about the existence of the narrow Δs.

4.4 A schematic model

The most evident question one might ask is: What could at all cause such narrow states against all intuition? One could search for a possibility by speculating about

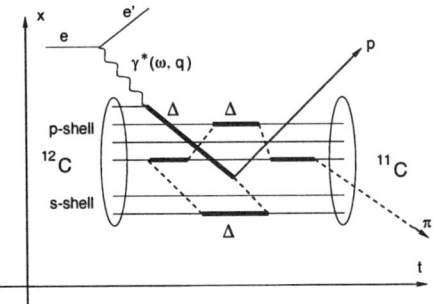 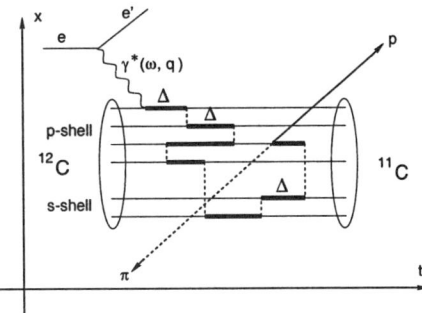

FIGURE 10. Amplitude for the "quasifree" production.

FIGURE 11. Amplitude for the production of a "bound Δ".

the change of the internal structure of the Δ through the mean nuclear field. There are no hints from other experiments that such an effect exists.

However, it is possible to construct a quenching mechanism of the Δ width in the frame work of classical physics [44]. In the model of ref. [44] the following schematic assumptions have been made:

1. The Δ moves as a single particle in the mean-field potential of the nucleus.

2. The large πN cross section causes a short mean free path of the decay π $\langle \lambda_{absorption} \rangle \approx 0.4 \; fm$ resulting in about 70 rescatters in ^{12}C. The wave length of the decay π is $\lambda_\pi = \hbar/k_\pi \approx 1 \; fm$ and, therefore, it will not propagate as an asymptotically free wave. It is assumed that the π propagates with "zero range", i.e. the usually taken π-N phase shifts are not applicable.

With these assumptions the two reaction mechanisms shown in figure 10 and figure 11 can be evaluated. The quasi-free Δ of figure 10 decays statistically along its path and all rescatter diagrams interfere destructively in the sum of final states so that this sum gives 1 and the Δ resonance curve stays unchanged. In contrast in figure 11 a bound Δ is the initial state which is not moving in space and does not pick up a statistical decay phase. The time dependence can be Fourier transformed to give the familiar Breit-Wigner amplitude:

$$\psi(\omega) = \psi_0 \frac{1}{\sqrt{2\pi}} \frac{-i\hbar}{(\omega - \omega_o) + i(\Gamma/2)} \qquad (1)$$

Since the rescattered πs are assumed to propagate with "zero range" only the phase advance between absorption and reemission given by eq. (1) is considered:

$$\phi(\omega) = arctan\left(\frac{\omega_0 - \omega}{M/2}\right) \qquad (2)$$

This means that at $\omega = \omega_0$ no phase advance occurs whereas above and below ω_0 the phase changes up to $\pm\pi/2$ for each rescatter. Therefore, the sum of all rescatter diagrams of figure 11 results in the diffraction amplitude of a normal optical grating and the Breit-Wigner decay curve is multiplied by the grating intensity distribution. For the 70 rescatters at ω_0 - above and below ω one has corresponding to the absorption cross section less rescatters - one gets a full width at half maximum of 6 MeV. The basic mechanism of the quenching is the destructive interference of the amplitudes left and right of the resonance maximum since there $|\phi(\omega)| > 0$. Of course, there are several damping mechanisms as an early p emission and a later π emission which have to be considered too. However, as argued in ref. [44], this does practically not change the basic result derived from the leading diagrams of figure 11.

5. CONCLUSIONS

Even after 70 years of nuclear physics one of the most important objects in nature, the atomic nucleus, is not understood on a fundamental level. For a long time one has concentrated on spectroscopy and its explanation in terms of the nucleonic compositeness of the nucleus. The investigations of non nucleonic degrees of freedom have mostly suffered from poor precision. With the new high precision facilities one can hope for a deeper penetrating to one of the most intriguing borderlines in physics: the transition from hadron to quark-gluon degrees of freedom. The nucleus offers observables for this study which are promising:

1. Short range correlations of nucleons in nuclei

2. Production of baryons in nuclei

3. Propagation of mesons in nuclei

As proven many times, the electromagnetic probe is well understood and promises to contribute particularly significant results.

REFERENCES

1. Fermi, E., *Nuclear Physics*, Chicago University Press (1949).
2. Povh, B., and Walcher, Th., *Comm. Nucl. Part. Phys.* **15**, 85 (1986); Walcher, Th., and Povh, B., *Naturwissenschaften* **74**, 468-473 (1987).
3. Faessler, A., *Progr. Part. Nucl. Phys.* **20**, 151 (1988).
4. see e.g. Lapikás, L., *Nucl. Phys.* **A553**, 297c (1993) and references therein.
5. Gottfried, K., *Nucl. Phys.* **5**, 557 (1958)
6. Wilhelm, P., et al., *Nucl. Phys.* **A597**, 613 (1996)
7. Onderwater, C.J.G., et al., *Phys. Rev. Lett.* **78**, 4893 (1997) *Phys. Rev. Lett.* **81**, 2213 (1998)
8. Blomqvist, K.I., et al., *Phys. Lett.* **B421**, 71 (1998)

9. Edelhoff, R., Doctoral thesis, Mainz University, in preparation
10. Rosner, G., International School on Nuclear Physics, "Electromagnetic Probes and the Structure of Hadrons and Nuclei" Erice/Sicily/Italy, September 17–25, 1999 to be published in *Progr. Part. Nucl. Phys.* **44** (2000)
11. Blomqvist, K.I., et al., *Nucl. Instr. and Meth.* **A403**, 263 (1998)
12. Ransome, R.D., et al., *Phys. Rev.* **C42**, 1500 (1990)
13. Enge, H.A. *Nucl. Instr. and Meth.* **162**, 161 (1979); Kowalski, S. and Enge, H.A. *Nucl. Instr. and Meth.* **A 258**, 407 (1987)
14. Bertozzi, W., et al., *Nucl. Instr. and Meth.* **162**, 211 (1979)
15. Ryckebusch, J., *Phys. Lett.* **B383**, 1 (1996)
16. Vanderhaeghen, M., et al., *Nucl. Phys.* **A580**, 551 (1994)
17. Ohmura, T., et al., *Prog. Theor. Phys.* **15**, 222 (1956)
18. Pieper, S., et al., *Phys. Rev.* **C46**, 1741 (1992)
19. Gearhart, C.C., PhD thesis, Washington University, 1994; Dickhoff, W., private communication
20. Arias de Saaverdy, F., et al., *Nucl. Phys.* **A605**, 359 (1996)
21. Kahrau, M., Doctoral thesis, Mainz University, in preparation
22. Ajzenberg-Selove, F., *Nucl. Phys.* **A523**, 1 (1991)
23. Geurts, W.J.W., et al., *Phys. Rev.* **C54**, 1144 (1996) *Phys. Rev.* **C53**, 2207 (1996)
24. Giusti, C., et al., *Phys. Rev.* **C57**, 1691 (1998)
25. Giusti, C. and Pacati, F.D. *Nucl. Phys.* **A615**, 373 (1996)
26. Wilhelm, P., et al., *Z. Physik* **A359**, 467 (1997)
27. Nadasen, A., et al., *Phys. Rev.* **C23**, 1023 (1981)
28. Ryckebusch, J., et al., *Nucl. Phys.* **A624**, 581 (1997)
29. Hulet, G.H., Hilfer, E.S., and Kleppner, D., *Phys. Rev. Lett.* **55**, 2137 (1985)
30. Povh, B., *Annu. Rev. Nucl. Part. Sci.* **28**, 1 (1978); *Progr. Part. Nucl. Phys.* **5**, 245 (1981)
31. Chrien, E.R., and Dover, C.B. *Annu. Rev. Nucl. Part. Sci.* **39**, 113 (1989)
32. Bertini, R., et al., *Phys. Lett.* **B90**, 375 (1980)
33. Dover, C.B., Gal, A., and Millener, D.J., *Phys. Lett.* **B 138**, 337 (1984)
34. Brockmann, R., and Oset, E., *Phys. Lett.* **B118**, 33 (1982)
35. Bertini, R., et al., *Phys. Lett.* **B136**, 29 (1984)
36. Povh, B., *Z. Physik* **A279**, 159 (1976)
37. Sawafta, R., et al., *Nucl. Phys.* **A585**, 103c (1995)
38. Nagae, T., et al., *Nucl. Phys.* **A631**, 363c (1998)
39. Hayano, R.S., et al., *Phys. Lett.* **B231**, 355 (1989)
40. Povh, B., et al., *SIN/PSI-proposal* (1985), unpublished
41. Brückner, W., et al., *Nucl. Phys.* **A469**, 617 (1987)
42. Bartsch, P., et al., *Eur. Phys. J.* **A4**, 209 (1999)
43. Lenz, F., and Moniz, E.J., *Comm. Nucl. Part. Phys.* **9**, 101 (1980)
44. Walcher, Th., submitted to *Eur. Phys. J. A*

II. STRUCTURE OF FEW-BODY NUCLEI

Chair: C.F. Williamson

Session II ....

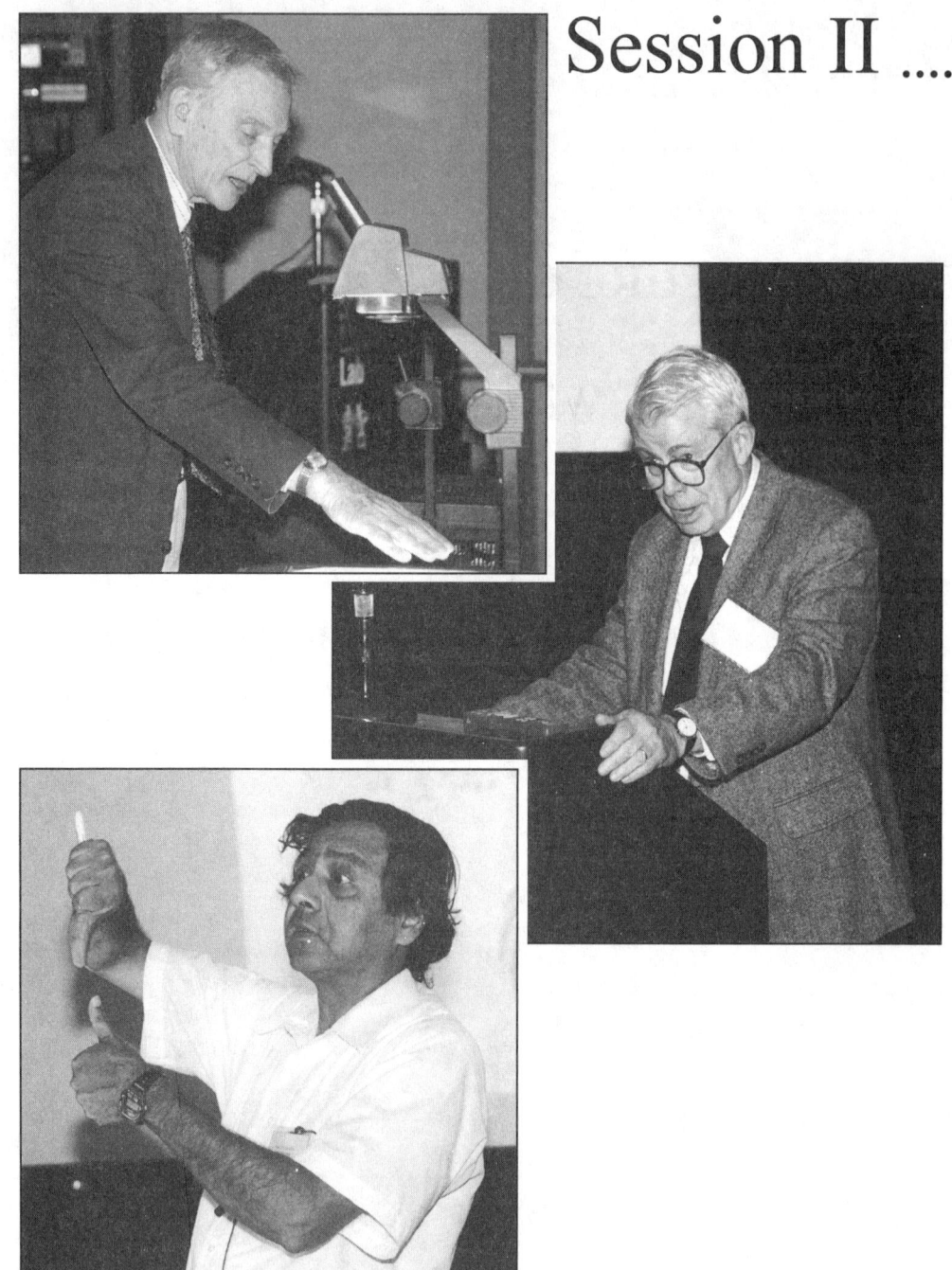

Nuclear Magnetism

Gerald A. Peterson

*Department of Physics and Astronomy, University of Massachusetts
Amherst, MA 01003, USA*

Abstract. The magnetic electron scattering program carried out at the Bates Linear Accelerator Center since 1977 is described. This is preceded by a brief discussion of some of the accelerator developments that made such experiments possible, and that have led to the present status of the Bates Laboratory. A short history of the very beginning of magnetic electron scattering, both elastic and inelastic, is also given.

INTRODUCTION

Nuclear magnetic dipole moments as deduced from atomic beam and nuclear magnetic resonance studies, as well as hyperfine optical studies, have been of great importance in understanding nuclear structure since the 1930's [1]. However, much more detailed information can be obtained by using electron scattering to probe the nuclear magnetization density. A sizeable program in magnetic electron scattering developed at the Bates Linear Accelerator Center starting in 1977, with many institutions participating. Before going to this topic, this 25th anniversary of the first physics experiments at Bates offers an opportunity to look back in time to consider the humble beginnings in our field. We also can recognize those efforts that made possible the magnificent tools and technical developments that are central to our studies today. The Bates accelerator complex is a superb facility in many respects. (i) The beam intensity is adequate for almost all experiments. (ii) The beam when recirculated through the linear accelerator has energies up to 1-GeV, suitable for many experiments. (iii) State of the art internal target experiments within the South Hall Ring are planned. And (iv) continuous duty factor experiments with the Out-Of-Plane Spectrometer system in the South Hall will be possible by using the extracted beams from the Ring. Let us first very briefly examine some of the advances that made possible the class of accelerator which includes that at Bates.

HISTORICAL COMMENTS

When I first entered academia 50 years ago, the study of nuclei by electron scattering was unheard of. At that time there were electron accelerators, mainly betatrons and synchrotrons, with beam energies up to about 300-MeV. The circulating electron beams of these accelerators could not be extracted from the accelerator donut-like vacuum chambers, but rather were used to produce bremsstrahlung beams by having the electron beams impinge upon high-Z targets contained within the vacuum chambers. Thus internal targets were used by necessity. Since the resulting bremsstrahlung spectra were continuous, the interpretation of results from the use of such bremsstrahlung beams was difficult. If we reflect today about the sophistication of the internal target facility for the Bates

BLAST facility [2], it is apparent that enormous progress has been made since the days when internal bremsstrahlung target facilities with all of their attendant problems were the only recourse.

The precursor of the Bates accelerator was the 17-MeV linear accelerator built on the MIT campus by the Laboratory for Nuclear Science and Engineering and the Research Laboratory for Electronics. [3]. In contrast to the early synchrotrons, an external electron beam could be easily extracted from a linear accelerator, but there were other problems. Each of the 21 resonant cavities of the MIT accelerator was driven by a self-excited tunable magnetron. It was difficult to lock the frequencies and phases of the magnetrons so that they would cooperatively accelerate the electrons. Nevertheless, valuable experience was gained, photo-neutron and photo-fission experiments were carried out for over a decade, and without question, that facility was the forerunner that made Bates a possibility.

A significant advance in microwave power sources was made in the early 1950's at Stanford University in the development of the high-powered klystron amplifier, e.g., as described by Chodorow et al. [4]. Their S-band klystron could achieve peak powers as high as 20 MW, and could drive a whole section of a linear accelerator section made up of many disk-loaded cavities. So the problem of locking frequencies and phases was eliminated, and external electron beams of high energy could be developed. The high-powered klystron was the key to the development of the high-energy room-temperature electron linear accelerator, such as Bates, with its easily extractable, focussed, high-current electron beam that could be used directly in fixed target experiments. However, the recent development of superconducting accelerator sections has rendered high power RF amplifiers less important.

EARLY EXPERIMENTS

Most of the very early work on electron scattering concentrated on elastic scattering from nuclear charge distributions [5]. But also high-energy electrons can undergo elastic magnetic scattering from nuclei to provide information about the spatial distributions of magnetism due to spin, convection, and exchange currents. The first observation of such scattering was by McAllister and Hofstadter in 1956 when they observed elastic magnetic scattering from the proton [6]. They used 188 MeV electrons from the Mark III accelerator [7] to scatter from hydrogen gas, and obtained results as shown in Fig. 1. Rosenbluth had predicted the scattering from the magnetic moment of the proton in 1950 for a point anomalous magnetic moment [8]. In Fig. 1 the experimental curve lies between Rosenbluth's curve and that calculated with a Dirac magnetic moment. The deviation from these curves in this seminal work represents the effect of a form factor and indicates structure within the proton. This measurement was carried out at a modest beam energy by Bates

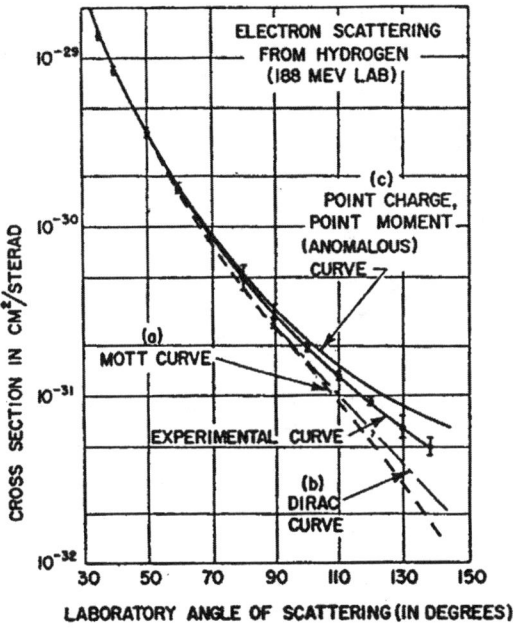

FIGURE 1. The electron scattering results of McAllister and Hofstadter showing that the proton had a magnetic structure, as is explained in the text.

standards, with an average beam current of 0.02 µA, a factor of several thousand smaller than beam currents that are routinely available at Bates.

Through the late 1950's, the use of electrons to observe nuclear excitations was relatively new, whereas there were many nuclear studies with real photons, especially of the electric dipole giant resonance. In an experiment by Barber and Gudden [9], in which the primary objective was to measure the E1 giant resonance of ^{12}C by electron scattering, the first observation of *inelastic* magnetic scattering was made. This involved the excitation of the now familiar T = 1, I = 1, 15.11-MeV excited state of ^{12}C. A short time later, Barber, Berthold, Gudden, and Fricke [10] also observed magnetic scattering from the known levels in ^6Li at 2.18- and 3.56-MeV, and in ^9Be at 2.43-MeV. Figures 2 and 3 show data taken at 132° and 160° scattering angles, respectively, for electrons of initial energy 42.5-MeV scattered from ^{12}C. Although the resolution is poor, the magnetic dipole excitation of the 15.11-MeV level is clearly evident. These results were interpreted by using a virtual photon theory [11] so that ground-state radiation widths could be compared with results from the many experiments using photons. It was hoped that from these measurements the excitation of the E1 giant resonance could be clearly seen, but the radiation tails were large compared to the giant resonance excitations. From Figs. 2 and 3, it is evident that going to the ultimate backward angles of 180° would reduce the radiation tails and give a relative enhancement of the excitations [12]. The decrease in the radiation tail is a consequence of the reduction of elastic scattering at 180°. In particular, for a nucleus with a spin-zero ground state, the Mott cross-section goes to zero because of helicity and angular momentum conservation [13].

Consequently the scattering from the nuclear charge is minimized at 180°, as is

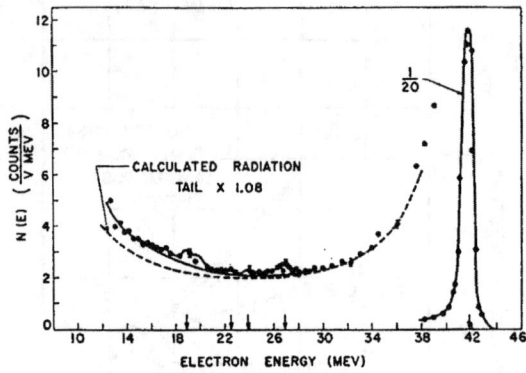

FIGURE 2. Energy spectrum of electrons initially of 41.5 MeV after scattering through 132° from a carbon target.

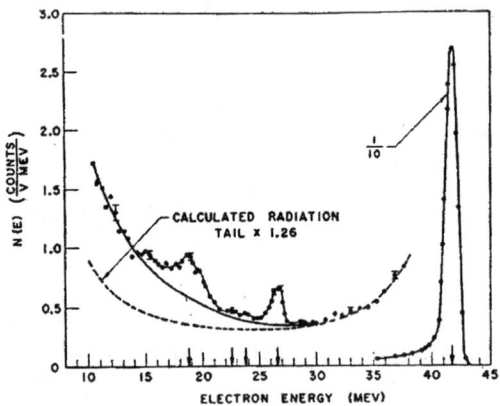

FIGURE 3. Energy spectrum of electrons initially of 41.5 MeV after scattering through 160° from a carbon target.

the attendant radiation tail. However, there is still a radiation tail due to large angle bremsstrahlung that rises gradually from zero at the position of the elastic peak to large values at high excitation energies.

In order to observe 180° scattering, a magnetic spectrometer cannot simply be moved back to that angle because the incident beam would hit the back of the spectrometer. So in this pioneering work [13], the spectrometer was set at 160° with respect to the incident beam, and a small dipole magnet was placed before the target, as shown in Fig. 4. This magnet deflected the incident beam through an angle δ_i before it hit the target. The back-scattered electrons returned through the magnet, and were deflected by an angle δ_f so that $\delta_i + \delta_f = 20°$. The magnetic field could be adjusted for inelastic scattering without moving the spectrometer, i.e., it stayed fixed at 160°, and the post-target beam was steered onto the beam catcher.

Figure 5 shows the first results for deuteron threshold electrodisintegration. The data were taken using the equipment shown in Fig. 4 [13]. Other work using

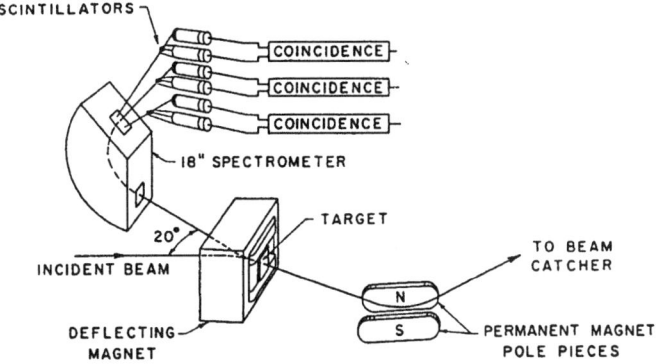

FIGURE 4. Experimental setup for the first 180° scattering experiments carried out on the Mark II accelerator at Stanford.

real photons could not easily separate the M1 breakup at threshold from the dominant p-wave E1 breakup [14]. By using electrons for which, in contrast to real photons, the energy transfer can be uncoupled from the momentum transfer, the M1 breakup could be measured. This measurement has been repeated many times at higher momentum transfers with much more intense beams than the 0.4 µA average current used in obtaining the data of Fig. 5, as will be discussed.

It was recognized that the magnetic elastic scattering, as seen for the proton in Fig. 1, should also be present for other nuclei possessing magnetic moments. This is illustrated in Fig. 6, where the 180° scattering results for the lithium isotopes are contrasted [15]. Since the results shown were taken at a very low momentum transfer, the ratio of the peaks are approximately proportional to the ratio of the square of the magnetic moments of the isotopes, i.e., $(\mu_7/\mu_6)^2 = (3.26/0.82)^2 = 15.7$.

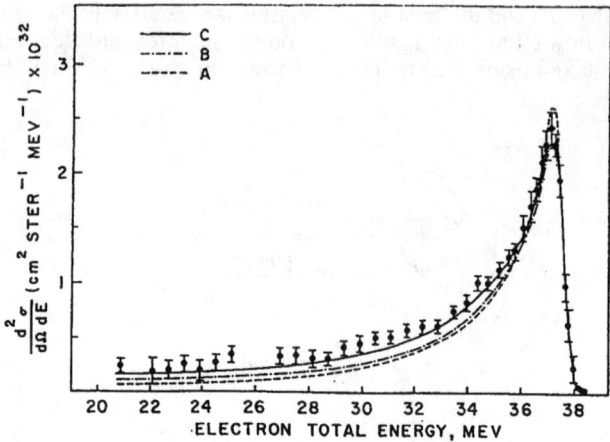

FIGURE 5. The first threshold M1 electrodisintegration cross-section measured for the deuteron for an incident beam energy of 41.5 MeV. The solid curve has the preferred radiative corrections.

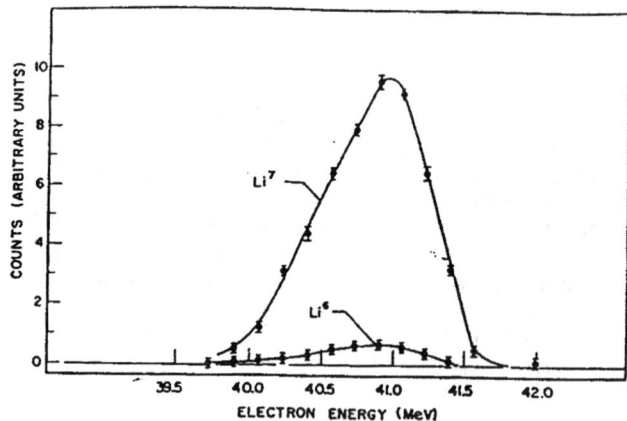

FIGURE 6. Elastic magnetic scattering from ^{6}Li and ^{7}Li using 41.5 MeV electrons.

This first measurement showed the potential for determining magnetization densities of nuclei heavier than the deuteron or the proton. Elastic magnetic scattering studies were pursued at higher momentum transfers at the Amsterdam EVA facility, as well as at several other laboratories [16].

MAGNETIC SCATTERING EXPERIMENTS AT BATES

The Bates Laboratory came into operation in 1974. About that time construction of a 180° scattering facility was authorized for use with the ELSSY spectrometer [17] in the Bates North Hall. A four-magnet chicane was built [18] in which the central two magnets could be moved over a wide vacuum chamber for inelastic scattering in order to keep the spectrometer angle fixed. This is shown in Fig. 7. The magnetic fields of all four magnets in the chicane were reduced for inelastic scattering in proportion to the energy of the back-scattered electrons to be received by the spectrometer. This ensured that the beam exiting the target would follow the same path to the beam dump along the incident beam direction. Since all electrons scattered from the target around 180° traveled through the same magnetic field shape, a constant solid angle was preserved. The system was used in the dispersion-matching (or energy-loss) mode [17] with the beam transport system and the spectrometer by dispersing the incident beam spot vertically on the targets. Resolutions less than 3×10^{-4} were obtained.

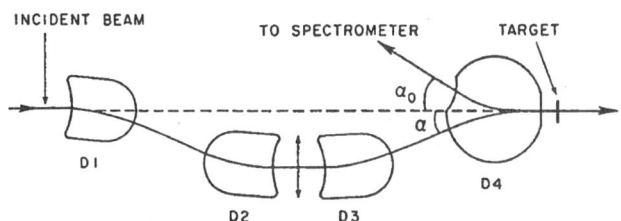

FIGURE 7. Schematic diagram of the Bates four-magnet chicane used for 180° scattering.

Figure 8 shows the first spectra obtained with the Bates 180° scattering facility. The rise and fall of the transverse excitation peaks in the 16-MeV region of ^{12}C for increasing beam energies is evident [19]. The 15.11-MeV M1 form factor goes through a diffraction minimum, the 16.11-MeV E2 is near a maximum, and the sum of the form factors of the complex of higher lying levels is increasing.

Figure 9 shows an example of elastic magnetic scattering from the ^{7}Li ground state doublet [the $J^{\pi} = (3/2)^{-}$ ground state and the $E_x = 478$ keV, $J^{\pi} = (1/2)^{-}$ first-excited state] taken with the Bates 180° scattering facility by Lichtenstadt et al. [20]. Much higher momentum transfers were obtained than in the initial work of Fig. 6, and correspondingly more detailed nuclear structure information was obtained. The ground state form factor has contributions from both M1 and M3 multipoles and the form factor for the first-excited state has both M1 and E2 (transverse electric quadrupole) contributions. These transverse form factors, in combination with longitudinal C0 and C2 form factors for the ground state and C2 form factor for the first-excited state, put severe constraints on nuclear models to be compared with the data.

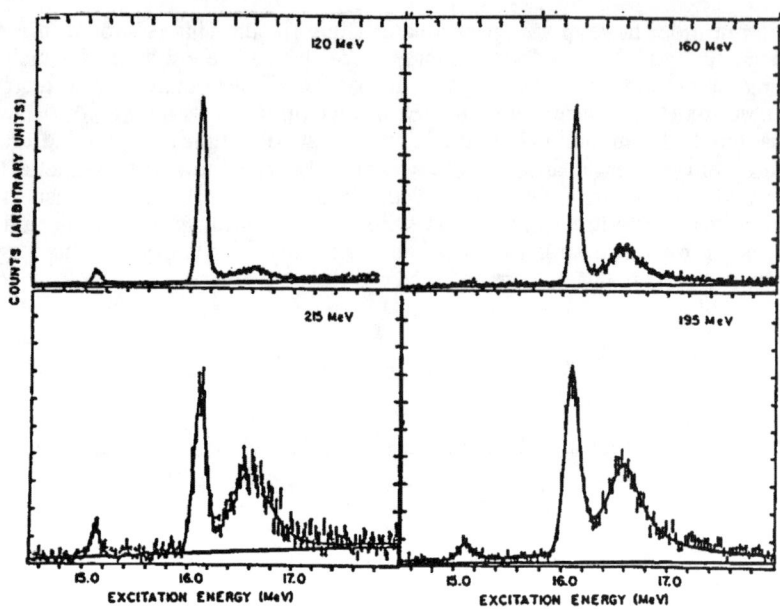

FIGURE 8. The 16 MeV region of ^{12}C, the first spectra taken with the Bates 180° scattering facility.

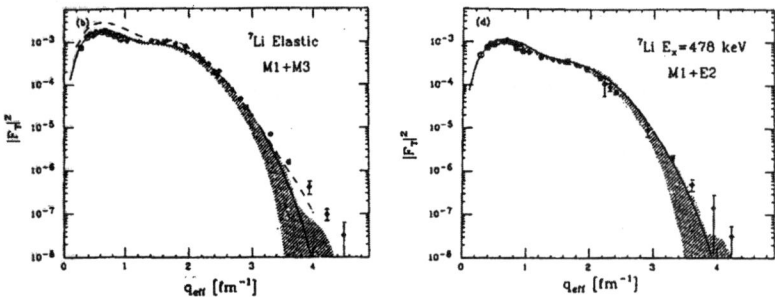

FIGURE 9. Magnetic form factors squared versus effective momentum transfer for the ground state and the first-excited state of ^7Li.

For photoreactions there is an ordering of the multipoles with E1 predominating. For electron scattering there is no ordering of the multipoles. An example of the participation of higher multipoles in the scattering is shown in the Bates data of Fig. 10 for ^{41}Ca taken in the momentum transfer range from 0.9 to 2.0 fm^{-1} [21] and Saclay data from about 1.8 to 3.3 fm^{-1}[22]. The charge scattering in the lower momentum transfer region taken at 180° is at least 400 times smaller than that in the 155° Saclay measurements taken in the same region. The Bates data showed that the M3 and M5 form factors were quenched relative to the shell model calculations, whereas the M7 multipole is in good accord with the model. Calculations indicated that the multiparticle-multihole configurations in the $1f_{7/2}$ and $1d_{3/2}$ subshells, first-order core polarization to higher excited orbitals, and meson exchange currents, give reasonable agreement with the data for all multipoles. The M7 multipole together with the other data permitted a determination of the rms radius of the $1f_{7/2}$ neutron orbit. After correcting for core polarization and meson exchange currents, the radius was found to be 3.96 ± 0.05 fm in agreement with mean-field Hartree-Fock-Bogolyubov calculations of Decharge and Gogny [23]. In Fig. 9 comparison is also made with the mean field calculations of Negele and Vautherin [24] and with the relativistic mean field calculations of Kim [25].

Elastic scattering from the deuteron involves isoscalar exchange currents that are of greater complexity than the predominant isovector pion exchange currents of the deuteron threshold electrodisintegration. Such deuteron meson exchange currents as well as relativistic effects are discussed by Van Order at this symposium [26]. Although for many nuclei 180° scattering is only important at modest momentum transfers up to about 3 fm^{-1} in order to reduce longitudinal contributions, there are cases where longitudinal contributions are large and transverse contributions are small even at the largest momentum transfers. The deuteron elastic magnetic form factor $B(Q^2)$ is an example. This was measured by Arnold et al. [27] at momentum transfers as large as large as $Q^2 = 2.77$ (GeV/c)^2 at the Stanford Linear Accelerator

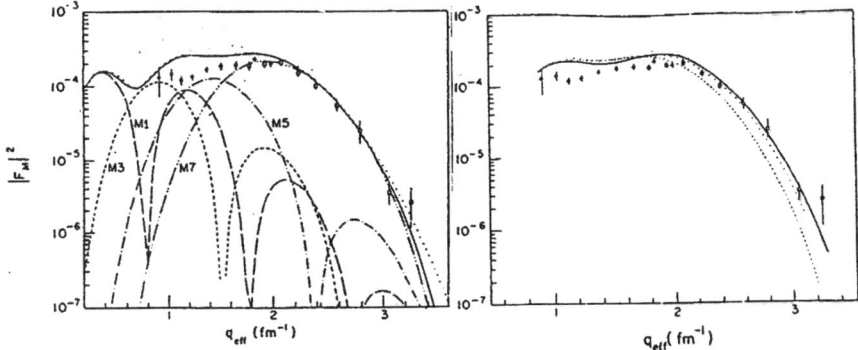

FIGURE 10. (left): Magnetic form factor data for ^{41}Ca vs. q_{ef}. The solid and dotted curves are for Woods-Saxon and harmonic-oscillator potentials, respectively. The various multipoles are shown for the Woods-Saxon potential. (right): Comparison of the measured magnetic for factor to mean-field predictions of Decharge and Gogny (Ref. 23) (solid line), Negele and Vautherin (Ref. 24) (dashed), and the relativistic calculation of Kim (Ref. 25) (dotted).

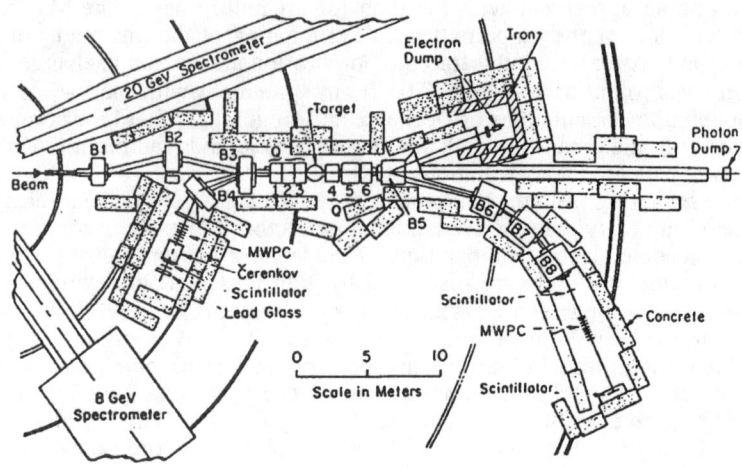

FIGURE 11. Experimental layout of the deuteron elastic magnetic scattering experiment at SLAC.

Center by the detection of electrons scattered at 180° in coincidence with recoiling deuterons at 0°. As shown in Fig. 11, a 180° chicane and the electron spectrometer are shown to the left, and the recoil spectrometer and the beam dump are shown to the right. Some of the cross sections measured in this experiment were extremely small, i.e., less than 10^{-41} cm²/sr. The theoretical comparisons with these cross sections are given in this symposium by Van Orden [26].

As was mentioned above, at large momentum transfers, e.g., above 3 fm^{-1}, the ratio of the Coulomb to the magnetic contributions to the cross sections generally decrease because of the fall off of the Coulomb form factor. Therefore corrections may be made for the Coulomb contributions and it is unnecessary to go to a 180° scattering angle to measure many cross sections. An example is the Bates measurement of deuteron electrodisintegration near threshold by Schmitt et al. [28], which was carried out at a 160° scattering angle. On the other hand, a recent 160° measurement at Bates of the elastic magnetic scattering from ³He by Nakagawa et al. [29] could have benefited from 180° scattering. For the data shown in Fig. 12, the Coulomb form factor reaches a sub-maximum near the first diffraction minimum of the magnetic scattering [30]. For this experiment, the Coulomb dominated the magnetic for the two new points shown in the vicinity of $Q^2 = 20$ fm^{-2} by factors as large as 6.

The diffraction minimum of the curve resulting from impulse approximation calculations, as shown in Fig. 12, is a consequence of S- and D-wave interference, similar to that which occurs in deuteron threshold electrodisintegration. These impulse approximation minima are entirely filled by exchange currents. These results for the ²H, ³He, and ³H [31] are the foremost evidence for exchange currents.

The two high q points shown in the Fig.12 have cross sections about 10^{-40} cm²/sr and do not agree well with the three theoretical curves shown. Because of these low cross sections, the target was operated at a temperature of 22 K and a pressure of 50 atmospheres so as to enhance the luminosity.

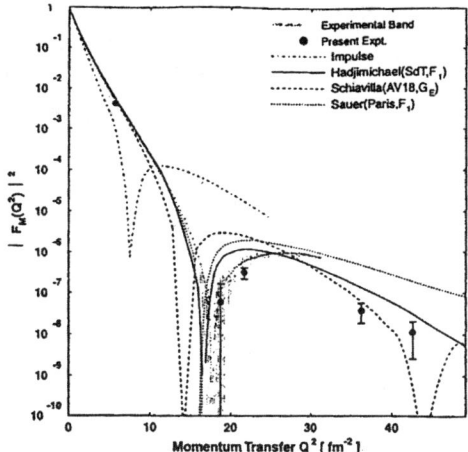

FIGURE 12. Elastic magnetic scattering data for ^{3}He. The experimental band is from previous experiments cited in Ref. 30.

FUTURE EXPERIMENTS

M. Petratos has recognized the need for going to far backward angles at the high momentum transfers available at the new Thomas Jefferson National Accelerator Facility [32]. Even for the lightest of nuclei, minimizing the longitudinal parts of

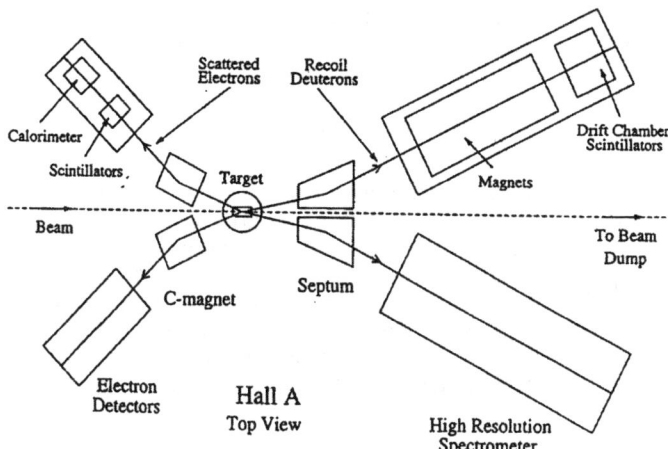

FIGURE 13. An experimental arrangement suggested by Petratos for measuring the deuteron magnetic form factor at the Thomas Jefferson National Accelerator Facility.

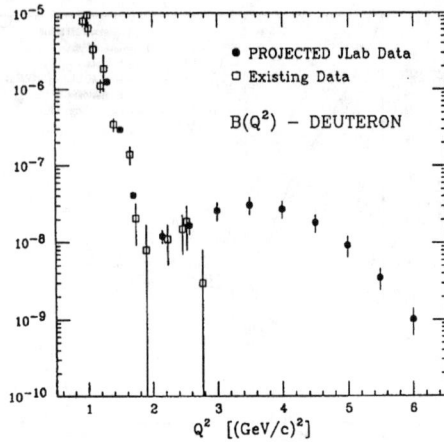

FIGURE 14. Projected elastic magnetic scattering form factor $B(Q^2)$ for the deuteron compared to the existing SLAC data from Ref. 27.

the cross sections, as well as eliminating the need for Rosenbluth separations of the longitudinal from the transverse, is important. Figure 13 shows a magnetic septum proposed by Petratos that allows the electrons scattered in the very far backward direction to be received by two magnetic spectrometers each of which is in coincidence with two existing spectrometers in Hall A that receive the recoiling nuclei of mass M. The electron spectrometers can be substantially smaller than those of the hadron arms because the recoiling electrons in the limit of infinite incident electron energy, the electrons, if scattered near 180°, will have an energy of only $Mc^2/2$. Figure 14 shows projected data together with the earlier SLAC data [27] obtained with the equipment shown in Fig. 11 and discussed previously.

CONCLUSIONS

The topics treated here represent only a tiny fraction of the work done in this area. Many other magnetic scattering studies carried out at Bates and elsewhere could have been discussed. Reflecting on these studies and other experiments at Bates over the past 25 years, and considering that our field of electron scattering was initiated less than 50 years ago, one can conclude that very rapid and substantial progress has been made. Most certainly the future bodes well for electromagnetic interaction nuclear physics, and significant advances can be expected at the powerful new facilities at Bates and elsewhere.

ACKNOWLEDGMENTS

The magnetic scattering program at Bates would not have been possible without the collaboration of many physicists from the following institutions who participated in the experiments: American University, University of Basel, California Institute of Technology, Catholic University, University of Colorado,

University of Glasgow, University of Illinois, Indiana University, Lawrence-Livermore National Laboratory, Los Alamos National Laboratory, University of Mainz. Massachusetts Institute of Technology, University of Massachusetts, Naval Research Laboratory, University of New Hampshire, NIKHEF-K, National Bureau of Standards. Oak Ridge National Laboratory, Saclay, Shizuoka University, Teikyo University, Tel Aviv University, Tohoku University, and the University of Virginia.

REFERENCES

[1] Rabi, I.I., Millman, S. Kusch, P., and Zacharias, J.R., *Phys. Rev.* **53**, 318 (1938); Bloch, F., Hansen, W.W., and Packard, M.E., *Phys. Rev.* **69**, 680 (1946); Purcell, E.M., Torrey, H.C., and Pound, R.V., *Phys. Rev.* **69**, 37-38 (1946).
[2] van den Brand, J., these proceedings.
[3] Demos, P.T., et al., *J. Appl. Phys.* **23**, 53-65 (1952)].
[4] Chodorow, M., Ginston, E.L., Neilsen, M., and Sonkin, P., *Proc. I.R.E.*, **41**, 1584-1596 (1953).].
[5] Hofstadter, R., Rev. Mod. Phys. **28**, 214-253 (1956)].
[6] McAllister, R.W., and Hofstadter, R., *Phys. Rev.* **102**, 851-856 (1956)].
[7] Chodorow, M., Ginzton, E.L., Hansen, W.W., Kyhl, R.L., Neal, R.B., Panofsky, W.K.H., and staff, *Rev. Sci. Instrum.* **26**, 134-204 (1955).
[8] Rosenbluth, M.N., *Phys. Rev.* **79**, 615-619 (1950).
[9] Barber, W.C., and Gudden, F.E., *Phys. Rev. Lett.* **3**, 219-221 (1959).
[10] Barber, W.C., Berthold, F., Fricke, G., and Gudden, F.E., *Phys. Rev.* **120**, 2081-2090 (1960).
[11] Dalitz, R.H., and Yennie, D.R., *Phys. Rev.* **105**, 1598-1615 (1957).
[12] Fricke, G. and Barber, W.C., private communications.
[13] Peterson, G.A., and Barber, W.C., *Proceedings of the Rutherford Jubilee International Conference*, edited by J.B. Birks, Academic Press, New York, 1961, p. 831, and *Phys. Rev.* **128**, 812-820 (1962).
[14] Fermi, E., *Nuclear Physics*, edited by J. Orear, A.H. Rosenfeld, and R.A. Schluter, The University of Chicago Press, Chicago, IL, 1950, pp. 175-177.
[15] Peterson, G.A., *Phys. Lett.* **2**, 162-163 (1962).
[16] deVries, C., and Bruinsma, P.J.T., *Nucl. Instrum. Methods* **74**, 5-12 (1969); van Niftrik, G.J.C., deVries, H., Lapikas, L., de Vries, C., ibid. **92**, 301- 309 (1971); de Jager, C.W., de Vries, H., and deVries, C., *At. Data Nucl. Data Tables* **14**, 479-508 (1974).
[17] Bertozzi, W., Hynes, M.V., Sargent, C.P., Turchinetz, W., and Williamson, C., *Nucl. Instrum. Methods* **162**, 211-237 (1982).
[18] Peterson, G.A., Flanz, J.B., Webb, D., de Vries, H., Williamson, C.F., *Nucl. Instrum. Methods* **160**, 375-381 (1979).
[19] Flanz, J.B., Ph.D. thesis, University of Massachusetts (1979).
[20] Lichtenstadt, J., et al., *Phys. Lett. B* **219**, 394-398 (1989).
[21] Baghaei, H., et al., *Phys. Rev. C* **42**, 2358-2366 (1990).
[22] Platchkov, S., et al., *Phys. Rev. Lett.* **61**, 1465-1468 (1988).
[23] Decharge, J. and Gogny, D., *Phys. Rev. C* **21**, 1568-1593 (1980).
[24] Negele, J.W., and Vautherin, D., *Phys. Rev. C* **5**, 1472-1493 (1972).
[25] Kim, L.J., Ph.D. thesis, Stanford University (1987).
[26] Van Orden, W., these proceedings.
[27] Arnold, R.G., et al., *Phys. Rev. Lett.* **58**, 1723-1726 (1987); Bosted, P.E., et al. *Phys. Rev. C* **42**, 38-64 (1990).

[28] Schmitt, W.M., et al., *Phys. Rev. C* **56**, 1687-1699 (1977).
[29] Nakagawa, I., et al., to be published; Nakagawa, I., Ph.D. thesis, Tohoku University (1999).
[30] Amroun, A., et al., *Nucl. Phys.* **A579**, 596-626 (1994).
[31] Beck, D., these proceedings.
[32] Petratos, M., private communications.

Quantum Monte Carlo Calculations of Nuclei

V. R. Pandharipande

Department of Physics, University of Illinois, Urbana, Illinois 61801

Abstract.
We review the recent progress in building realistic models of nuclear forces and the solving the resultant nuclear Schrödinger equation with quantum Monte Carlo methods. The present models of nuclear electroweak current operators and the transition rates they predict are also discussed.

I INTRODUCTION

The many-body theory of nuclei [1] and neutron stars [2] is based on the Hamiltonian:

$$H = \sum_i -\frac{\hbar^2}{2m_i}\nabla_i^2 + \sum_{i<j} v_{ij} + \sum_{i<j<k} V_{ijk}, \qquad (1)$$

containing kinetic, and two- and three-nucleon interaction energies. In several recent calculations the mass difference between the proton and the neutron is taken into account, isovector and isotensor terms are included in the v_{ij}, and leading relativistic corrections are considered. The theory of strong interactions has not yet progressed enough to permit calculations of v_{ij} and V_{ijk} with the accuracy required to calculate nuclear binding energies. We model them with guidance from theory, and determine the parameters in the model from available data. Their present models are discussed in sections II to IV.

The many-body Schrödinger equation, $H\Psi_i = E_i\Psi_i$, has to be solved to calculate the bound and continuum states of this Hamiltonian. In the eighties this equation could be solved only for the states of two- and three-nucleons. In the nineties it was possible to solve it with useful accuracy for all the bound and quasi-bound states of up to eight nucleons by building upon the variational Monte Carlo (VMC) methods developed by Wiringa [3] and the Greens function Monte Carlo (GFMC) methods developed by Carlson [4]. Fortunately nuclei with $A \leq 8$ dominate solar and primordial nuclear reactions, and have been extensively studied in laboratories. Moreover, in the coming years it will be possible to extend the mass limit up to $A = 12$ with the present methods discussed in section V, and new methods are being developed to calculate ground states of heavier nuclei, nuclear and neutron star matter as outlined in section VI.

Knowledge of the nuclear electroweak current operator is necessary to calculate natural processes like radiative capture reactions that take place during stellar evolution, as well as electron-nucleus scattering experiments conducted in laboratories, such as Bates, to probe various aspects of nuclear physics. We have mostly used theoretical models of the electroweak current, based in part on the nucleon-nucleon (NN) interaction v_{ij}, developed by Riska and Schiavilla [5–7]. The results obtained with these models are reviewed in sections VII and VIII. The outstanding problems are summarized in last section IX. Many details of the work carried out before 1997 can be found in the excellent review by Carlson and Schiavilla [8].

II MODELS OF TWO-NUCLEON INTERACTION

Accurate models of the NN interaction v_{ij} are essential for *ab initio* calculations of nuclear binding energies, spectra, transition and reaction rates. In the early 1990's the Nijmegen group [9] carefully examined all the data on NN scattering at energies below 350 MeV published between 1955 and 1992. They extracted 1787 proton-proton and 2514 proton-neutron "reliable" data, and showed that these could determine all NN scattering phase shifts and mixing parameters quite accurately. NN interaction models which fit this Nijmegen database with a $\chi^2/N_{data} \sim 1$ are called "modern". These include the Nijmegen models [10] called Nijmegen I, II and Reid 93, the Argonne v_{18} (A18) [11] and CD-Bonn [12]. In order to fit both the proton-proton and neutron-proton scattering data simultaneously and accurately, these models include a detailed description of the electromagnetic interactions and terms that violate the isospin symmetry of the strong interaction.

All realistic models of the NN interaction have the one pion-exchange potential (OPEP) as the long range part. The five modern models use different parameterizations of the phenomenological short- and intermediate-range parts, and the Nijmegen-I and CD-Bonn models include nonlocalities suggested by boson-exchange representations. Thus, like the older models, they make different predictions for the many-body systems. However, the differences in their predictions are much smaller than those between older models, presumably because they accurately fit the same large data set.

The interaction in the spin-isospin $S, T = 1, 0$ has the largest model dependence. Fortunately the deuteron structure provides significant information on this interaction. Fig.1 shows the deuteron wave functions obtained with the five modern potentials [13]. The three potentials, Reid 93, Nijmegen II and A18, which are local in each NN partial wave, give essentially the same deuteron wave function. We expect that they will give rather similar correlations in all nuclei. The Nijmegen I and CD-Bonn potentials have momentum dependent terms associated with heavy meson exchange. These two potentials give larger $^3S_1$ wave functions at $r < 0.8$ fm, because they have softer repulsive cores than the local models; however, this effect is not very large as can be seen from Fig.1. At $r > 0.8$ fm only the CD-Bonn predictions differ from the rest. This is because CD-Bonn has a strongly nonlocal OPEP as suggested by pseudoscalar pion-nucleon coupling. It predicts a smaller D-wave in the deuteron. The triton ground state energies predicted by

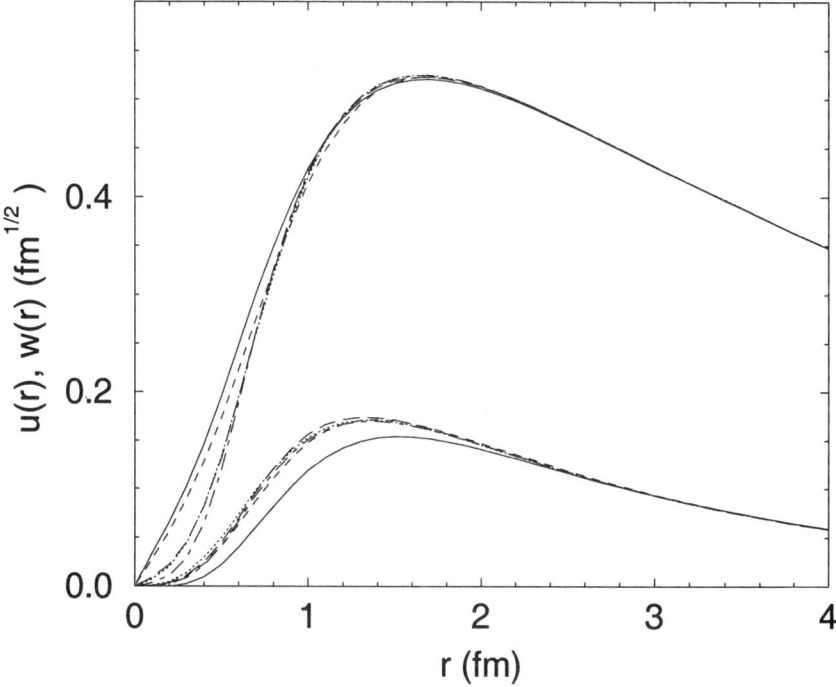

FIGURE 1. Deuteron wave functions: large curves, $u(r)$; small curves, $w(r)$. The solid, dashed, dash-dotted, dotted and long-dashed curves are generated from the CD-Bonn, Nijm-I, Nijm-II, Reid93 and A18 potentials respectively.

the modern Nijmegen and Argonne models are between -7.62 to -7.72 MeV [14], while that of CD-Bonn is -8.00 MeV [12]. This difference is also due to the nonlocal OPEP in CD-Bonn.

These results indicate that the main uncertainty in the preset models of v_{ij} is from the assumed nonlocality in OPEP. We first note that the deuteron wave functions predicted by the local models give form factors that are in excellent agreement with the data, including the new data from Jefferson Lab [15–17]. Second, effective chiral lagrangians do not favor pseudoscalar pion nucleon coupling. These two facts suggest that potentials with local OPEP may be more realistic. However, relativistic field theories permit use of OPEP with different nonlocalities [18,19] related with the Dyson transformation. The three-nucleon interaction (TNI) and the pair current operators depend upon the choice of OPEP, and the final results obtained after including them should be independent of the choice. Therefore, if relativistic field theories can be used to describe pion exchange forces, one

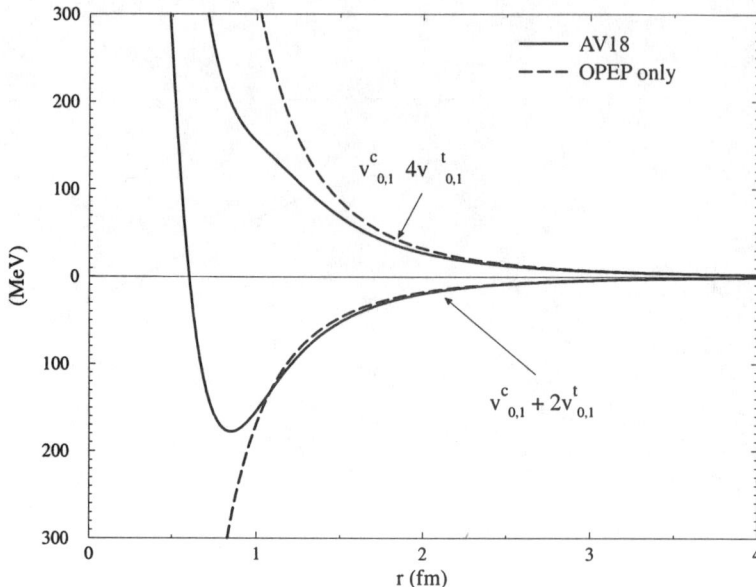

FIGURE 2. The static part of the A18 interaction (without spin-orbit terms), and the OPEP in $S, M_S, T = 1, 0, 0$ two-nucleon states at $\theta = \pi/2$ (lower curves) and $\theta = 0$ (upper curves).

can use either the local OPEP in A18, or the nonlocal one in CD-Bonn. In this case it is obviously better to use the local representation because accurate many-body calculations can be carried out with it. Most of the results discussed in this talk are obtained with the A18 interaction.

The OPEP has a dominant tensor force which depends upon the angle θ made by the vector $\mathbf{r}_i - \mathbf{r}_j$ with the spin directions [20]. The local potential, (*i.e.* without the spin-orbit term), in the $S, T = 1, 0$ is shown in Fig.2 for spin projection $M_S = 0$. It is very attractive when $\theta = \pi/2$, *i.e.* when the $\mathbf{r}_i - \mathbf{r}_j$ is perpendicular to the z-axis, while it is repulsive when $\theta = 0$. This anisotropy of the v_{ij} in $S, T = 1, 0$ is similar to that of the interaction between magnetic dipoles, and is reflected in the structure of the deuteron and all nuclei. Fig.2 also shows the OPEP for comparison with the A18 model. The OPEP dominates the v_{ij} at $r_{ij} > 1$ fm in all states except those with $S, T = 0, 1$, where the OPEP is the smallest.

The nucleon density distribution of the deuteron with spin projection $M = 0$, in its center of mass frame, calculated from A18, is shown in Fig.3. This distribution is axially symmetric under rotations about the z'-axis, and the figure shows the cross section of the $\rho(\mathbf{r}')$ in the x'-z' plane. Here $\mathbf{r}' = \pm \mathbf{r}/2$ are the nucleon coordinates in the center of mass frame. The two-peaks at z'=0, x'$\sim \pm 0.5$ fm correspond to the two nucleons being close to the minimum of the potential shown in Fig.2. The density is large along the z'=0 ($\theta = \pi/2$) line where the interaction is attractive. In contrast the density is very small along the x'=0 ($\theta = 0$) line where the v_{ij} is repulsive. The elastic form factors of the

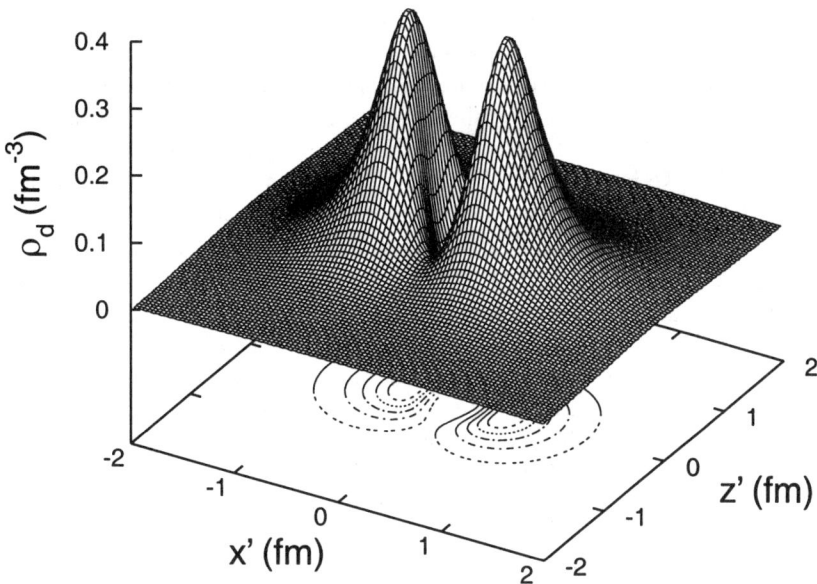

FIGURE 3. The nucleon density distribution in the deuteron in spin projection $M = 0$ state, in the x'-z' plane of the center of mass frame.

deuteron are sensitive to the distance between the peaks of $\rho(\mathbf{r})$, and the width of these peaks [20], and the experimental data [15–17] confirms the predicted distance between the peaks, and supports the predicted width.

The two-nucleon distribution functions in nuclei [20] and nuclear matter [21], in the $S, T = 1, 0$ NN state have very similar shapes, as in the deuteron, up to $r \sim 1.2$ fm. The ratio of the peak hight of the distribution function in the nucleus to that of the deuteron, gives most accurate predictions for the Bethe-Levinger factor of the nucleus. The calculated values of this factor are in good agreement with the observed ratios of $\sim$ 100 MeV pion and photon absorption cross sections [20].

III MODELS OF THREE-NUCLEON INTERACTION

Nuclear Hamiltonians containing only NN interactions underbind the triton [12,14], and give too large equilibrium density for symmetric nuclear matter [21,22]. It is therefore necessary to include TNI in the nuclear Hamiltonian. The presence of two-pion exchange TNI, $V_{ijk}^{2\pi}$, was pointed out in 1957 by Fujita and Miyazawa [23]. More recent models of $V_{ijk}^{2\pi}$ are based on chiral perturbation theory [24,25]. The $V_{ijk}^{2\pi}$ is attractive, and either its short-range cutoff, or its strength can be adjusted to reproduce the triton binding energy. However, inclusion of the attractive $V_{ijk}^{2\pi}$ in the Hamiltonian further increases the saturation density of nuclear matter away from the empirical value of $\rho_0 \sim 0.16$ fm^{-3}. It is thus necessary to add repulsive terms in V_{ijk} to obtain reasonable values of ρ_0.

In Urbana models, the V_{ijk} is expressed as sum of $V_{ijk}^{2\pi}$ and V_{ijk}^R. The V_{ijk}^R is assumed to be spin-isospin independent, and its spatial dependence is taken as $T_\pi^2(r_{ij})T_\pi^2(r_{jk})$, where $T_\pi(r)$ gives the radial dependence of the dominant tensor force in OPEP. This term was meant to simulate the quenching of the attractive two-pion exchange interaction present in the v_{ij}, which has approximately the radial shape of $T_\pi^2(r_{ij})$, by pion exchanges with the third nucleon k. However, $\sim$ 37% of it seems to simulate known relativistic effects discussed in the next section. The Urbana model U-IX assumed that the $V_{ijk}^{2\pi}$ has the form suggested by Fujita and Miyazawa, and adjusted its strength along with that of V_{ijk}^R to reproduce the binding energy of ^{3}H calculated essentially exactly, and the nuclear matter ρ_0 calculated variationally using hypernetted chain summation methods. The short-range cutoffs in V_{ijk} are assumed to be the same as those in A18 v_{ij} used in the calculations. The two-parameter U-IX model gave rather good predictions for nuclei with $A \leq 6$ [26], and properties of nucleon matter other than the equilibrium binding energy E_0 whose empirical value is ~ -16 MeV per nucleon. The variational calculations give only ~ -12 MeV, however much of the underbinding is presumably due to missing many-body correlations in the variational wave function of nuclear matter [21].

In the past few years calculations with errors less than a percent or two, of all bound and narrow resonance states of up to eight nucleons have become possible [1,27] with the GFMC method discussed in the section V. Their results show that the U-IX interaction, determined from the energy of ^{3}H and nuclear matter ρ_0, underbinds $A = 8$ nuclei, and since ^{8}He is more underbound than ^{8}Be, it misrepresents the isospin dependence of V_{ijk}. The new Illinois models of V_{ijk} [28] resolve this problem by including the leading three-pion exchange term, $V_{ijk}^{3\pi}$, that is attractive in triplets having isospin $T = 3/2$, but has little effect on the $T = 1/2$ triplets in ^{3}H and ^{4}He. A much improved fit to the observed energies is obtained by adjusting the strength of $V_{ijk}^{3\pi}$. Pieper has developed two Illinois models called IL-2R and IL-3H. They can explain the energies of up to 8 nucleon states with errors less than $\sim$ 2% as shown in Fig.4. The three parameters in IL-2R model are the strengths of the $V_{ijk}^{2\pi}$ (P-wave part), V_{ijk}^R and $V_{ijk}^{3\pi}$. The chiral model containing the S-wave and P-wave parts of $V_{ijk}^{2\pi}$ [25] is used along with the cutoffs of A18 v_{ij} and the theoretical value for the strength of the small S-wave part. In the IL-3H model the strengths of both S- and P-wave parts of $V_{ijk}^{2\pi}$ are taken from theory, and the three parameters are the short range cutoff and strengths of V_{ijk}^R and $V_{ijk}^{3\pi}$. Both models seem to be equally successful in light nuclei; their predictions for nuclear and neutron star matter are being calculated.

The OPEP gives a large contribution to the interaction energies of nuclei. The expectation values of v_{ij}^π and $v_{ij} - v_{ij}^\pi$ in ^{8}Be (^{8}He) are respectively $\sim$ -238 (-162) and -70 (-66) MeV. The expectation values of $V_{ijk}^{2\pi}$, V_{ijk}^R and $V_{ijk}^{3\pi}$ of IL-2R model, in ^{8}Be (^{8}He) are respectively $\sim$ -38 (-27), +19 (+14) and -2 (-5) MeV. The small binding energies of these nuclei are mostly because the kinetic energy cancels a large fraction of the attraction from v_{ij}. The V_{ijk} presumably has additional smaller terms, but it is difficult to determine their strengths from the nuclear spectra that can be calculated accurately from bare forces at present.

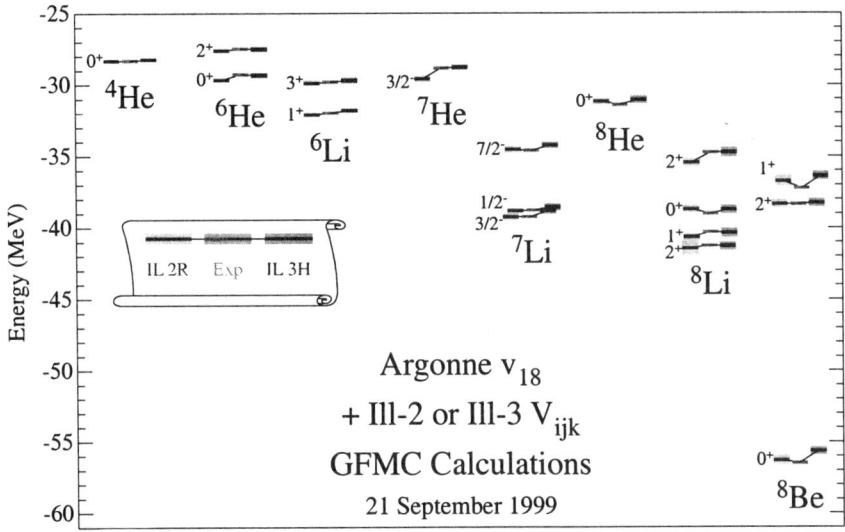

FIGURE 4. Energies calculated with A18 NN and IL-2R and IL-3H TNI models are compared with the experimental data.

IV FOUR-NUCLEON INTERACTIONS AND RELATIVISTIC EFFECTS

Since TNI are needed to reproduce the observed nuclear spectra, it is natural to ask if four nucleon interactions (FNI) are indeed negligible. Since we do not have *ab initio* models of FNI, it is not possible to calculate the expectation value of a realistic FNI, though that can be done with quantum Monte Carlo methods. Thus the estimates of FNI contributions are more indirect. All the models of V_{ijk} we have studied reproduce the energy of ^{4}He with an error of less than 0.5 %, after fitting the energy of ^{3}H. The total contribution of V_{ijk} is much smaller than that of the v_{ij} in all nuclei computed. For example, in ^{8}Be the expectation values of the v_{ij} and V_{ijk} are -308 and -21 MeV, and by comparison with experiment we estimate that the possible contribution of the neglected FNI is less than $\sim$ 1 MeV in magnitude. The binding energy of ^{8}Be has little influence on the three parameters of the IL-2R model; they are primarily determined by the energies of ^{3}H, ^{8}He and nuclear matter ρ_0.

Our Hamiltonian is nonrelativistic, thus we should expect relativistic corrections to it. However, when nonrelativistic potentials are fit to the experimental data, relativistic effects present in the data are automatically buried in these potentials. In order to study the magnitude of a chosen relativistic effect, such as that of the kinetic energy or the boost interaction [29], or those making OPEP nonlocal [30], it is necessary to refit the same data set fitted to obtain the nonrelativistic Hamiltonian, and then study the differences. Such comparisons [30] indicate that the relativistic effects associated with kinetic energies and nonlocalities of on shell OPEP are small, whereas the boost corrections which give the dependence of the NN interaction on the total momentum of the interacting pair, are significant. This is not surprising since the boost interaction is totally omitted from the conventional nonrelativistic nuclear Hamiltonian. Its contributions account for about 37 % of that of V_{ijk}^R, *i.e.* when boost interactions are included in the Hamiltonian the strength of V_{ijk}^R must be reduced by $\sim$ 37 % to reproduce the energies of light nuclei. After this adjustment the results for nuclear matter at $\rho \leq \rho_0$ are also not much changed. However at high densities the effects are noticeable. The maximum mass of neutron stars predicted by the nonrelativistic Hamiltonian in which boost interactions are buried in V_{ijk}^R, is larger by $\sim$ 8 % than that by the semi-relativistic Hamiltonian in which they are correctly included [2].

V QUANTUM MONTE CARLO CALCULATIONS

Until mid eighties only the ground state of ^{3}H could be calculated by solving the Faddeev equations. In the past ten years Quantum Monte Carlo (QMC) methods have been developed so that solutions of nuclear Schrödinger equation are now available for all the bound and narrow states of up to eight nucleons, and the prospects for applications to larger p-shell nuclei are excellent. Accurate calculations of the bound states of many-body systems in which the interactions do not depend upon the spins of the particles have been carried out in the past. In these systems the spin degrees of freedom do not play a role, and one can work with a wave function $\Phi(\mathbf{R})$ that depends only upon the positions of all the particles represented by the configuration vector $\mathbf{R} = \mathbf{r}_1, \mathbf{r}_2, ...\mathbf{r}_A$. The main difficulty in applying QMC methods to nuclear problems is that nuclear forces change spins and isospins of the interacting nucleons, and thus nuclear wave functions contain superpositions of all possible spin-isospin states of A-nucleons. Since the spin of each nucleon can be up or down, there are 2^A spin-states, and since any Z of the A nucleons can be protons, the number of isospin states is $A!/[Z!(A-Z)!]$. The total number, $2^A A!/[Z!(A-Z)!]$, of these states increases very rapidly with A. The isospin factor in this number can be reduced somewhat by assuming that the total isospin is a good quantum number [1], but the problem persists.

A Variational Monte Carlo

VMC calculations constitute the first step of QMC. The variational wave functions used for nuclei with $A \leq 8$ are discussed in refs. [1] and [27]. They have a shell model part multiplied by pair and triplet correlation operators. The wave function is represented as $\Psi(\mathbf{R}) = \sum_\alpha \psi_\alpha(\mathbf{R})|\alpha\rangle$, where $|\alpha\rangle$ are the all possible spin-isospin states, and the correlation operators as well as the interactions are matrix functions of $\mathbf{R}$.

The energies obtained for the $A = 3, 4$ nuclei from VMC calculations are above the exact E_0 only by 2-3 %. However the error increases as we go from $A = 4$ to 8. This is partly because the binding energies of these nuclei are only $\sim 15\%$ of the interaction energy, hence difficult to calculate. However, the structure of the nuclei seems to be rather well described by the present variational wave functions. For example, the pair distribution functions obtained from VMC are fairly close to those from GFMC.

B Greens Function Monte Carlo

In the GFMC method one operates on the variational wave function with the imaginary time evolution operator $exp(-[H - E_0]\tau)$ to project out the ground state for the chosen spin-parity. All the bound or narrow states of light nuclei can be studied with GFMC because those in the same nucleus have different spin-parities.

GFMC calculations of Fermion systems suffer form the "Fermion sign problem". The real wave functions of simple Fermi systems have nodal surfaces because the $\Phi(\mathbf{R})$ must equal $-\Phi(\mathbf{R}')$ when the configurations $\mathbf{R}$ and $\mathbf{R}'$ are related by the exchange of a pair of particles. GFMC configurations which diffuse across nodal surfaces, as the system evolves in imaginary time, increase the variance of the calculated quantity. The variance increases rapidly as the imaginary time τ increases and more configurations diffuse across nodal boundaries, making unconstrained propagation impractical. In the fixed node method [31] for simple systems this growth of variance is eliminated by restricting the configurations to domains enclosed by the nodal surfaces of the variational wave function. Such calculations give the exact result only when the variational wave function has the correct nodal structure; generally they have an error due to imperfections in that structure. We can make this error small by choosing a good variational wave function.

A similar problem comes in nuclear GFMC with the additional complexity due to nuclear wave functions having many spin-isospin components, each with a different nodal structure. The growth of the variance with τ depends upon the number of nucleons; it is tolerable for $A \leq 7$, so that energies with better than 1% accuracy can be obtained without constraints to reduce the variance. When $A \geq 8$ it is necessary to use constrained path methods [32] to control the variance. The constraint can be removed at large τ to test if it influenced the calculated energy significantly. It appears that calculations with $\sim$ 1% accuracy in the binding energy are possible in this way. They have been carried out for nuclei having A up to 8, and are being attempted for the $A = 9$ nuclei. The results of these calculations have been used to develop the Illinois models of V_{ijk} discussed in section III.

VI AUXILIARY FIELD DIFFUSION MONTE CARLO

Auxiliary field diffusion Monte Carlo (AFDMC) [33] seems to be the long-sought breakthrough needed to eliminate the exponential (2^A) growth of spin states in GFMC calculations of neutron matter. It can presumably be applied to study large nuclei and nuclear matter. By keeping the successful elements of GFMC, also called diffusion Monte

Carlo, for the spin-isospin independent parts, and applying an auxiliary field to the spin-isospin dependent parts of the Hamiltonian, the method combines two major themes in QMC. Auxiliary-field methods are used in the shell model Monte Carlo calculations [34], and several condensed matter systems in which the continuous spatial degrees of freedom have been eliminated.

In the approach developed by Schmidt and Fantoni [33], the spatial parts, (*i.e.* kinetic energy and spin-independent interactions), of the Hamiltonian are propagated as in GFMC and the spin-dependent interactions between neutrons are replaced by interactions of neutrons with auxiliary fields. Integrating over the auxiliary fields reproduces the original spin-dependent interaction. The method consists of a Monte Carlo sampling of the auxiliary fields and then propagating the spin variables at the sampled values of the auxiliary fields. This propagation is essentially a rotation of each particle's spin. In addition they introduce a constraint analogous to the fixed-node approximation in GFMC by requiring that the real part of the overlap with a trial function remains positive.

The AFDMC method is described in more detail in ref. [33], where results for neutrons interacting with a semi-realistic v_{ij} are reported. Since then Schmidt and Fantoni [35] have carried out calculations with a more realistic Hamiltonian consisting of Argonne v'_8 NN interaction used in the GFMC calculations [1], and the U-IX TNI. Results of calculations with 38 neutrons in a periodic box with finite size corrections have been obtained. They are in good agreement with the results obtained earlier with variational methods using hypernetted chain summation (VCS) techniques [21,2]. For example, the AFDMC and VCS energies for Argonne v'_8 and U-IX interactions are 21.8 (65.5) and 23.2 (68.6) MeV per neutron at $\rho = 0.2$ (0.4) fm^{-3}. The statistical errors in these AFDMC results are very small (~ 0.1) MeV.

The trial functions used to constrain the present AFDMC calculations are rather simple without any spin correlations. In contrast it is possible to use more accurate variational wave functions with spin correlations to constrain the GFMC calculations. Carlson [36] has compared AFDMC and GFMC results for 14 neutrons in a periodic box. At $\rho = 0.15$ fm^{-3} the GFMC energy (220 ± 1) is about 8 % below the AFDMC result of 236.4 ± 1.5 MeV. The accuracy of AFDMC can presumably be improved by using more accurate constrains. In order to study nuclei and nuclear matter with the AFDMC method, additional auxiliary fields need to be introduced to replace isospin dependent interactions.

VII NUCLEAR ELECTROMAGNETIC CURRENT

The present models of nuclear electroweak current operators are discussed in the review by Carlson and Schiavilla [8]. The electric charge and current operators are expressed as:

$$\rho(\mathbf{q}) = \sum_i \rho_i^{(1)}(\mathbf{q}) + \sum_{i<j} \rho_{ij}^{(2)}(\mathbf{q}) + \dots , \qquad (2)$$

$$\mathbf{j}(\mathbf{q}) = \sum_i \mathbf{j}_i^{(1)}(\mathbf{q}) + \sum_{i<j} \mathbf{j}_{ij}^{(2)}(\mathbf{q}) + \dots . \qquad (3)$$

TABLE 1. Thermal neutron capture cross sections in mb

Target	$\mathbf{j}^{(1)}$	$\mathbf{j}^{(1+2,MI)}$	$\mathbf{j}^{(1+2,MI+MD)}$	Expt.
^{1}H	304.1	326.9	331.4	334.2(5)
^{2}H	0.38	0.48	0.58	0.508(15)
^{3}He	0.006	0.073	0.086	0.055(3)

The $\rho_i^{(1)}$ and $\mathbf{j}_i^{(1)}$ are calculated from Dirac spinors with momenta $\mathbf{p}$ and $\mathbf{p}' = \mathbf{p} + \mathbf{q}$, and the experimental nucleon form factors. In most calculations they are expanded in powers of $1/m$ and terms up to $1/m^2$ are retained; however, in some of the recent calculations by Forest and Schiavilla this expansion is avoided (see the paper by van Orden in this symposium).

Following Riska's notation [5,6], the pair terms $\rho_{ij}^{(2)}$ and $\mathbf{j}_{ij}^{(2)}$ are divided in to two parts:

$$\rho_{ij}^{(2)} = \rho_{ij}^{(2,MI)} + \rho_{ij}^{(2,MD)}, \qquad (4)$$

$$\mathbf{j}_{ij}^{(2)} = \mathbf{j}_{ij}^{(2,MI)} + \mathbf{j}_{ij}^{(2,MD)}. \qquad (5)$$

The model independent parts, denoted by superscripts MI, are related to the NN interaction v_{ij}, and are free of parameters. These parts also ensure current conservation. The model dependent parts, with superscript MD, contain terms in which the photon changes a hadron. The MD terms in most QMC calculations include $N\Delta\gamma$, $\rho\pi\gamma$ and $\omega\pi\gamma$ couplings. The $N\Delta\gamma$ coupling gives large contributions to radiative capture reactions, and it is treated nonperturbatively with correlation operators in recent calculations [37,38].

This model is quite successful in explaining the charge form factors of $A \leq 6$ nuclei; those of ^{7}Li are yet to be calculated. However, its predictions for the ^{3}He magnetic form factor are lower than the data at $q > 3$ fm^{-1} [38]. Many more results can be found in the Carlson Schiavilla review [8]. The results for thermal neutron radiative capture cross section are summarized in Table 1. The computational methods used in the calculations of n-p and n-d capture cross sections are essentially exact; thus the difference with experiment is due to imperfections of the model. The n-^{3}He capture cross section is calculated with VMC wavefunctions in which the coupling between the incoming n-^{3}He and p-^{3}H channels is neglected.

The n-p capture proceeds mostly via the better known one-body current, and its cross section is predicted with an error of only 1 %. The parameters in the MD pair current can be tuned to obtain the observed cross section if so desired. In contrast, the n-d and n-^{3}He captures are suppressed by the pseudo orthogonality of the continuum state with the bound. The one-body currents give much smaller contributions, by a few orders of magnitude, to these reactions, and the pair currents give significant and dominant contributions respectively. The present theoretical predictions for their cross sections are larger than observed presumably due errors in the pair current operator. Accurate reproduction of these three observed cross sections provides a stringent test of the pair current operator, and indirectly of the interaction operators also, however, the later are constrained by other data such as NN scattering and nuclear binding energies, charge form factors, etc.

VIII NUCLEAR WEAK CURRENT OPERATOR

The observed beta decay transition rates in $A \leq 8$ nuclei provide tests for the nuclear weak current operator. The Fermi matrix elements in beta decay of ^{3}H and ^{7}Be are given by those of the isospin raising and lowering operators. They thus test isospin purity of these states. Accurate calculations of the trinucleon ground states have been carried out by the Pisa group including Coulomb and other isospin symmetry breaking interactions. The calculated matrix element of the T_+ operator is 0.9993, against 1 for pure $T = 1/2$ partners [13]. The present correction, -0.0007, is about twice that obtained with the older interaction models [39].

The Gamow-Teller (GT) matrix elements of the leading one-body term of the axial current, $\sum_i \vec{\sigma}_i \tau_{i,\pm}$, are influenced by the tensor correlations in nuclear ground states. For a pure S wave function the ^{3}H GT matrix element is $\sqrt{3} \times 1$. In the following we renormalize it by a factor $1/\sqrt{3}$ for clarity, so that the renormalized GT (RGT) matrix element is one for pure S wave function. Its values predicted by the modern Argonne and Nijmegen v_{ij} with the Tucson-Melbourne (TM) [24] V_{ijk}, with cutoff adjusted to give the observed ^{3}H binding energy, are between 0.925 and 0.927, while the CD-Bonn+TM give 0.937. These results are from Faddeev calculations by the Bochum group [13]. The A18+U-IX interactions give 0.922 from correlated hyperspherical harmonics calculations by the Pisa group, and the observed decay rate gives 0.957 ± 0.003. Thus one-body axial currents underestimate the decay rate by a few percent.

Models of the axial pair current operator were constructed by Chemtob and Rho [40]. The largest contributions seem to come from the weak excitation of nucleons to Δ. The $N\Delta$ weak axial coupling constant is not known. Using the quark model value $g_{N\Delta} = (6\sqrt{2}/5)g_A$, the total one- plus two-body RGT matrix element is found to be 0.964 for the A18+U-IX interactions [13]. The experimental value of 0.957 is obtained by reducing the quark model value of $g_{N\Delta}$ by ~ 20 %.

The resulting model of nuclear weak current has been used to predict the weak p-p and ^{3}He-p capture rates of interest in solar physics. Like the beta decays of ^{3}H, ^{6}He and ^{7}Be, the p-p capture is super-allowed, and proceeds mostly via the one-body axial current, with pair terms contributing only ~ 1 % to the capture cross section. In contrast the ^{3}He-p capture is suppressed by pseudo orthogonality as is the ^{3}He-n electromagnetic. In this case the axial pair current seems to cancel the small one-body current matrix element and drastically reduce the cross section [37,42]. Bahcall and Krastev [41] have argued that the solar neutrino spectrum observed by Super Kamiokande indicates a much larger ^{3}He-p capture rate than predicted by the present models. New calculations [42] indicate that the first-forbidden capture from negative parity incident waves is comparable to the capture via the suppressed GT, and may increase the predicted rate, though not by as much as needed to explain the Super Kamiokande data.

IX CONCLUSIONS

Accurate calculations, with errors less than a percent, are essential to extract information on nuclear forces and currents from observed data. In many cases, the contributions of three-body forces as well as pair currents are small, and therefore the leading terms must be accurately computed. For the three- and four-body nuclei this is now possible with the Faddeev, correlated hyperspherical harmonics and GFMC methods. Bound states of nuclei with $A \leq 8$ can now be computed with the GFMC method. Seemingly realistic models of V_{ijk} have been constructed using this ability. However, the real test of these models will come when we can apply them to heavier nuclei. Calculations based on these models indicate that the better known pion exchange forces give the largest contributions to nuclear binding; however, the phenomenological parts in the force models are quite significant.

We can also compute elastic scattering form factors, Euclidean responses [8], and a few radiative capture reactions accurately. The present results indicate that our theoretical models of the pair electromagnetic current are not accurate enough. For example, the n-d radiative capture cross section and the ^{3}He magnetic form factor are computed essentially exactly, but the results are in significant disagreement with experiment. A major challenge is to develop exact methods to compute the initial state in ^{3}He-n capture with its coupling to the ^{3}H-p channel. Attempts to calculate the α-d, α-t and ^{7}Be-p captures are in progress.

Accurate calculations of ^{3}H beta decay with all modern interactions confirm the need of axial pair currents. These are to be expected because there must be a weak axial N-Δ coupling. The beta decays studied so far with VMC wave functions are superallowed, and have very small pair-current contributions. Significant pair-current contributions are expected for the beta decays of the $J^\pi = 2^+$ ground states of ^{8}Li and ^{8}B to the 2^+ resonance in ^{8}Be. The main challenge here is to compute the resonant final state.

It is necessary to extend the domain of accurate computations beyond $A = 8$. Wiringa has started doing $A = 9$ VMC calculations, and the present GFMC can be pushed up to $A = 12$ in the next ~ 5 years. The data on $9 \leq A \leq 12$ nuclei will provide valuable information on nuclear forces and currents. A significant computational challenge is to apply the new AFDMC method, or any other exact method, to study larger nuclei like ^{16}O, ^{40,48}Ca, and nuclear matter.

ACKNOWLEDGMENTS

It is a pleasure to thank my colleagues Ana Arriaga, Arya Akmal, Joe Carlson, Stefano Fantoni, Jun Forest, Jim Friar, Walter Glöckle, Laura Marcucci, Steve Pieper, Brian Pudliner, Geoff Ravenhall, Dan Riska, Sergio Rosati, Rocco Schiavilla, Kevin Schmidt and Robert Wiringa for their extensive contributions to this work, and for numerous discussions. This work has been supported in part by the U.S. National Science Foundation via grant PHY 98-00978.

REFERENCES

1. B. S. Pudliner, V. R. Pandharipande, J. Carlson, S. C. Pieper and R. B. Wiringa, Phys. Rev. C **56**, 1720 (1997).
2. A. Akmal, V. R. Pandharipande and D. G. Ravenhall, Phys. Rev. C **58**, 1804 (1998).
3. R. B. Wiringa, Phys. Rev. C **43**, 1585 (1991).
4. J. Carlson, Nucl. Phys. A **522**, 185c (1991).
5. R. Schiavilla, V. R. Pandharipande and D. O. Riska, Phys. Rev. C **40**, 2294 (1989).

6. R. Schiavilla, V. R. Pandharipande and D. O. Riska, Phys. Rev. C **41**, 309 (1990).
7. J. Carlson, D. O. Riska, R. Schiavilla and R. B. Wiringa, Phys. Rev. C **44**, 619 (1991).
8. J. Carlson and R. Schiavilla, Rev. Mod. Phys. **70**, 743 (1998).
9. V. G. J. Stoks, R. A. M. Klomp, M. C. M. Rentmeester and J. J. de Swart, Phys. Rev. C **48**, 792 (1993).
10. V. G. J. Stoks, R. A. M. Klomp, C. P. F. Terheggen and J. J. de Swart, Phys. Rev. C **49**, 2950 (1994).
11. R. B. Wiringa, V. G. J. Stoks and R. Schiavilla, Phys. Rev. C **51** 38 (1995).
12. R. Machleidt, F. Sammarruca and Y. Song, Phys. Rev. C **53**, R1483 (1996).
13. R. Schiavilla et.al., Phys. Rev. C **58**, 1263 (1998).
14. J. L. Friar, G. L. Payne, V. G. J. Stoks and J. J. de Swart, Phys. Lett. B **311**, 4 (1993).
15. L. C. Alexa et. al. Phys. Rev. Lett. **82**, 1374 (1999).
16. D. Abbot et. al. Phys. Rev. Lett. **82**, 1383 (1999).
17. E. Beise, private communication (1999).
18. S. A. Coon and J. L. Friar, Phys. Rev. C **34**, 1060 (1986); and J. L. Friar preprint LA-UR-99-296.
19. J. L. Forest, Phys. Rev. C (2000), to be published.
20. J. L. Forest, V. R. Pandharipande, S. C. Pieper, R. B. Wiringa, R. Schiavilla and A. Arriaga, Phys. Rev. C **54**, 646 (1996).
21. A. Akmal and V. R. Pandharipande, Phys. Rev. C **56**, 2261 (1997).
22. L. Engvik, M. Hjorth-Jensen, R. Machleidt, H. Muther, and A. Polls, Nucl. Phys. **A627**, 85 (1997).
23. J. I. Fujita and H. Miyazawa, Prog. Theor. Phys. **17**, 360 (1957).
24. S. A. Coon, M. D. Scadron, P. C. McNamee, B. R. Barrett, D. W. E. Blatt and B. H. J. McKeller, Nucl. Phys. A **317**, 242 (1979).
25. J. L. Friar, D. Huber and U. van Kolck, Phys. Rev. C **59**, 53 (1999).
26. B. S. Pudliner, V. R. Pandharipande, J. Carlson, and R. B. Wiringa, Phys. Rev. Lett. **74**, 4396, (1995).
27. S. C. Pieper, R. B. Wiringa, J. Carlson and V. R. Pandharipande, to be published, (2000).
28. J. Carlson, S. C. Pieper, R. B. Wiringa and V. R. Pandharipande, to be published, (2000).
29. J. L. Forest, V. R. Pandharipande, and J. L. Friar, Phys. Rev. C **52**, 568 (1995).
30. J. L. Forest, V. R. Pandharipande, and A. Arriaga, Phys. Rev. C **60** 014002 (1999).
31. J. B. Anderson, J. Chem. Phys. **65**, 4122 (1976).
32. S. Zhang, J. Carlson and J. E. Gubernatis, Phys. Rev. Lett. **74**, 3652 (1995).
33. K. E. Schmidt and S. Fantoni, Phys. Lett. **B446**, 99 (1999).
34. S. E. Koonin, D. J. Dean, and K. Langanke, Phys. Rep. **278**, 1 (1997).
35. K. E. Schmidt and S. Fantoni, Private Communication (1999).
36. J. Carlson, Private Communication (1999).
37. R. Schiavilla, R. B. Wiringa, V. R. Pandharipande and J. Carlson, Phys. Rev. C **45**, 2628 (1992).
38. L. Marcucci, D. O. Riska and R. Schiavilla, Phys. Rev. **C58**, 3069 (1998).
39. T. -Y. Saito, Y. Wu, S. Ishikawa and T. Sassakawa, Phys. Lett. B **242**, 12 (1990).
40. M. Chemtob and M. Rho, Nucl. Phys. A **163**, 1 (1971).
41. J. N. Bahcall and P. I. Krastev, Phys. Lett. B **436**, 243 (1998).
42. L. Marcucci, private communication (1999).

Elastic Scattering Studies of Deuterium

Michel Garçon

DAPNIA/SPhN, CEA-Saclay, 91191 Gif-sur-Yvette, France.

Abstract. The experimental information concerning elastic electron-deuteron scattering is reviewed, with emphasis on latest results at intermediate to high momentum transfers. The Bates Linear Accelerator Center played a key role in initiating and developing the necessary measurements of polarization observables, and will continue to do so with the BLAST project. The interpretation of observables and form factors is illustrated with some recent theoretical calculations, while their asymptotic behaviour is discussed in the light of perturbative quantum chromo-dynamics. In connection with the search for quark effects in nuclei, a qualitative description of the proton charge distribution is attempted.

INTRODUCTION

Elastic electron-deuteron (*e-d*) scattering has been for over fifty years the reaction of choice to study the electromagnetic structure of the simplest compound nucleus. The deuteron, as a two-nucleon system, is in effect the best testing ground for precise descriptions of the nucleon-nucleon interaction, for the investigation of the role of non-nucleonic degrees of freedom in nuclei, as well as the role of relativity in nuclear structure.

As the four-momentum Q transferred to the deuteron in elastic *e-d* scattering is increased, one explores the structure of this two-nucleon system at a shorter relative distance. A breakdown of the non-relativistic description of two interacting nucleons should occur at some distance scale. Furthermore, when this scale becomes comparable to, or smaller than, the size of the nucleons, we would expect, from a classical geometrical picture of overlapping spheres, that the deuteron structure be sensitive to details of the nucleon structure or excitations. This scale would set a limit to our conventional description of nuclei.

THE NON-RELATIVISTIC PICTURE OF THE DEUTERON

We review briefly the connection between the nucleon-nucleon potential the deuteron wave function and its shape.

The only well established part of the nucleon-nucleon interaction is the long range one pion exchange, which has a tensor part (in configuration ⊗ spin space). For internucleon distances of the order of 2 fm, the attractive interaction is described in terms of the exchange of two pions and/or heavier mesons, including sometimes the hypothetical σ meson. At still shorter distances (1 fm or less), the interaction becomes repulsive. All potentials used in non-relativistic calculations include a purely phenomenological part to describe the short range repulsion, and part of the medium range attraction. Potentials used in field theory based relativistic calculations are entirely built as a superposition of one boson exchanges (OBE).

Given the spin-parity $J^P = 1^+$ of the deuteron, its wave function is obtained by coupling its spin $S = 1$ to an orbital angular momentum $L = 0$ or 2. Each component in a given substate $|J = 1, M >$ can be written as

$$\Psi^M = \frac{1}{r}[u(r) + \frac{1}{\sqrt{8}}w(r)S_{12}] \cdot \chi_1^M \tag{1}$$

where χ_1^M is the three-component spin 1 spinor. The tensor operator S_{12} introduces a mixing between the $^3S_1$ and $^3D_1$ waves (u and w). The Schrödinger equation in presence of a tensor interaction leads to a set of coupled equations for u and w [1].

The shape of the deuteron is then characterized by the nucleon densities in given substates: $|\Psi^0|^2$ and $|\Psi^1|^2 = |\Psi^{-1}|^2$. These are illustrated in Fig. 1.

Electron elastic scattering off the deuterium determines form factors (see below) which, in the non-relativistic impulse approximation, are Fourier transforms of these densities (or of combinations of them). For example a recent re-analysis of low Q e-d data determines precisely the charge radius (which results from the convolution of the illustrated density with the nucleon intrinsic charge distribution): $< r^2 >^{\frac{1}{2}} = 2.130 \pm 0.012$ fm [5].

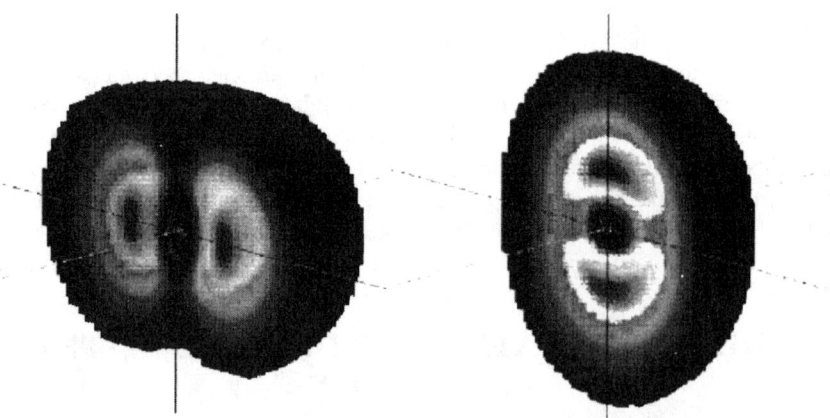

FIGURE 1. Deuteron densities in $M = 0$ (left) and $M = 1$ (right) states, for the Paris potential [2]. Outer surface is for 10% of maximal density. See [3,4] for equivalent representations.

OBSERVABLES AND EXPERIMENTS

All observables of elastic e-d scattering are quadratic combinations of the three electromagnetic form factors that characterize a spin 1 nucleus: charge monopole $G_C(Q)$, magnetic dipole $G_M(Q)$ and charge quadrupole $G_Q(Q)$. The differential cross section differs from the Mott scattering by a factor $S \equiv A + B\tan^2\frac{\theta_e}{2}$. Its measurements at forward and backward angles θ_e yield the elastic structure functions $A(Q)$ and $B(Q)$:

$$A = G_C^2 + \frac{8}{9}\eta^2 G_Q^2 + \frac{2}{3}\eta G_M^2, \qquad (2)$$

$$B = \frac{4}{3}\eta(1+\eta)G_M^2, \qquad (3)$$

with $\eta = Q^2/4M_d^2$. The separate determination of G_C and G_Q necessitates the measurement of at least one polarization observable. The recoil deuteron tensor polarization (t_{2q}, with $q = 0, 1, 2$), or alternatively the tensor analyzing powers T_{2q} measured via scattering off a tensor polarized target, yield other combinations of the form factors, with the following structure:

$$t_{20} = \tilde{t}_{20} + \mathcal{O}\left(\frac{B}{A}, \frac{B\tan^2\frac{\theta_e}{2}}{A}\right), \text{ with } \tilde{t}_{20} = f\left(\frac{2\eta G_Q}{3G_C}\right), \qquad (4)$$

$$t_{21} = \frac{G_M G_Q}{S} g(\eta, \theta_e), \qquad (5)$$

$$t_{22} = \frac{G_M^2}{S} h(\eta) \propto \frac{B}{A}. \qquad (6)$$

The exact expressions can be found e.g. in [6]. t_{20} is the most useful quantity to separate G_C and G_Q. It is a measure of the relative probabilities of finding the deuteron in substates $M = -1, 0$ or 1 after the scattering. Using a longitudinally polarized electron beam, the recoil deuteron can also acquire a vector polarization (alternatively, the scattering will depend on the vector polarization of the target): these observables, it_{11} or iT_{11}, were never measured (see discussion in [7]).

The structure function A has been measured recently at Jefferson Lab up to a very high momentum transfer, 6 (GeV/c)^2 or 12.4 fm^{-1} [8]. An independent measurement was also carried out at intermediate values of Q [9]. The two measurements resolve in part some discrepancies between older data sets [10–12]. The remaining discrepancies are of the order of 12% for $4.5 < Q < 6.8$ fm^{-1}.

Since the observation of a minimum in the structure function B (node of G_M) around 7 fm^{-1} [13], there are no new published measurements concerning the magnetic form factor. A recent experiment at Jefferson Lab [14], between 4.2 and 5.9 fm^{-1}, seems to confirm earlier data [13,15]. An extension of this experiment up to about 9 fm^{-1} is being envisaged [14].

For an exhaustive list of A and B measurements, see [5,16]. As for the measurements of polarization observables in e-d scattering, they are all mentioned below.

The Bates Linear Accelerator Center was the first to measure t_{20} [17], benefiting from the development of a low energy deuteron polarimeter [18]. After that, recoil polarization techniques were developed for higher energy deuterons, mostly at the synchrotron SATURNE [19], applied once more at Bates [6] with the polarimeter AHEAD [20] and recently at Jefferson Lab [21] (POLDER polarimeter [22]) in double scattering experiments. The alternative measurement of the tensor analyzing power T_{20}, using a polarized target [23] intercepting a stored electron beam, was pioneered at Novosibirsk/VEPP-2 [24,25], and then gradually improved at VEPP-3 [26] and NIKHEF/AmPS [27,28]. An attempt to use a solid cryogenic polarized target (ND_3) in an external electron beam at Bonn was not very successful [29].

The competition between these techniques has up to now turned to the advantage of the recoil polarization measurement, simply because it allows to reach higher momentum transfers. From the point of view of statistics, the internal polarized deuterium targets are not dense enough to reach equivalent luminosities. A

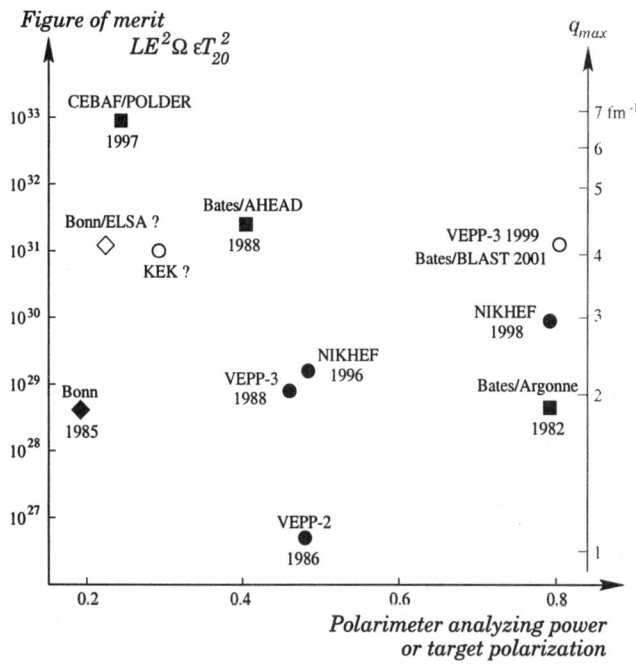

FIGURE 2. Figure of merit for all existing (full symbols) and planned (open symbols) polarization experiments. Squares are for t_{20} measurements (recoil polarimetry), circles for T_{20} measurements using internal polarized targets, and diamonds for T_{20} with external polarized target. See references in text. The dates correspond to the data taking periods.

quantitative comparison of the two techniques is presented in Fig. 2, using a figure of merit defined as the product luminosity (L) × solid angle (Ω) × efficiencies (ε) × (polarimeter analyzing power or target polarization)^2 × (beam energy E)^2 [7]. The last factor accounts for the approximate dependence of the elastic cross section for a fixed value of Q. But the methods should also be compared from the point of view of systematic errors, which are quite different. The recoil polarimetry necessitates a separate calibration experiment and absolute measurements (see however [6,7]) which render the data analysis very delicate. In contradistinction, the polarized target experiments are conceptually simpler, with a more straightforward data analysis, the most difficult issues being the measurement of the target polarization and, at the highest energies, the elimination of background.

We will witness in the next few years the continuous improvements of internal polarized targets experiments, both at Bates [30] and at Novosibirsk [31]. In addition, the new detection capabilities at small angles and for recoil particles of the HERMES detector [32] might make high Q measurements of T_{20} possible at DESY. This should be investigated. As for external polarized targets, the use of ^{6}LiD now allows to withstand beam intensities of the order of 80 nA [33], but there are no foreseen e-d experiments using this technique.

The polarimetry method, on the other hand, seems difficult to extend to still higher momentum transfers: not only there is no established way to measure the tensor polarization of 500 MeV to 1 GeV deuterons (corresponding to $7 < Q < 10$ fm^{-1}), but a double scattering experiment becomes very hard at high Q with the small solid angles associated to the necessary magnetic spectrometers (and the increasing kinematical mismatch between them).

LATEST RESULTS AND THEIR SIGNIFICANCE

The latest experimental results (T_{20} from NIKHEF, t_{20} and A from Jefferson Lab) deal mostly with charge scattering. Only t_{21}, proportional to the magnetic form factor, helps to confirm a node of this form factor around 7 fm^{-1}, although with limited accuracy [21]. So our discussion will focus mostly on charge scattering. Nevertheless, it should be kept in mind that the magnetic form factor is very sensitive to various theoretical assumptions or ingredients.

The meso-nucleonic description of the deuteron

In the non-relativistic impulse approximation (NRIA), each form factor is a product of the isoscalar nucleon form factor by an integral over a function of u^2, uw and w^2. Nucleon-nucleon potentials such as Paris, Argonne v18, Nijmegen, generate very similar deuteron wave functions, so that their predictions for the e-d observables do not differ very much. The various versions of the Bonn potential can give significantly different values of G_C and therefore of $\tilde{t}_{20}$ already at the NRIA level. In addition, uncertainties in the nucleon charge form factors should also be considered [34,35]: precise direct determinations of the neutron charge form factor G_E^n are still in their infancy [36], while recent measurements of the proton charge form factor G_E^p [37] could lead to a reexamination of our empirical knowledge of the nucleon form factors. We will have a remark connected to this at the end of

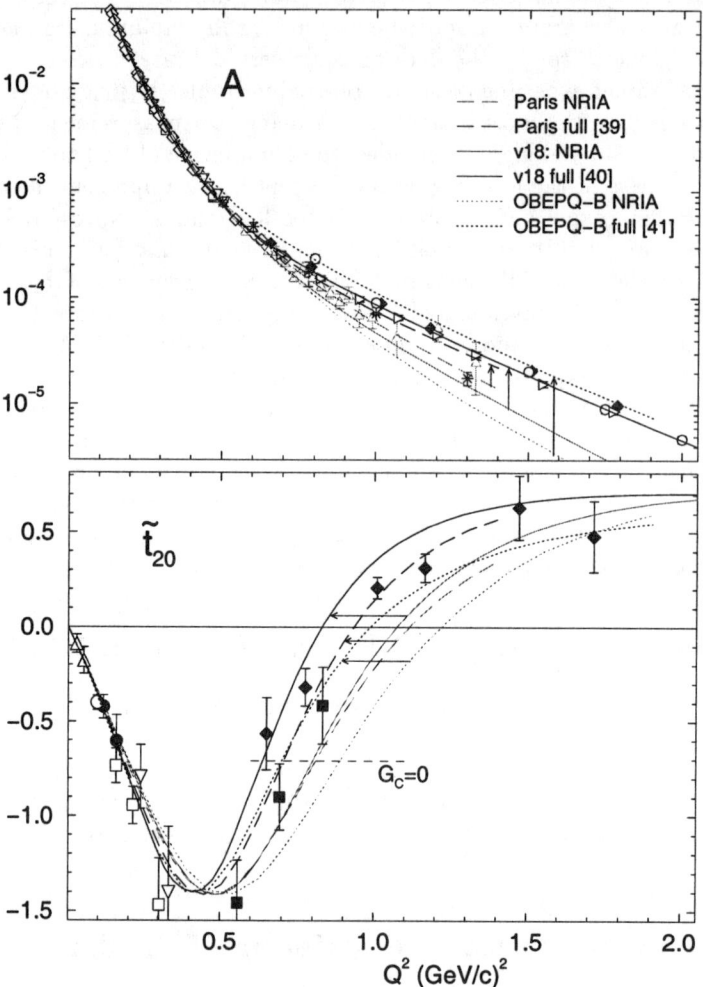

FIGURE 3. A and $\tilde{t}_{20}$ as a function of Q^2. The A data are from Refs [8–12,43–45]. The t_{20} experiments performed at Bates are indicated by solid circles [17] and squares [6]. On the $\tilde{t}_{20}$ graph, full symbols are associated with double scattering experiments [6,17,21], while open symbols are for internal polarized target experiments [24–28]. The value $\tilde{t}_{20} = -\sqrt{2}/2$ (horizontal dashed line) corresponds to the node of G_C. The theory curves are associated two by two by an arrow showing the change from a NRIA calculation to a full (NRIA + RC + MEC) calculation.

this paper. These uncertainties affect the calculation of A, but not $\tilde{t}_{20}$, since in this ratio (and still in the NRIA), the nucleon form factors are eliminated.

Relativistic corrections (RC) and meson exchange currents (MEC) contributions to the deuteron form factors can be implemented in a systematic way [38], but in practice, various calculations can differ significantly from each other. The isoscalar MEC are not as well determined as their isovector counterpart. For the later only, the direct coupling of the virtual photon to a pion in flight is allowed, leading to large contributions to isovector transitions such as $ed \to e(pn)$ at threshold. The isoscalar MEC, in contradistinction, are significantly smaller and dominated by the pion pair term in e-d elastic scattering. This term always shifts the node of the charge form factor G_C to smaller values of Q, and this is reflected in the shift in $\tilde{t}_{20}$ illustrated in Fig. 3. The $\rho\pi\gamma$ MEC also plays a role in all observables, but the exact magnitude of this contribution is model dependent. Different calculations of MEC's differ by the meson-nucleon-nucleon (and $\rho\pi\gamma$) vertex form factors or coupling constants. Some calculations do not include the proper Lorentz boost of the wave functions (see discussion in [38,41]).

A test of our understanding of the isoscalar MEC is provided by a comparison of the isoscalar charge form factors for the A=2 (deuteron) and A=3 (^{3}H/^{3}He) systems [42]. A fit to the world e-d data indicates a position of the node of G_C around $Q = 4.17$ fm^{-1} [16], quite compatible with expectations and with the position of the node of the isoscalar ^{3}H/^{3}He system [42]. The heights of the secondary maximum of $|G_C|$, however, do not seem to be compatible in both systems, as illustrated in Fig. 4. In this figure, the Argonne $v18$ calculation [46] is not correlated with the calculations performed with other potentials by the Hannover group [42]: this is probably due to a different approach of the MEC calculation and confirms the need to establish a "standard model" of MEC.

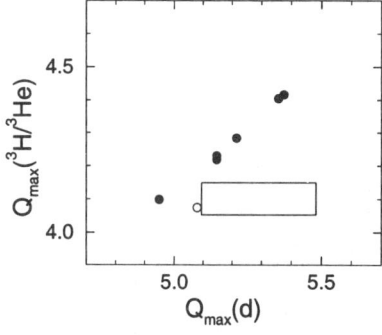
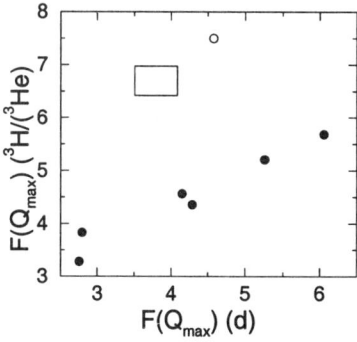

FIGURE 4. Position Q_{max} (fm^{-1}) and values ($\times 10^3$) of secondary maximum of the charge form factor, in the isoscalar A=3 system (F = $\frac{1}{2}|2F_C(^3\text{He})+F_C(^3\text{H})|$) vs the deuteron (F = $|G_C|$). The full circles are from [42] and correspond to the potentials RSC, Bonn C, Paris, Bonn B, Bonn OBEPQ, Bonn A. The open circle is from [40,46], with Argonne $v18$ potential. All calculations are NRIA + RC + MEC. The boxes correspond to experiment. The figure is adapted from [42].

The non relativistic description of the deuteron form factors should break down as Q/M_N becomes comparable to or higher than 1. Some authors indeed caution about the use of their non relativistic calculations beyond 5 fm^{-1} [41]). Amazingly enough, such calculations seem to describe the data much beyond this value.

Purely relativistic calculations have made significant progress in the past decade. They are reviewed in these proceedings by W. van Orden. Using One Boson Exhange nucleon-nucleon potentials consistently with quantum field theory, it has now become possible to include MEC contributions to form factors calculated in

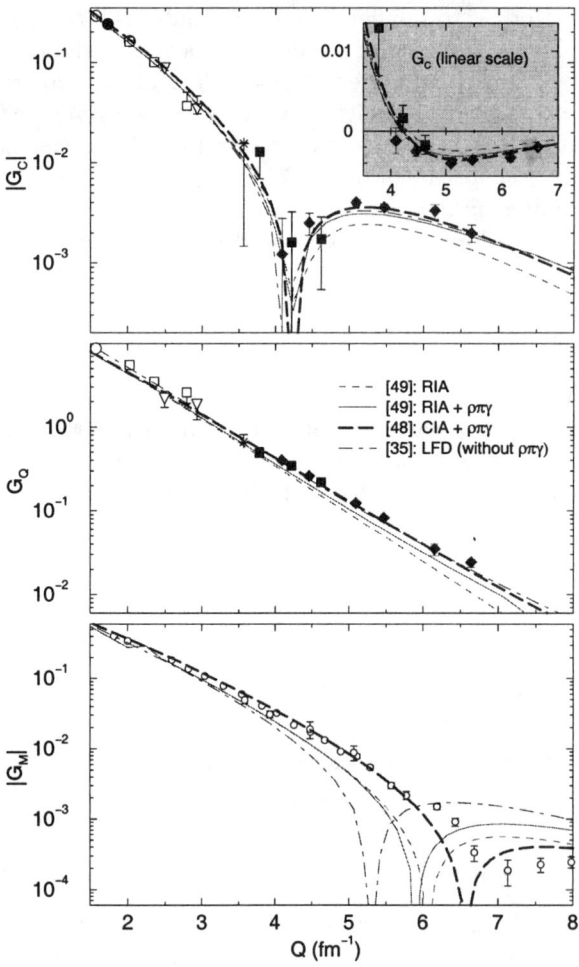

FIGURE 5. Relativistic calculations of the three deuteron electromagnetic form factors. The G_C and G_Q data are extracted from t_{20} measurements [16], while G_M is readily calculable from B measurements.

various 3-D reductions of the Bethe-Salpeter equation [47–49] or in covariant Light-Front Dynamics [35,50]. These are illustrated in Fig. 5 where, instead of plotting e-d observables, we choose to plot the three form factors (the extraction of G_C and G_Q from A, B and t_{20} is rather straightforward, but with some subtleties [16]). The magnetic form factor is very sensitive to small components of the deuteron wave function, of purely relativistic origin. The excellent agreement obtained in [48] is due in part to the use of a very soft $\rho\pi\gamma$ form factor.

Isobar and quark cluster components in the deuteron

As illustrated above, a description of the nucleon in a meso-nucleonic picture, whether non relativistic or relativistic, can reproduce the main trend of the data. Small deviations from experiment could still be due to uncertainties in these calculations. In particular, we pointed out some differences between different non-relativistic calculations.

Alternatively, one can go beyond the picture of two nucleons exchanging mesons. The nucleons can be excited into baryonic resonances, Δ or N^*, and these may be explicitly considered in the deuteron wave function. In the light of such calculations [51,52], the new Jefferson Lab t_{20} data does not seem to accomodate a percentage of $\Delta\Delta$ components in the deuteron much higher than 0.5%. Quark clusters in the deuteron have also been considered in [53] and earlier work. These configurations do not necessarily lead to predictions much different from the conventional calculations. The precision of the new e-d data at intermediate and high Q probably makes it worthwhile to reexamine and update these calculations.

On the relevance of perturbative QCD

The asymptotic behaviour of electromagnetic form factors can be inferred from quark counting rules and perturbative quantum chromo-dynamics (PQCD). The Q-scale at which this asymptotic behaviour becomes applicable is however the subject of much theoretical debate [54–56]. If PQCD becomes applicable, it would certainly mean that the electron scatters from a six-quark object which does not bear much ressemblance with our conventional deuteron. But the reverse is not necessarily true: quark effects in nuclei, though still elusive, could manifest themselves before PQCD becomes applicable.

There are three levels of predictions:

i) Dimensional scaling [57]: the leading form factor of a hadron composed of n quarks should behave asymptotically as $Q^{-2(n-1)}$. For the deuteron, n=6 and one expects $A \sim Q^{-20}$.

ii) Logarithmic corrections to the leading amplitude, together with the assumption of the factorization of the nucleon form factors, yield [58]:

$$\sqrt{A} \sim F_N^2 \left(\frac{Q^2}{4}\right) \cdot \frac{1}{Q^2} \cdot \left(\ln \frac{Q^2}{\Lambda^2}\right)^{-1+\epsilon} \tag{7}$$

where F_N is the nucleon form factor and Λ is an energy scale characteristic of QCD. The absolute normalization of A is not predicted.

iii) Helicity conservation at the quark level, at each photon/gluon-quark vertex, implies helicity conservation for the deuteron between the initial and final states [54]. In the light-cone frame (LCF), the form factors are best expressed in an helicity basis [56]: helicity conservation is then equivalent to the prediction that G_{00}^+, the "+" component of the current matrix element, is the leading amplitude. Other components G_{+0}^+ (single helicity flip) and G_{+-}^+ (double helicity flip) should be smaller by respectively one or two powers of Q. Note that point ii) above supposes that the scattering is dominated by the non-helicity flip amplitude. Since the usual form factors are linear combinations of the $G_{hh'}^+$'s, it is straightforward to calculate ratios such as B/A, t_{20} and t_{21}. Again, the absolute scale of these amplitudes is not known.

What do elastic e-d scattering experiments tell us in this respect? The new A data [8] extends to very high Q and is certainly suggestive of a behaviour in Q^{-20}, although a Q^{-16} behaviour is still not excluded (see Fig 6). Furthermore, the logarithmic corrections to the Q^{-20} behaviour, mostly due to the QCD running coupling constant, yield the observed behaviour of A between 2 and 6 $(GeV/c)^2$. Considering the helicity amplitudes, the minimal ansatz (which ignores logarithmic corrections and does not constitute a systematic expansion in powers of $1/Q$) is:

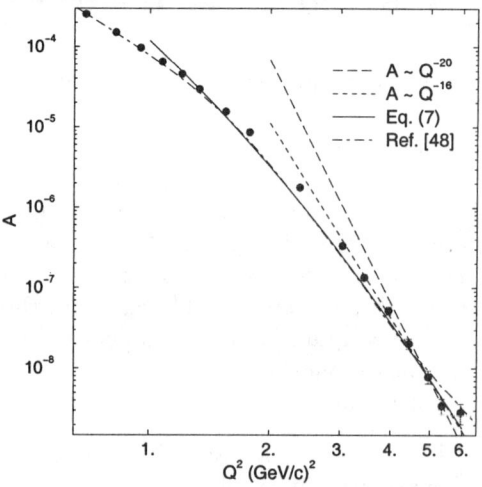

FIGURE 6. $A(Q^2)$, in a Log-Log scale. Data from Jefferson Lab./Hall A [8]. All curves are normalized to the point at 4.95 $(GeV/c)^2$, except for the calculation of Ref. [48] which illustrates that nucleon-nucleon dynamics also accounts for the high Q data.

$$\frac{G^+_{+0}}{G^+_{00}} = a\left(\frac{\Lambda}{Q}\right) \quad \text{and} \quad \frac{G^+_{+-}}{G^+_{00}} = b\left(\frac{\Lambda}{Q}\right)^2. \tag{8}$$

The pure dominance of G^+_{00} ($a = b = 0$) [56] fails to reproduce the existing data below 2 (GeV/c)^2 (see Fig. 7). The prescription $a = 5$ and $b = 0$ was built to accomodate the existence of a node in G_M [59], but it fails to reproduce the new t_{20} data. The most economical way to reproduce the trend of the data in the 1-2 (GeV/c)^2 range is to fix a and b by the position of the nodes of G_C and G_M [60]. All data is then accounted for, at least qualitatively (this is not a fit). Note that the prescriptions [56,61] both give the wrong relative sign of G_Q and G_M, since they lead to $t_{21} < 0$. With $\Lambda = 200$ MeV and $b \simeq 38$, our prescription implies $G^+_{+-}/G^+_{00} \simeq 1.5$ to 0.75, which means that the double helicity flip amplitude is as large as the non helicity flip amplitude in the momentum transfer range considered. The single helicity flip amplitude seems to be smaller

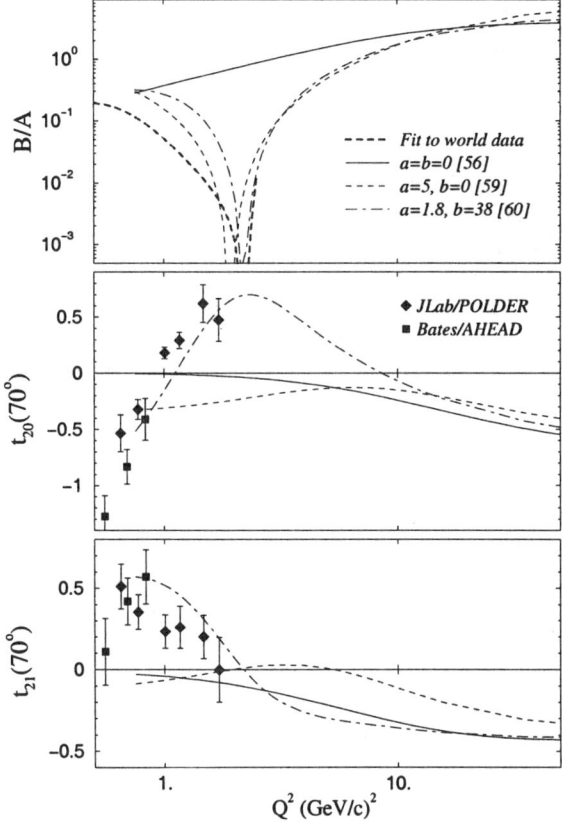

FIGURE 7. Ratios B/A, t_{20} and t_{21} from light-cone frame helicity amplitudes. See text for definition of parameters a and b. Note the logarithmic scale in Q^2.

($a \simeq 1.8 \Rightarrow G^+_{0+}/G^+_{00} \simeq 0.4$ to 0.25). A fit to the data using reduced LCF helicity amplitudes leads to the same conclusion [16,61].

From this discussion, we infer that PQCD can also account for the Q-dependence of A above 2 (GeV/c)^2, but that polarization data up to this value implies that the non helicity flip amplitude is not the dominant one. Thus there seems to be a contradiction around 2 (GeV/c)^2 between these two statements. They could be reconciled if the double helicity flip amplitude were to decrease very quickly beyond the point where it is now determined. It would certainly constitute a big progress if the absolute normalization of these amplitudes could be evaluated within PQCD.

A REMARK ABOUT THE NUCLEON STRUCTURE

We do not have yet a description of the deuteron form factors that prevails over all other ones. Still, we can come back to our introductory question: at high Q, do the nucleons overlap so much that their internal structure has to play a role in the process? Around 2 (GeV/c)^2, we should be sensitive to internucleon distances of the order of 0.3 fm. This is certainly smaller than the charge radius of a proton (~ 0.8 fm).

Let us in this respect make a conjecture based on a qualitative interpretation of recent measurements of the proton charge form factor: the ratio of charge to magnetic form factors, G^p_E/G^p_M, was recently measured with significantly increased accuracy [37]. Using the experimentally established $G^p_M \simeq \mu_p/(1.+Q^2/0.71)^2$, G^p_E can be readily obtained. The Fourier transform of a dipole form factor is an exponential density. But it would be very strange if the proton had exactly such a density, down to its center. Intuitively, one would expect the density to "level off" at some value r_0 of the radius, which could be represented by:

$$\begin{aligned}\rho(r) &= \rho_0 \text{ if } r < r_0, \\ \rho(r) &= \rho_0 e^{-\alpha(r-r_0)} \text{ if } r > r_0,\end{aligned} \quad (9)$$

with α fixed as a function of r_0 by the proton charge radius and ρ_0 by the integral charge. Such a schematic density with $r_0 = 0.24$ fm (see Fig. 8) does correspond, by Fourier transform, to a form factor with the measured behaviour. This reasoning can only be qualitative, since the correspondance between form factor and static density, without taking into account any relativistic contraction, is correct only non relativistically. One can certainly object to its application beyond $Q^2 \simeq 1$ (GeV/c)^2 [62]. Nevertheless, the picture may tell us that the proton has a small core (quarks only?) and an exponential tail (the pion cloud?). This is also what many theoreticians have been proposing for a decade or so.

It may just be that the quark core of the nucleon is very small. This could explain why quark degrees of freedom do not play an explicit role in the structure of nuclei.

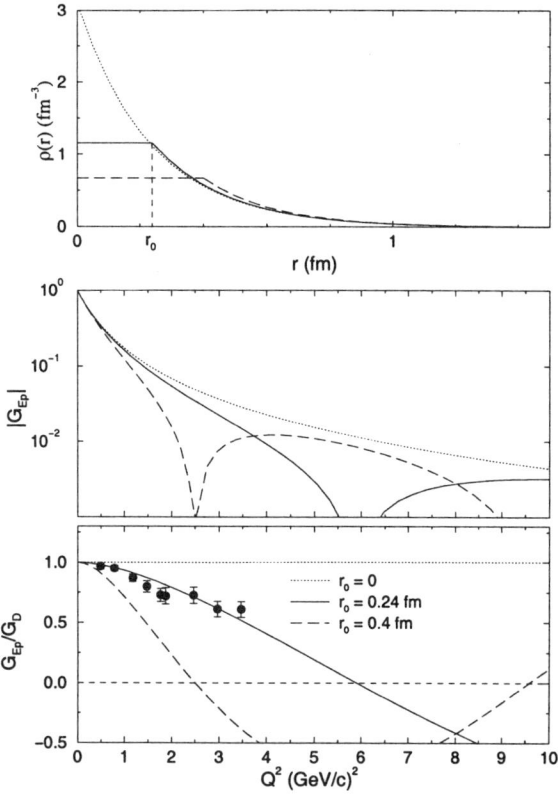

FIGURE 8. Proton charge density and its Fourier transform, according to Eq.(9), for 3 values of r_0. The data are from Ref. [37], with the additional assumption that $G^p_M = \mu_p G_D$.

ACKNOWLEDGEMENTS

It should be acknowledged here that the initiative and dedication of the Argonne group (led by R. Holt) and of the Alberta group at Edmonton (J. Cameron) were determining factors in the success of the two t_{20} experiments at Bates. It was a great experience for me to contribute to the Bates scientific program (through the $^{16}O(\gamma, n)^{15}O$ and AHEAD t_{20} experiments) and a pleasure to work with people like E. Moniz, R. Redwine, W. Turchinetz and many others at MIT. A good part of the work presented in this talk was done within the JLab/POLDER collaboration (see *http://t20.jlab.org*). Finally, I would like to acknowledge stimulating discussions with my students K. Hafidi and D. Pitz, and I thank H. Arenhövel for communicating the results of his calculations prior to publication.

REFERENCES

1. G.E. Brown and A.D. Jackson, *The nucleon-nucleon interaction*, North-Holland, Amsterdam, 1976.
2. M. Lacombe et al., *Phys. Lett.* **101B**, 139 (1981).
3. T. Ericson and W. Weise, *Mesons and Nuclei*, Clarendon Press, Oxford, 1988, p. 49.
4. J.L. Forest et al., *Phys. Rev. C* **54**, 646 (1996).
5. I. Sick and D. Trautman, *Nucl. Phys.* **A456**, 559 (1998).
6. M. Garçon et al., *Phys. Rev. C* **49**, 2516 (1994); I. The et al., *Phys. Rev. Lett.* **67**, 173 (1991).
7. M. Garçon, *Nucl. Phys.* **A508**, 445c (1990).
8. L.C. Alexa et al. (Jefferson Lab Hall A collaboration), *Phys. Rev. Lett.* **82**, 1379 (1999).
9. D. Abbott et al. (Jefferson Lab t_{20} collaboration), *Phys. Rev. Lett.* **82**, 1374 (1999).
10. J.E. Elias et al., *Phys. Rev.* **9**, 521 (1969).
11. R.G. Arnold et al., *Phys. Rev. Lett.* **35**, 776 (1975).
12. R. Cramer et al., *Z. Phys.* **C29**, 513 (1985).
13. P.E. Bosted et al., *Phys. Rev. C* **42**, 38 (1990).
14. M. Petratos, private communication.
15. S. Auffret et al., *Phys. Rev. Lett.* **54**, 649 (1985).
16. D. Abbott et al. (Jefferson Lab t_{20} collaboration), Report DAPNIA/SPhN-99-76, to be submitted.
17. M.E. Schulze et al., *Phys. Rev. Lett.* **52**, 597 (1984).
18. E.J. Stephenson et al., *Nucl. Instr. Meth.* **178**, 345 (1980).
19. M. Garçon, to be published in *The 20 years of SATURNE-2*, edited by A. Boudard and P.-A. Chamouard, World Scientific, Singapore, 2000.
20. J. M. Cameron et al., *Nucl. Instr. Meth. Phys. Res.* **A305**, 257 (1991). .
21. D. Abbott et al. (Jefferson Lab t_{20} collaboration), Report ISN-2000-01, to be submitted to *Phys. Rev. Lett.*
22. S. Kox et al., *Nucl. Instr. Meth. Phys. Res.* **A346**, 527 (1994).
23. J.F.J. van der Brand et al., *Phys. Rev. Lett.* **78**, 1235 (1997); and references therein.
24. V.F. Dmitriev et al., *Phys. Lett.* **157B**, 143 (1985).
25. B.B. Voĭtsekhovskiĭ et al., *JETP. Lett.* **43**, 733 (1986).
26. R. Gilman et al., *Phys. Rev. Lett.* **65**, 1733 (1990).
27. M. Ferro-Luzzi et al., *Phys. Rev. Lett.* **77**, 2630 (1996).
28. M. Bouwhuis et al., *Phys. Rev. Lett.* **82**, 3755 (1999).
29. B. Boden et al., *Z. Phys.* **C49**, 175 (1991).
30. W. Turchinetz et al., Proc. 2nd Workshop on Electronuclear Physics with Internal Targets and the Bates Large Acceptance Toroid (BLAST), 1998.
31. I. Rachek, private communication.
32. K. Rith, Workshop on new detectors and facilities, Santorini 1999, unpublished.
33. S. Bültmann et al., *Nucl. Instr. Meth. Phys. Res.* **A425**, 23 (1999).
34. V. Christian et al., *Few Body Syst. Suppl.* **9**, 429 (1995).
35. J. Carbonell and V.A. Karmanov, *Eur. Phys. J.* A **6**, 9 (1999).

36. H. Gao, these proceedings; H. Schmieden, these proceedings.
37. G. Quéméner et al., *Nucl. Phys.* **A654**, 469c (1999); K. De Jager, these proceedings; M.K. Jones et al., to be published in *Phys. Rev. Lett.*
38. W. van Orden, these proceedings.
39. B. Mosconi and P. Ricci, *Few-Body Syst.* **6** 63, (1989); *Erratum*, **8**, 159 (1990).
40. R.B. Wiringa et al., *Phys. Rev.* C **51**, 38 (1995).
41. H. Arenhövel et al., Preprint nucl-th/9910009.
42. H. Henning et al., *Phys. Rev.* C **52**, R471 (1995).
43. C.D. Buchanan and R. Yearian, *Phys. Rev. Lett.* **15**, 303 (1965).
44. S. Galster et al.,*Nucl. Phys.* **B32**, 221 (1971).
45. S. Platchkov et al., *Nucl. Phys.* **A510**, 740 (1990).
46. J. Carlson and R. Schiavilla, *Rev. Mod. Phys.* **70**, 743 (1998).
47. E. Hummel and J. Tjon, *Phys. Rev.* C **42**, 423 (1990).
48. W. van Orden et al., *Phys. Rev. Lett.* **75**, 4369 (1995); and private communication.
49. D.R. Phillips et al., Preprint nucl-th/9906086.
50. K. Hafidi, Thesis, Univ. Paris-Sud/Orsay (1999), Report DAPNIA/SPhN-99-05T.
51. P.G. Blunden et al., *Phys. Rev.* C **40**, 1541 (1989).
52. R. Dymarz and F.C. Khanna, *Phys. Rev.* C **41**, 2438 (1990).
53. A. Buchmann et al., *Nucl. Phys.* **A496**, 621 (1989).
54. C.E. Carlson and F. Gross, *Phys. Rev. Lett.* **53**, 127 (1984).
55. N. Isgur et al., *Phys. Rev. Lett.* **52**, 1080 (1984).
56. S.J. Brodsky and J. Hiller, *Phys. Rev D* **46**, 2141 (1992).
57. S.J. Brodsky and G.R. Farrar, *Phys. Rev. D* **11**, 1309 (1975); and references therein.
58. S.J. Brodsky et al., *Phys. Rev. Lett.* **51**, 83 (1983).
59. A. Kobushkin and A. Syamtomov, *Phys. Rev D* **49**, 1637 (1994).
60. M. Garçon et al. (Jefferson Lab t_{20} collaboration), *Nucl. Phys.* **A654**, 493c (1999).
61. A. Kobushkin and A. Syamtomov, *Phys. At. Nucl.* **58**, 1477 (1995).
62. P.A.M. Guichon, private communication.

MEC and Relativistic Effects in the Deuteron

J. W. Van Orden

Department of Physics, Old Dominion University
Norfolk, Virginia 23529
and
Jefferson Lab, 12000 Jefferson Ave.
Newport News, Virginia 23606

Abstract. This talk shows recent results for the contributions of Meson Exchange Currents (MEC) and relativistic effects on the elastic electron scattering structure functions of the deuteron in the context of several recent models. These calculations show that these effects are substantial at 1 GeV^2 and above and that care must be taken in consistently treating these effects within the context of a given model. The current data suggest that the contributions to the structure functions from the $\rho\pi\gamma$ meson exchange current must be small.

INTRODUCTION

For some time measurements of elastic electron scattering from the deuteron have been made at momentum transfers greater that the mass of the nucleon. Under these circumstances, it is necessary that any attempt to provide a quantitative description of these experiments must consider the possible effects of special relativity on the observables. For example, a nonrelativistic expansion of the current matrix elements that appear in this process shows that the leading order meson-exchange-current (MEC) correction to the charge density is on the order of $(v/c)^2$. Therefore, it is necessary that all possible contributions to the current matrix elements of this order must also be considered.

The construction of models that produce relativistically covariant matrix elements is considerably more complicated than in the usual nonrelativistic case of two nucleons interacting through a simple potential. It is possible to realize the Poincaré algebra in a remarkable number of ways, to construct potential models with a simple well defined Hilbert space, to use field theory as a basis for the description of matrix elements, use a "time-ordered" or Feynman perturbation expansion for the multiple scattering series, and the system in all cases can be quantized on a variety of space-time surfaces such as with equal time or on the light cone. (A sample of the various approaches is included in Refs. [1]- [34].) While in principle all of these approaches should yield the same results for a fixed dynamical input, the dynamical content of the models may be obscured by the need for imposing phenomenological constraints on the model and by the practical necessity of approximations such as the truncation of the sets of infinite sums of contributions to the multiple scattering series. It should also be noted that although the matrix elements must satisfy covariance, the identification of the current operators and wave functions that constitute the matrix elements depend upon the method of organization and may not themselves be individually covariant. Dynamical contributions that may appear as contributions to the current operator in one case may appear in the wave functions in another.

The purpose of this presentation is to examine variations on two common approaches to this problem in order to obtain some understanding of the basic requirements of relativity and to see if it is possible to extract some common trends concerning the dynamical input for the theories. In all cases it is assumed either explicitly or implicitly that the basic degrees of freedom in the models are nucleons and mesons. The comparison is made through calculation of the elastic electron-deuteron structure functions $A(Q^2)$, $B(Q^2)$ and $t_{20}(Q^2)$.

All of these organizations of the problem must, however, satisfy a similar set of constraints. On the simplest level, the model must provide a reasonable description of the nucleon-nucleon scattering data (at least in the np channel) and of the deuteron bound state. The models must be Lorentz covariant and the electromagnetic current for each model must be conserved. However, in cases where a well defined expansion process such as an expansion in v/c or in coupling constants are used, it may not be necessary to satisfy these constraints exactly but only up to a consistent order in the expansion, provided that it can be shown that the expansion is convergent to that order. Without Lorentz covariance and current conservation, it is not possible to uniquely extract the deuteron form factors from the current matrix elements.

To see how the current conservation and Lorentz covariance requirements are related consider the current matrix element

$$\mathcal{J}^\mu = <\psi_f(P')|J^\mu(q)|\psi_i(P)> \quad (1)$$

where $\psi_i(P)$ and $\psi_f(P')$ are the initial and final states of the two-nucleon system with four-momenta P and P'. The greater the momentum transfer, the greater the difference between the rest frames of the two-body relative wave functions.

Current conservation requires that the current operator $J^\mu(q)$ satisfy a commutation relation similar to the Ward-Takahashi [35] identity of the Bethe-Salpeter equation [3]

$$q_\mu J^\mu = \left[e_1(q) + e_2(q), G^{-1}\right] \quad (2)$$

where here $e_i(q)$ is a combination of the charge operator for particle i and a four-momentum shift operator, and G^{-1} is the fully interacting two-body propagator for the system. This commutation relation imposes a constraint on the current operator which relates it to the interaction (potential) between the two nucleons. Since the wave equation can be written as

$$G^{-1}|\psi> = 0, \tag{3}$$

(2) and (3) imply that

$$q_\mu <\psi_f|J^\mu|\psi_i> = 0 . \tag{4}$$

So the current matrix element satisfies the continuity equation as required. Since the inverse propagators act on states which are defined with different rest frames, this also requires a consistent treatment of the Lorentz boosts between the various frames appearing in the matrix element.

It is convenient to classify contributions to the current operator in as one-body currents, longitudinal two-body exchange currents contrained by (2), and transverse two-body exchange currents where $q_\mu J^\mu_{\text{transverse}} = 0$ and are thus not constrained by (2). In all of the cases that are discussed here, the only transverse current will be the $\rho\pi\gamma$ exchange current.

It must be emphasized here that although all models must obey a commutator constraint similar to (2), that the separation of the dynamical constituents of the matrix element into wave functions and current operators is not unique and is determined by the chosen approach to the inclusion of relativity. That is an effect that may show up as a longitudinal exchange current in one approach may be subsumed into the wave function in another approach. Therefore, in all cases it is necessary that all model calculations use a consistent organizational framework.

Four models of the deuteron elastic structure functions are presented here in order to illustrate the importance of consistency and to determine whether these approaches provide any insight into the dynamical content of this process. In all cases the single-nucleon electromagnetic form factors are either Höhler 8.2 or the form factor of Mergell, Meisner and Drechsel which are very similar over the range considered here.

RELATIVISTIC EXPANSION

This approach assumes that the basic dynamical content of the deuteron is nonrelativistic and that the necessary relativistic effects can be described as corrections to the nonrelativistic current matrix elements as an expansion in v/c. This approach is based on the work of J. L. Friar [30].

The nucleon-nucleon interaction is taken to be a standard nonrelativict potential with paramenters determined by fitting to the nucleon-nucleon scattering data and to give the deuteron bound state. It is assumed that the potential is, at least in part, represented by a one-meson-exchange model, since the meson degrees of freedom are necessary to constructing two-body exchange currents from simple

Feynman diagrams and for constructing corrections due to retardation of meson propagators.

Relativity is imposed by requiring that the currents and interactions are consistent with operator commutators of Poincaré invariance to some order in v/c. This approach guarantees that the interaction model can be very well constrained by data but its application can become technically complicated. In addition, the expansion in v/c must fail at some value of momentum transfer.

The calculations shown below are by Jiří Adam [36,33] and are based on a version of this technique described in papers by Adam and Arenhövel [32]. The interaction model is chosen to be based on the Paris potential although a recent preprint [34] gives similar results using the Bonn B potential.

Figure 1 shows the structure functions $A(Q^2)$, $B(Q^2)$ and $t_{20}(Q^2)$ for this model. The relative importance to the various corrections to the nonrelativistic impulse approximation (NRIA) are shown by adding the contributions successively to the NRIA.

First consider the structure function $A(Q^2)$. The NRIA (represented by the dashed-triple-dotted line) represents the data well up to about 1 GeV^2 were it then falls below the data. The addition of relativistic corrections (RC, dashed-dotted line) (ie. v^2/c^2 corrections to the one-body currents and boost corrections) increases the curvature of the structure function decreasing its magnitude below about 2 GeV^2 and increasing it considerably above this value. The addition of the pion contact exchange current (pi cont., short-dashed line) then increases the value of the structure function everywhere below about 5 GeV^2. Inclusion of corrections associated with retardation of the pion (pi ret., long-dashed line) decreases the structure function everywhere. Finally, the $\rho\pi\gamma$ exchange current (rpg, solid line) increases the value of the structure function everywhere bringing the calculation into reasonable agreement with the data, although it is slightly larger than the data for intermediate momentum transfers.

Here it is clear that the various corrections to the NRIA are not small, they tend to be of similar magnitude, and they contribute to the deuteron form factors with varying signs. As a result, any calculations that includes some, but not all of these corrections is seriously flawed. Furthermore, the data suggest that the $\rho\pi\gamma$ (transverse) exchange current is necessary to give reasonable agreement with the data.

In the case of the magnetic structure function $B(Q^2)$, this calculation contains only the NRIA and the $\rho\pi\gamma$ exchange currents. The number of corrections to order v^2/c^2 is large for the magnetic form factor and these have not been calculated in this example. Here the NRIA has a diffraction minimum at a Q^2 below the position suggested by the data. This is remedied by including the $\rho\pi\gamma$ exchange current which pushes the position of the minimum to larger Q^2 in good agreement with the data. The corrections that have not been included in this calculation are examined in the recent work of Arenhövel, ritz and Wilbois. These corrections are substantial and lead to a considerable overestimation of the data.

The polarization structure function $t_{20}(Q^2)$ again contains the full range of correc-

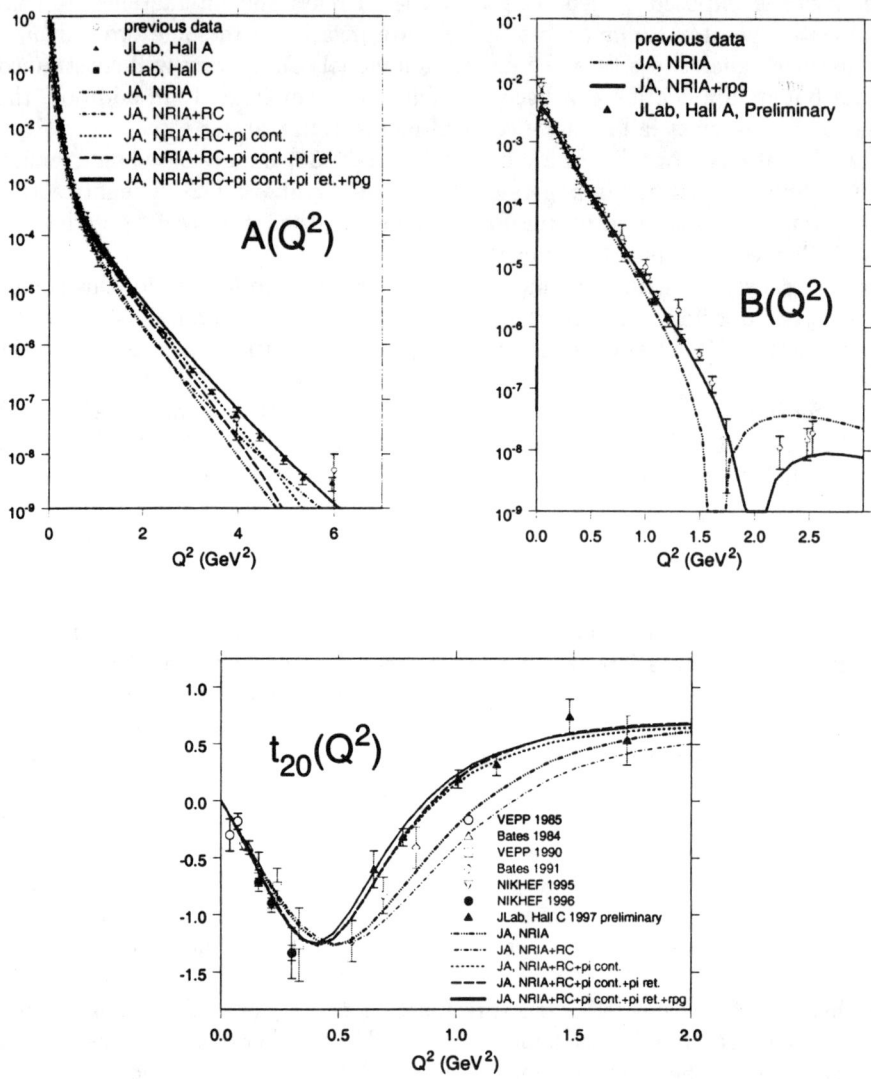

FIGURE 1. Deüteron structure functions $A(Q^2)$, $B(Q^2)$ and $t_{20}(Q^2)$ calculated by Jiří Adam (JA) for the Paris potential. Contributions for the nonrelativistic impulse approximation (NRIA), relativistic corrections (RC), pion-contact meson exchange current (pi cont), retardation corrections (pi ret.) and $\rho\pi\gamma$ exchange current (rpg) are displayed. New Jefferson Lab data for $A(Q^2)$ are from [37] (Hall A) and [38] (Hall C), for $B(Q^2)$ from [39] (Hall A), and for $t_{20}(Q^2)$ from [40] (Hall C).

tions through the charge and quadrupole form factors. Here the NRIA is substantially below the data in the region above the minimum. Inclusion of the relativistic corrections further decreases the structure function in this region. Adding the pion contact exchange current brings the calculation into reasonable agreement with the data. The effects of including retardation corrections and the $\rho\pi\gamma$ exchange current are relatively small.

RELATIVISTIC HAMILTONIAN MODEL OF FOREST AND SCHIAVILLA

The relativistic Hamiltonian model of Forest and Schiavilla [41] is carried out according to much the same philosophy as the previous example but with the goal of minimizing recourse to the expansion in v/c. Here the wave equation is chosen to be a relativised Schrödinger equation where the nonrelativistic kinetic energy has been replaced by the relativistic onshell energy operator. The potential is similar to the Argonne V18 potential but with the one-pion-exchange part of the potential replaced by a form containing the full relativistic form of the vertices without any expansion in v/c. The shorter range parts of the potential are exactly like those of the AV18 potential. The parameters of the new potential are used with the new wave equation and are adjusted to fit the NN scattering data and deuteron properties. The Poincaré commutators are then imposed on the current while avoiding and expansion in v/c where possible.

The structure functions for this calculation are shown in Fig. 2. For $A(Q^2)$, The relativistic impulse approximation (RIA) is below the data above approximately 1 GeV. The inclusion of the longitudinal MEC (LMEC) brings the calculation into much better agreement with the data although it is above the data for intermediate values of Q^2. The inclusion of the of the $\rho\pi\gamma$ exchange current further increases the size of the calculation and further degrades the agreement with the data.

For $B(Q^2)$ there are no longitudinal MEC contributions in this model. The RIA is in good agreement with the data in the region between about 1 and 2 GeV^2. The inclusion of the $\rho\pi\gamma$ exchange current decreases the size of the calculation and brings it below with the new data from Hall A at Jefferson Lab. Note that the effect of the $\rho\pi\gamma$ exchange current in this case is to move the diffraction minimum to the left. This is in contrast to the case of the v/c expansion where this current moves the minimum to the right. This is due to the tensor coupling of the ρ to the nucleon. In a strict v/c expansion this term is of higher order than the vector coupling. However, since the tensor coupling constant is large, neglecting the tensor coupling produces the wrong sign for this contribution and results in moving the minimum to the left rather than to the right as is the case in all of the relativistic calculations that include this coupling.

For $t_{20}(Q^2)$ the RIA is very much like the sum of the NRIA and RC for the v/c calculation. The inclusion of the LMEC brings the calculation into very good

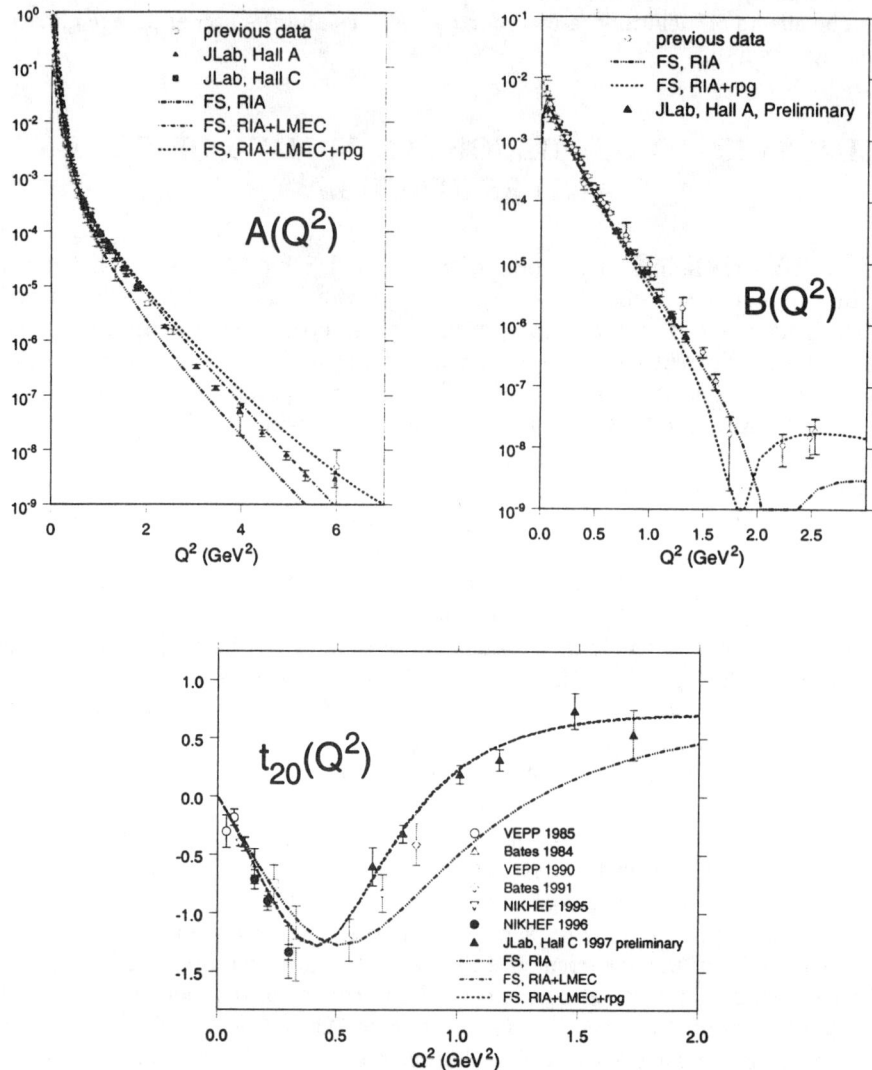

FIGURE 2. Deuteron structure functions $A(Q^2)$, $B(Q^2)$ and $t_{20}(Q^2)$ according to Forest and Schiavilla (FS). Contributions for the relativistic impulse approximation (RIA), RIA with longitudinal meson exchange currents (RIA+LMEC) and RIA with LMEC and $\rho\pi\gamma$ exchange current (RIA+LMEC+rpg) are displayed

agreement with the new data from Hall C at Jefferson Lab. The inclusion of the $\rho\pi\gamma$ exchange current has an extremely small effect here.

BETHE-SALPETER-LIKE MODELS

The Bethe-Salpeter like approaches [3]–[24] are motivated by a field-theoretical description of the two-body problem. The Bethe-Salpeter [3] equation represents a summation of the multiple scattering series written in terms of Feynman diagrams. The resummation involves organizing all two-particle-irreducible diagrams into the interaction kernel and then constructing the Bethe-Salpeter integral equation to produce the complete sum. Since the Feynman series is organized such that all of the individual propagators and vertices are covariant with respect to the free-particle Lorentz transformations, the sum is manifestly covariant as well. The currents can be constructed by combining the free one-nucleon currents with exchange currents obtained by attaching a photon to every possible place within the irreducible interaction kernel. These currents will then satisfy the Ward-Takahashi identity (2). As a practical matter, calculations of these type usually use a kernel truncated at the one-meson-exchange level.

The Bethe-Salpeter equation is a four-dimensional integral equation. As such it is much more difficult to solve numerically than the comparable nonrelativistic three-dimensional Lippmann-Schwinger equation. A class of approaches to simplify the solution of the relativistic equations is the infinite class of quasipotential equations [7]–[23]. The quasipotential equations resum the multiple scattering series in terms of a simplified propagator that has a delta-functions constraint on either the energy or time in the four-dimensional loop integrals. The resulting sum can then be made equivalent to the original Bethe-Salpeter equation by introducing a new kernel called the quasipotential. The wave or t-matrix equation is now a three-dimensional integral equation, while the quasipotential is given by a four-dimensional integral equation in terms of the original Bethe-Salpeter kernel. Since the kernel of the Bethe-Salpeter equation is usually truncated, a similar truncation of the quasipotential kernel reduces the equation for the quasipotential to a quadrature over the truncated Bethe-Salpeter kernel. This truncation, however, eliminates the formal equivalence to the Bethe-Salpeter equation and can in general introduce problems maintaining Lorentz covariance and current conservation. The problem with lack of equivalence to the Bethe-Salpeter equation is not as serious as it would naively appear since both the truncated Bethe-Salpeter and the truncated quasipotential equations both represent approximations to the full Bethe-Salpeter equation and there is considerable indication that the quasipotentials may converge more rapidly to the full result than does the truncated Bethe-Salpeter equation.

The calculations shown here are those of Hummel and Tjon using the Bethe-Salpeter equation [21], and the Van Orden, Gross and Devine [18–20], using the Gross or spectator equation, one of the infinite class of quasipotential equations. The Gross equation has the desirable properties that it is always Lorentz covariant and that it has been shown to conserve current for a variety of different truncations schemes [42,43].

The results of these two calculations are shown in Fig. 3. In each case the calculations contain a relativistic impulse approximation and a $\rho\pi\gamma$ exchange current. For $A(Q^2)$, the RIA of both Van Orden, Gross and Devine (VOGD) and Hummel and Tjon (HT) fall below the data above about 1 GeV^2 although the VOGD result tends to be higher throughout most of this region. Addition of the $\rho\pi\gamma$ exchange current to the HT calculation places the calculated result substantially above the new data from Jefferson Lab. This is the result of using a vector-meson-dominance model (VMD) of the $\rho\pi\gamma$ form factor along with the rather hard meson-nucleon form factors typical of a relativistic one-meson-exchange model. This produces a very large $\rho\pi\gamma$ exchange current which is inconsistent with the data. For this reason the VOGD calculation uses a much softer $\rho\pi\gamma$ form factor giving a more reasonable representation of the data. This, however, involves making an arbitrary choice that is not constrained by any data for this form factor and therefore represents an unwanted ambiguity in the calculation. The calculation of Adam used an overall factor of $F_1^s(Q^2)$ for the $\rho\pi\gamma$ exchange current while that of Forest and Schiavilla uses the VMD form factor with much softer meson-nucleon form factors. In both cases the $\rho\pi\gamma$ exchange current has a smaller effect than that of Hummel and Tjon.

For $B(Q^2)$ the HT RIA clearly disagrees with the new Jefferson Lab data while that of VOGD is in reasonably good agreement with this data. A consequence of this is that the minimum in the structure function for HT is lower in Q^2 than that of VOGD. This difference has been shown to be the result of a very small p-wave contribution of relativistic dynamical origin. These small p-waves which contribute less than 0.01% to the normalization of the wave functions differ in sign between the two calculations. It is this sign change in an interference term between the large s and d waves and the p waves that results in this shift in the position of the minimum. This is clearly a case where an apparently negligible contribution can have substantial effect when the large contributions are interfering to give zero. This clearly indicates that a great deal of care must be made in approximating contributions to the magnetic form factor of the deuteron.

For $t_{20}(Q^2)$, the HT and VOGD calculations are fairly similar and tend to reproduce the new Jefferson Lab data reasonably well while being two large at lower Q^2. This problem at lower momentum transfer may be associated with an underprediction of the deuteron quadrupole moment by these models and is clearly something that requires further investigation. For both calculations, the effects of the $\rho\pi\gamma$ exchange currents are relatively small, but help the agreement of the calculations with the Jefferson Lab data.

SUMMARY

All of the calculations described above are summarized in Figs. 4 and 5. Figure 4 shows the four calculations with all contributions except the $\rho\pi\gamma$ exchange

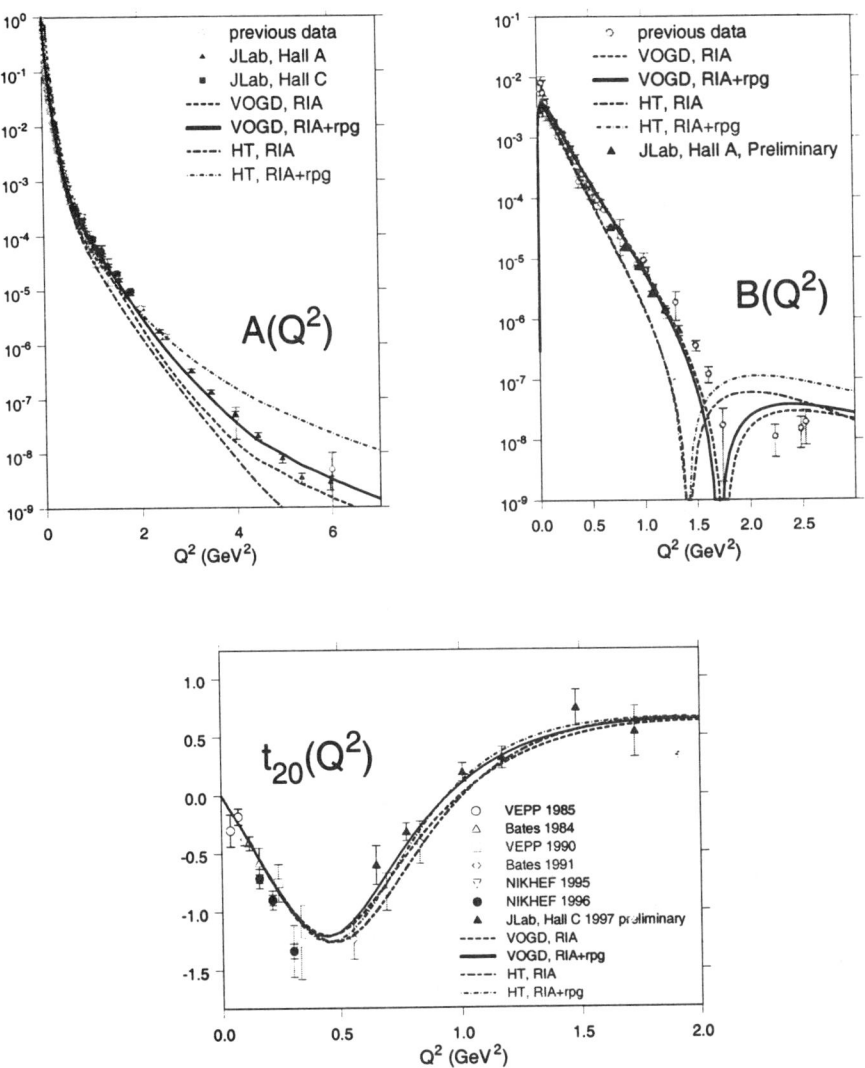

FIGURE 3. Calculations of Hummel and Tjon (HT), and Van Orden, Gross and Devine (VOGD) for relativistic impulse approximation (RIA) and $\rho\pi\gamma$ exchange currents (rpg).

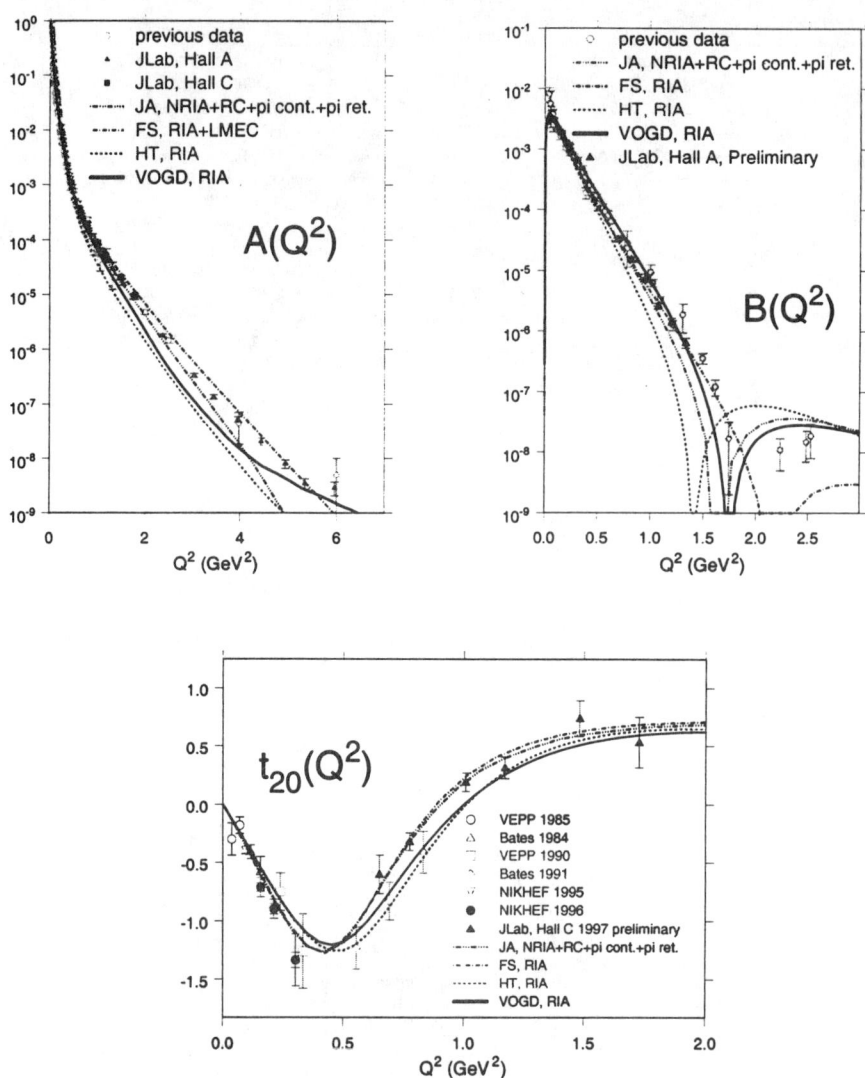

FIGURE 4. Calculations of Jiři Adam (JA), Forrest and Schiavilla (FS), Hummel and Tjon (HT), and Van Orden, Gross and Devine (VOGD) with all contributions except the $\rho\pi\gamma$ exchange current.

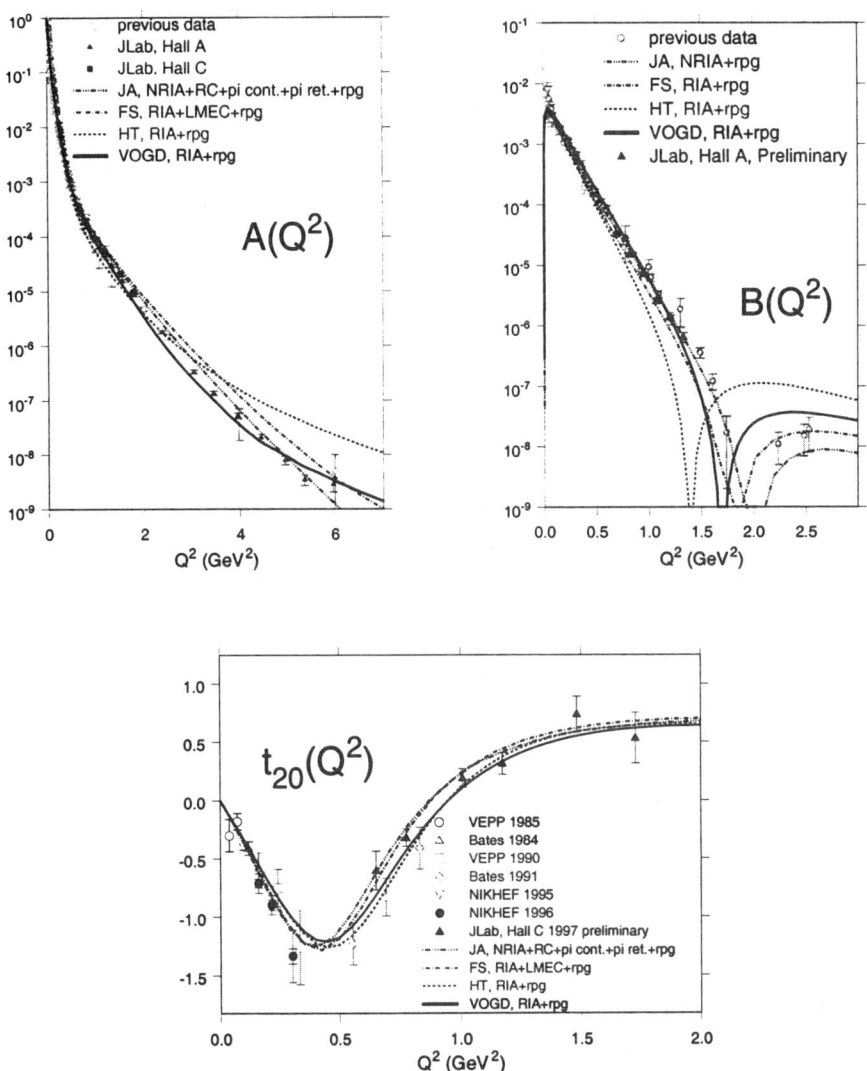

FIGURE 5. Calculations of Jiří Adam (JA), Forrest and Schiavilla (FS), Hummel and Tjon (HT), and Van Orden, Gross and Devine (VOGD) with all contributions including the $\rho\pi\gamma$ exchange current.

current, while Fig. 5 includes the $\rho\pi\gamma$ exchange current. Clearly there is still a substantial amount of variation among these various models. This is particularly apparent in $A(Q^2)$ and $B(Q^2)$ while the variation in $t_{20}(Q^2)$ is considerably smaller. The underlying dynamical reasons for this variation is not clear and needs to be addressed by a detailed comparison of the calculations to determine whether the variation results from the quality of the various interaction models or is due to the truncations and approximations required by each model.

The one dynamical element that can be easily identified in each case is the $\rho\pi\gamma$ exchange current. All of the models except that of Forest and Schiavilla seem to require the presence of this current in order to describe the data for the structure functions. However, it is clear that this contribution must be relatively small in order to avoid problems with $A(Q^2)$ in particular. The model of Forest and Schiavilla appears to work best with no contributions from this current.

Note that the full calculations of Forest and Schiavilla, and Van Orden, Gross and Devine provide a reasonable representation of the new Jefferson Lab data for $B(Q^2)$. However, they give quite different predictions as to the location of the diffraction minimum. The existing SLAC data seem to favor the calculation of Forest and Schiavilla. It would obviously be interesting if new data with smaller errors could be obtained closer to the diffraction minimum.

The full calculations of $t20(Q^2)$ lie in a relatively narrow band reflecting the insensitivity of this structure function to the $\rho\pi\gamma$ exchange current. All of the calculations do a good job or reproducing the four highest Q^2 data from Jefferson Lab, but the lower Q^2 data from Jefferson Lab and NIKHEF favor models which place them minimum in $t_{20}(Q^2)$ at lower values of Q^2.

Ambiguous elements of the various calculations must be better constrained either by experiment, such as experimental determination of the $\rho\pi\gamma$ form factor, or by reference to underlying degrees of freedom of the problem possibly through effective field theory.

ACKNOWLEDGEMENTS

The author would like to thank Jun Forrest, Rocco Schiavilla and Jiři Adam for providing previously unpublished calculations for use here.

REFERENCES

1. B. D. Keister and W. N. Polyzou, Adv. Nucl. Phys. **20**, 225 (1961).
2. P. L. Chung, F. Coester, B. D. Keister, and W. N. Polyzou, Phys. Rev. C **37**, 2000 (1988).
3. E. E. Salpeter and H. A. Bethe, Phys. Rev. **84**, 1232 (1951).
4. J. Fleischer and J. A. Tjon, Nucl. Phys. **B84**, 375 (1975); Phys. Rev. D **15**, 2537 (1977); **21**, 87 (1980). E. van Faassen and J. A. Tjon, Phys. Rev. C **28**,2354 (1983); **30**, 285 (1984); **33**, 2105 (1986).

5. M. J. Zuilhof and J. A. Tjon, Phys. Rev. C **22**, 2369 (1980); **24**, 736 (1981).
6. A. Yu. Umnikov and F. C. Khanna, Phys. Rev. C **49** 2311 (1994).
7. R. Blankenbecler and R. Sugar, Phys. Rev. **142**, 1051 (1966); A. A. Logunov and A. N. Tavkhelidze, Nuovo Cimento **29**, 380 (1963).
8. R. H. Thompson, Phys. Rev. D **1**, 1738 (1970).
9. I. T. Todorov, Phys. Rev. D **10**, 2351 (1971).
10. K. Erkelenz and K. Holide, Nucl. Phys. **A194**, 161 (1972).
11. V. G. Kadychevsky, Nucl. Phys. **B6**, 125 (1968).
12. F. Gross, Phys. Rev. **186**, 1448 (1969); Phys. Rev. D **10**, 223 (1974).
13. F. Gross, Phys. Rev. C **26**, 2203 (1982).
14. F. Gross, J. W. Van Orden and K. Holinde, Phys. Rev. C **41**, R1909 (1990).
15. F. Gross, J. W. Van Orden and K. Holinde, Phys. Rev. C **45**, 2094 (1992).
16. W. W. Buck and F. Gross, Phys. Rev. D **20**, 2361
17. R. E. Arnold, C. E. Carlson, and F. Gross, Phys. Rev. C **21**, 2361 (1980).
18. J. W. Van Orden, Czech. J. Phys. **45**, 181 (1995).
19. J. W. Van Orden, N. Devine and F. Gross, Phys. Rev. Lett. **75**, 4369 (1995).
20. J. W. Van Orden, N. Devine and F. Gross, Few-body Systems Suppl. **9**, 415 (1995).
21. E. Hummel and J. A. Tjon, Phys. Rev. Lett. **63**, 1788 (1989).
22. S. J. Wallace and V. B. Mandelzweig, Nucl. Phys. **A503**, 673 (1989).
23. N. K. Devine and S. J. Wallace, Phys. Rev. C **48**, 973 (1993).
24. X. Zhu, R. Gourishankar, F. C. Khanna, G. Y. Leung, and N. Mobed, Phys. Rev. C **45**, 959 (1992).
25. M. G. Fuda, Phys. Rev. D **44**, 1880 (1991).
26. V. A. Karmanov and A. V. Smirnov, Nucl. Phys. **A546**, 691 (1992).
27. J. Carbonell, B. Desplanques, V. A. Karmanov and J.-F. Mathiot, Phys. Rept. **300**, 215 (1998).
28. J. Carbonell and V. A. Karmanov, Few Body Syst. Suppl. 10, 427 (1999).
29. J. Carbonell and V. A. Karmanov, Eur. Phys. J, **A6**, 9 (1999).
30. J. L. Friar, Phys. Rev. C**12**, 695 (1975).
31. R. Schiavilla and D. O. Riska, Phys. Rev. C **43**, 437 (1991).
32. J. Adam, Jr. and H. Arenhövel, Nucl. Phys. **A614** 289 (1997).
33. H. Henning, J. Adam, Jr., P. U. Sauer and A. Stadler, Phys. Rev. C**52**, R471 (1995).
34. H. Arenhövel, F. Ritz and T. Wilbois, nucl-th/9910009.
35. J. C. Ward, Phys. Rev. **78**, 182 (1950); Y. Takahashi, Nuovo Cimento **6**, 371 (1957).
36. J. Adam, priviate communication.
37. L. C. Alexa, et al. (Jefferson Lab Hall A Collaboration), Phys. Rev. Lett. **82**, 1374 (1999).
38. D. Abbott, et al. (Jefferson Lab t_{20} Collaboration, Phys. Rev. Lett. **82**, 1379 (1999).
39. K. McCormick, Ph.D. Dissertation, Old Dominion University (1999).
40. K. Hafidi (Jefferson Lab t_{20} collaboration), Contr. 15th Int. Conf. Particles and Nuclei (PANIC-99), Uppsala, Report DAPNIA/SPhN-99-50.
41. Jun Forest and Rocco Schiavilla, private communication.
42. F. Gross and D. O. Riska, Phys. Rev. C **36**, 1928 (1987).
43. J. Adam, J. W. Van Orden and F. Gross, Nucl.Phys. **A640**, 391(1998).

Unpolarized Inclusive Studies of A=3 Nuclei

Douglas H. Beck

Department of Physics
University of Illinois at Urbana-Champaign
1110 West Green Street
Urbana, IL 61801-3080

Abstract. A review of the MIT-Bates and Saclay studies of inclusive scattering from ^{3}H and ^{3}He is presented. Elastic, threshold inelastic, quasi-elastic and Δ production are considered

INTRODUCTION

During the 1980's, programs of inclusive electron scattering from ^{3}H and ^{3}He were undertaken at both MIT-Bates and Saclay. These measurements presented challenges of target technology, measurements of small cross sections, measurements at large energy loss as well as accurate longitudinal/transverse separations. Theoretical work aimed at understanding these results continues to the present - indeed significant progress has been made since the last experiments. Discussions are presently underway regarding use of the three-body systems as a good *nuclear* isospin doublet in new experiments at Jefferson Lab.

Few-body nuclei are the most important testing ground for microscopic nuclear models because both ground state and continuum wave functions can be calculated starting from the microscopic input of NN interactions. The electromagnetic current of these systems directly reflects their underlying structure, although not without some complications. The one-body current of the nucleons is certainly representative of the nuclear wave function; however, meson exchange currents (MEC), Δ effects (both in the wave function and in the current), relativistic effects and the particular effects of the three-body force also play a role. This requires that a broad range of measurements be made against which models can systematically be tested to determine their suitability.

The cross sections discussed herein are characterized in terms of form factors F_C and F_M for the elastic measurements and response functions R_L and R_T for the inelastic measurements defined as follows

A recent review of few-body systems in general may be found in Ref. [1].

$$\frac{d\sigma}{d\Omega} = \sigma_M \frac{1}{f_{rec}} \frac{Q^4}{q^4} Z^2 F_C^2 + \left[\frac{q^2}{2M^2}\frac{-Q^2}{2q^2} + 2\tan^2\theta/2\right]\mu^2 F_M^2$$

$$\frac{d\sigma}{d\Omega dE} = \sigma_M \frac{Q^4}{q^4} R_L(\omega,q) + \left[\frac{-Q^2}{2q^2} + \tan^2\theta/2\right] R_T(\omega,q)$$

EXPERIMENTS

Saclay

Four separate experiments were run at the ALS during the late 70's and early 80's at Saclay. The ALS was a pulsed electron linac with 10 μs pulses at a rate of 500 Hz. Energies from 189 - 689 MeV were utilized for these experiments. The ^{3}He runs for forward and backward elastic scattering as well as for the quasi-elastic were run at different times. The entire ^{3}H elastic experiment (both forward and backward angles) was run in the same period [2]. Measurements were made with the "900" spectrometer with an angular range of 25-155°.

The ^{3}He target for these experiments consisted of a loop of flowing ^{3}He gas cooled with a liquid hydrogen heat exchanger. This target operated at 20 K with pressures of 12-14 atm yielding a target density of about 50 mg/cm^2.

The problem of providing a high density ^{3}H target could be approached in one of two ways. At Saclay a small liquid ^{3}H target was chosen in order to reduce the inventory ^{3}H as much as possible. A cylindrical target cell 10 mm in diameter and 40 mm long was used in conjunction with a small expansion volume maintained at a higher temperature. This system yielded a target thickness of about 1 g/cm^2; the total inventory was 10 kCi. Again, liquid hydrogen was used as a coolant (although the ^{3}H did not flow as it did in the ^{3}He target). A doubly nested containment system (with thin windows for the scattered electrons) enclosed the cell during normal operation. A heavier container was closed around this system while the experiment was not taking data.

MIT-Bates

At MIT-Bates, the ^{3}H/^{3}He experiments were performed at the same time [3]. Fixed volume (i.e. not flowing) gas targets (~10 cm diameter x 4 cm high) were chosen in order to make the two targets as similar as possible. They operated at about 45 K and 15 atm. yielding target thicknesses of 80 and 35 mg/cm^2 for ^{3}H and ^{3}He, respectively. The targets were used for both elastic and inelastic measurements accommodated by the range of kinematics available: beam energies from 65 - 790 MeV and scattering angles of 54 and 135°. The Bates Linear Accelerator was used providing beam pulses of 15 μs duration at a maximum repetition rate of 600 Hz with currents up to 25 μA. For all measurements the ELSSY (energy loss spectrometer system) was used; detection included vertical drift chambers, plastic scintillator trigger counters and a Freon 12 Cherenkov detector.

Use of such a gas target for the ^{3}H experiment required a total inventory of about 140 kCi or roughly 2 mol, of which 115 kCi was in the target during the measurement. For safety reasons the ^{3}H was stored in uranium as U^{3}H$_3$ during the times when it was not needed in the target. The uranium - hydrogen system affords the capability to essentially build a pump with no moving parts. At room temperature the "vapor pressure" of hydrogen over pure uranium is a small fraction of 1 torr; at 450 °C, the operating temperature of our oven, the pressure is 2.5 atm. Operationally, in order to move the ^{3}H to the target, the U^{3}H$_3$ was first heated to 450 °C and the tritium liquified in the target cell. The oven was then valved off and the target slowly heated to vaporize the liquid and reach the operating pressure of 15 atm. A total of some 50 Ci of gaseous ^{3}H were released into the atmosphere in planned pump-outs of the system.

After some weeks of operation, decay ^{3}He, which would otherwise blanket the U^{3}H$_3$ and greatly slow the rate of ^{3}H absorption, was from the system.

ELASTIC SCATTERING

Low Momentum Transfer

One of the important goals of the Bates measurement was to carefully compare the ^{3}H and ^{3}He elastic scattering at low momentum transfer by measuring concurrently using very similar apparatus [4]. The ^{3}He results were checked carefully against the world's data and agreed at about the 3% level.

An important comparison of the form factors is expressed in the isoscalar and isovector form factors defined as

$$F^S = \frac{F(^3He) + F(^3H)}{2}, \quad F^V = \frac{F(^3He) - F(^3H)}{2}$$

As shown in Figure 1, the meson exchange current effects (MEC) were already

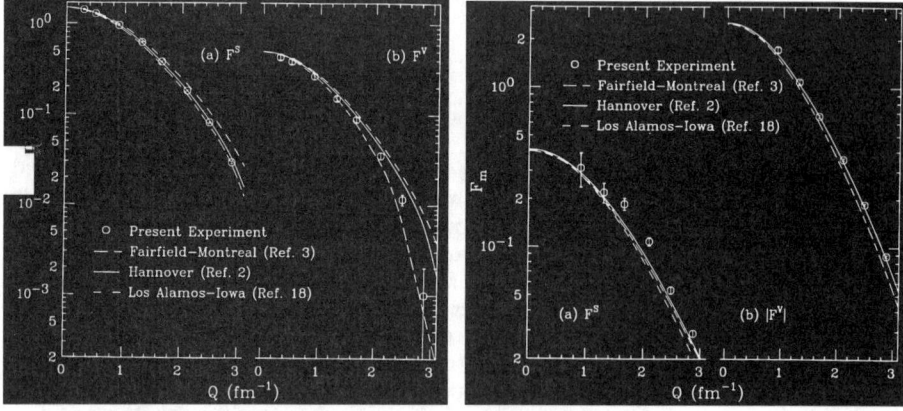

FIGURE 1. Low momentum transfer elastic scattering results from Bates. The isoscalar, F^S, and isovector, F^V form factors are shown.

apparent in the data: compare the Los Alamos-Iowa calculation [5] without MEC (but including the 3-body force) to those of the Hannover [6] (MEC plus coupled channel calculation of $\Delta(1232)$ effects) and Fairfield-Montreal [7] groups (MEC, Δ current contributions and 3-body force). Even in the cases where these then state-of-the-art calculations showed small effects (e.g. F_C^S), the calculations including MEC improved the agreement. We note that the impulse calculations of F_M^V, where MEC come in to lowest order, greatly underpredict the data.

High Momentum Transfer

The Saclay group performed an analysis of the complete elastic scattering data sets for the ^{3}H and ^{3}He, again expressing the results in terms of isoscalar and isovector form factors [2,8] again showing the dominance of MEC in the magnetic isovector form factor. Figure 2 shows this compilation of data (this time form factors of the two nuclei) with the state of the art calculation of Marcucci, et al. [9]. These calculations are based on "converged" Faddeev ground state wave functions and include the most consistent modern inclusion of MEC, the effects of the Δ in the wave function and in the current as well as the Urbana three-body force. The one remaining problem is in the ^{3}He magnetic form factor where the first diffraction minimum is at too low a momentum transfer. The experimental situation in this case has been reinvestigated;

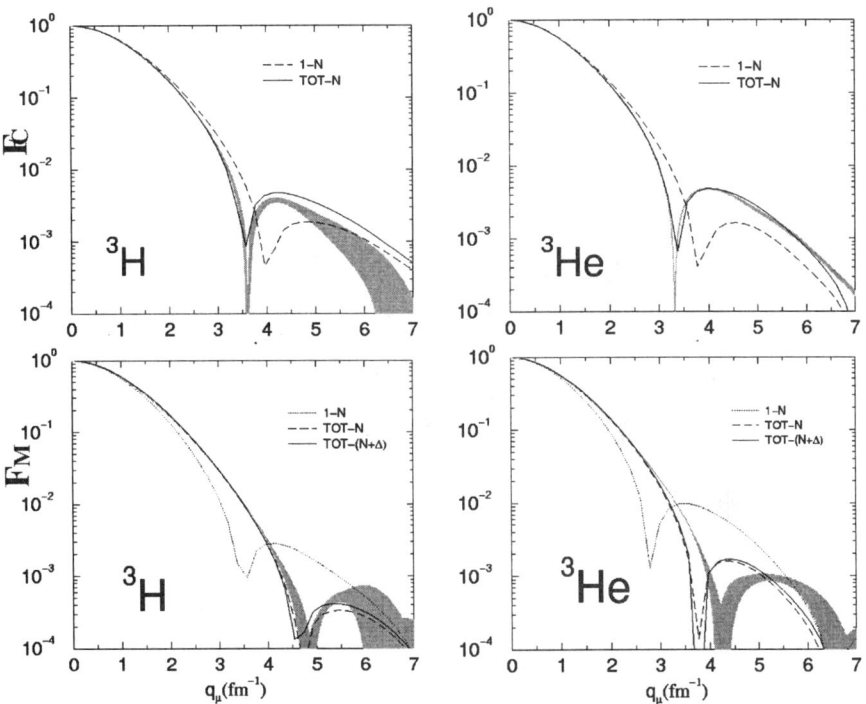

FIGURE 2. High momentum transfer elastic scattering form factors. Data are from the compilation of Amroun, et al. [2,8].

however, the preliminary results reported at this symposium by the UMass group confirm the earlier measurements (with smaller uncertainties) [10].

INELASTIC SCATTERING

Threshold Inelastic

In the Bates experiment, the comparison of ^{3}H and ^{3}He extended through the inelastic region including quasi-free D production. By the time of these experiments, there was significant evidence that for light nuclei simple plane wave descriptions of the final states was inadequate to describe especially the longitudinal response. This is illustrated strikingly by comparing the threshold inelastic response in ^{3}H and ^{3}He as shown in Figure 3 [11]. In this case, there is a strong continuum "state" just above threshold in the two-body channel. For the electrodisintegration the ^{2}S - ^{2}S monopole (C0) transition proceeds with the virtual photon coupling to the proton in the case of ^{3}He but the neutron in the case of ^{3}H. Typical plane wave calculations do not reproduce the ^{3}He data well (and do not include the two-body breakup of ^{3}H at all). Continuum wave functions (Schiavilla, et al.) that are orthogonal to the ground state [12] do a better job; the continuum Faddeev calculations with a simple NN potential of van Meijgaard and Tjon [13] do a very good job for the longitudinal response, although they appear to consistently underpredict the ^{3}H transverse response.

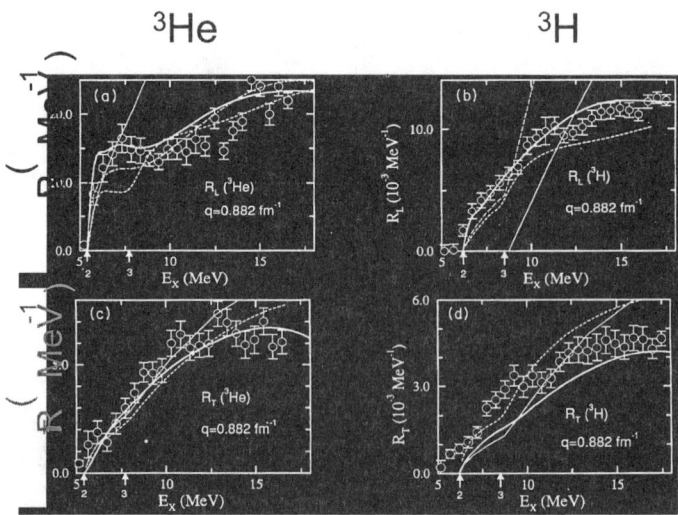

FIGURE 3. Threshold inelastic response of ^{3}H and ^{3}He at a momentum transfer of 0.9 fm^{-1}.

Quasi-free scattering

Separation of the quasi-elastic responses of ^{3}He into longitudinal and transverse components was performed at Saclay (C. Marchand, et al. [14]) and extended in the Bates measurements to include ^{3}H (K. Dow, et al. [15]). The separated ^{3}He responses generally show good agreement, though in the transverse response the Saclay points are generally lower at the peak (by several %) and higher in the tail region.

The longitudinal response, often a difficult characteristic of nuclei to understand, shows remarkably good agreement between experiment and modern theory as shown in Figure 4. The tour-de-force continuum Faddeev calculations of the Bochum group (Golak, et al. [16]) show excellent agreement with the data for both ^{3}H and ^{3}He over the range of momentum transfers measured. An integral transform calculation of the longitudinal response by Martinelli, et al. [17] wherein only the ground state wave functions are utilized to generate the continuum also shows good agreement with the data for both nuclei.

Not surprisingly, the Coulomb sum rule for ^{3}H and ^{3}He is now also well

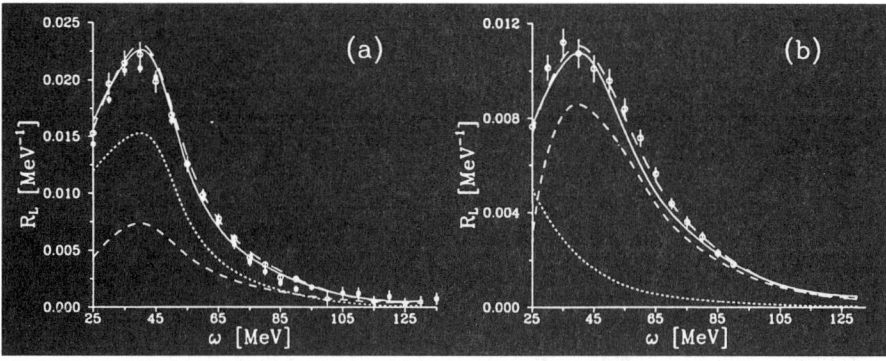

FIGURE 4. Longitudinal quasi-elastic response for ^{3}H and ^{3}He at a momentum transfer of 300 MeV.

understood. Integrating the longitudinal response over all excitation energies (dividing out the square of the nucleon form factor) leaves two contributions - essentially the elastic form factor of the nucleus and a two-body correlation function. This correlation function describes the probability of finding a second proton some distance from the first (actually the Fourier transform of this distribution is measured). As shown in Figure 5 (D. Beck [18]), there is a minimum in this function at only 1.7 fm^{-1} (c.f. minima of the elastic form factors). We note that despite initial appearances, the minimum is actually located quite accurately because both the magnitude and the sign of the correlation function is determined by the data. These data are well accounted for in a calculation (R. Schiavilla, et al. [19]) which includes, in particular, the effect of the relativistic Darwin Foldy term. This observable is one of the few that apparently show substantial relativistic effects at such a low momentum transfer. Similar effects are seen in ^{4}He [19].

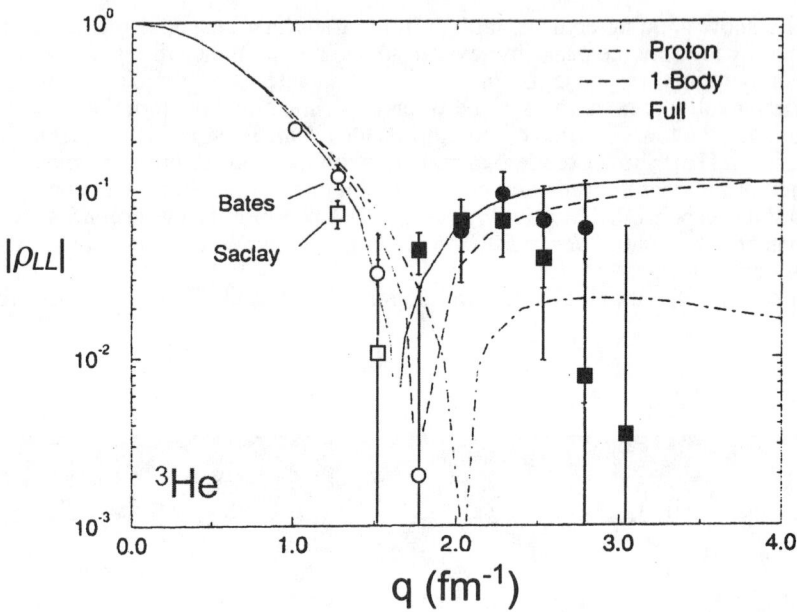

FIGURE 5. Two-body correlation function derived from the Coulomb sum for ^{3}He.

The transverse response in the three-body system is reasonably well reproduced by the modern continuum calculations [16], especially in terms of the shape. However, as shown in Figure 6, the calculations are systematically 10-15% below the data, with the larger discrepancy evident in the ^{3}H response. This is undoubtedly as a result of the absence of MEC effects. This effect is also seen in the Euclidean response calculated by J. Carlson and R. Schiavilla [20]. Including MEC in these calculations is a difficult task, but we look forward to the possibility of completing our

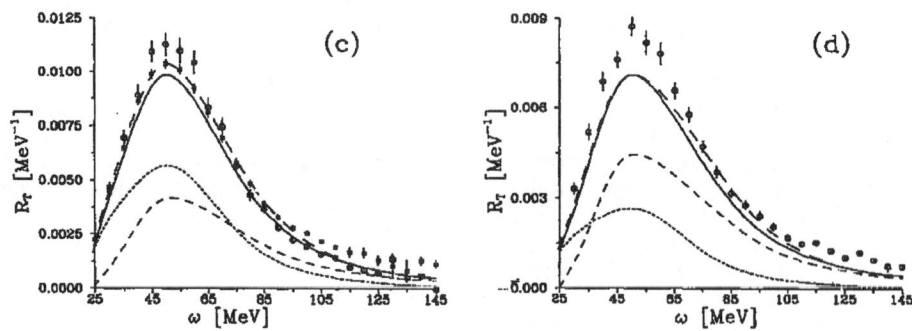

FIGURE 6. Transverse quasi-elastic response for ^{3}H and ^{3}He at a momentum transfer of 300 MeV.

understanding of the A=3 quasi-elastic responses.

Measurements were also made at Bates in the region of quasi-free $\Delta(1232)$ production for both ^{3}H and ^{3}He as shown in Figure 7 (T. Ueng [21]). As expected, and in contrast to the quasi-free nucleon knockout region, the difference between the ^{3}H and ^{3}He responses in the region of the Δ is perhaps 5% at most. The existing PWIA calculation, however, apparently shows either a downward shift of the peak relative to Δ production on the nucleon, or the again the strong effects of MEC.

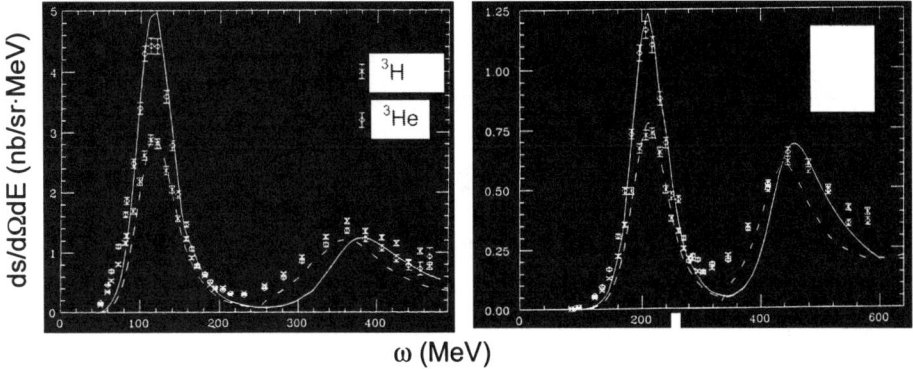

FIGURE 7. Inelastic spectra for ^{3}H and ^{3}He taken at 54° in the lab and for incident energies of 558 MeV (left) and 790 MeV (right).

CONCLUSIONS AND OUTLOOK

There exists a broad data set for the inclusive scattering from three-body nuclei. These data are complemented by many extensive calculations. Taken as a whole, the most recent of these, which have been completed only in the past couple of years, describe the data with satisfying precision. The two dominant themes that have developed with improved calculations are the strong roles that exact wave functions and meson exchange currents play. In the former case, the effect is perhaps the most striking for the continuum wave functions, where previous impulse approximation and various distorted wave calculations did not reproduce the main features of the data. There is also strong evidence of relativistic effects (Darwin-Foldy term) in the longitudinal two-body correlation function. The three-body force seems to be unimportant in the description of these data; perhaps there is some indication of its presence in new data taken at IUCF (R. Cadman, et al. [23]).

Despite the difficulties of dealing with ^{3}H targets, there is renewed interest at JLab [24] in again taking advantage of this unique nuclear isospin pair. In addition to extending the elastic form factors to momentum transfers comparable to those for which ^{2}H data now exist, several important issues involving comparison of the proton and neutron are being considered. These include the ratio and difference of the deep-inelastic F_2^p and F_2^n as well as the Gerasimov-Drell-Hearn sum rule. A precision measurement of the "original" EMC effect F_2^A/F_2^N is also possible with this mirror pair. Finally, there is the possibility of extending the current program of polarized ^{3}He measurements to ^{3}H using a target based on a laser-driven source.

ACKNOWLEDGMENTS

It is a pleasure to thank the organizers, especially W. Donnelly and W. Turchinetz, for encouraging the very stimulating atmosphere of the symposium - this was a particularly enjoyable few days.

REFERENCES

1. Carlson, J., and Schiavilla, R., Rev. Mod. Phys. 70, 743 (1998).
2. Amroun, A., et al., Nucl. Phys. A579, 596 (1994).
3. Beck, D. H., et al. Nucl. Instrum. & Meth. A277, 323 (1989).
4. Beck, D. H., et al. Phys. Rev. Lett. 59, 1537 (1987).
5. Friar, J., et al., Phys. Rev. C 34, 1463 (1986).
6. Strueve W., et al., Nucl. Phys. A405, 620 (1983).
7. Hadjimichael, E., et al., Phys. Rev. C27, 851 (1983).
8. Amroun, A., et al., Phys. Rev. Lett. 69, 253 (1992).
9. Marcucci, et al., Phys. Rev. C58, 3069 (1998).
10. Peterson, G. contribution to this symposium.
11. Retzlaff, G. A., et al., Phys. Rev. C 49, 1263 (1994).
12. Schiavilla, R., et al., Phys. Lett. B218, 1 (1989).
13. van Meijgaard and Tjon, J., Phys. Rev. C 45, 1463 (1992).
14. Marchand, C. et al., Phys. Lett. B153, 29 (1985).
15. Dow, K. et al. Phys. Rev. Lett. 61, 1706 (1988).
16. Golak, et al., Phys. Rev. C 52, 1216 (1995).
17. Martinelli, et al., Phys. Rev. C 52, 1778 (1995).
18. Beck, D., Phys. Rev. Lett. 64, 268 (1990).
19. Schiavilla, R., et al. Phys. Rev. Lett. 70, 385 (1993).
20. Carlson, J., and Schiavilla, R., Phys. Rev. Lett. 68, 3682 (1992).
21. Ueng, T. S., Ph. D. thesis, University of Virginia, unpublished (1988).
22. Meier-Hajduk, H. et al. Nucl. Phys. A499, 637 (1989).
23. Cadman, R. V., et al., to be published.
24. Jones, C., Calarco, J. and Ransome, R. private communication.

III. FEW-BODY NUCLEI AND NUCLEON STRUCTURE

Chair: J.L. Matthews

Session III ....

Recent Results On Spin Dependent Scattering from Few Body Systems Obtained in Amsterdam

J.F.J. van den Brand

Vrije Universiteit, de Boelelaan 1081, 1081 hv Amsterdam , The Netherlands
E-mail: jo@nat.vu.nl

Abstract. We report on aspects of the physics program with polarized electrons and polarized few-body systems in Amsterdam. Specifically, we focus on the first measurement of spin-correlation parameters in quasifree electron scattering from vector-polarized deuterium. Polarized electrons were injected into the AmPS electron storage ring at a beam energy of 720 MeV. A Siberian snake was employed to preserve longitudinal polarization at the interaction point. Vector-polarized deuterium was produced by an atomic beam source and injected into an open-ended cylindrical cell, internal to the electron storage ring. Spin correlation parameters were measured for the $(e,e'p)$ and $(e,e'n)$ reactions at a four-momentum transfer squared of 0.21 $(GeV/c)^2$. The $(e,e'p)$ data give detailed information about the spin structure of the deuteron, and are in fair agreement with the results of microscopic calculations that include various spin-dependent reaction mechanism effects. The $(e,e'n)$ data provide new information on the charge form factor of the neutron.

INTRODUCTION

The deuteron has been used extensively in the past as a testing ground for nuclear models. As it is the simplest nucleus, only consisting of a proton and neutron, accurate microscopic calculations can be made and data can be interpreted without much model dependence. The ground state wave function is dominated by a central interaction, which results in the fact that the proton and neutron are mainly in a relative S wave. The tensor component in the nucleon-nucleon interaction induces a D-wave component in the wave function. The S- and D-wave components depend on the momentum of the proton (p) and neutron ($-p$) inside the deuteron. The D wave scales with $r\times p$, so for $p = 0$, the contribution of the D wave vanishes, and the wave function consists of a pure S wave. This S wave crosses zero for an internal momentum in the range 250 to 350 MeV/c, due to the short-range interaction of the nucleon-nucleon potential. The wave function has been investigated with elastic electron-deuteron scattering, where one measures the two structure functions, A and B, of the deuteron. The electromagnetic properties of a deuteron cannot be described completely with these structure functions, because for a spin-1 particle at least three observables are required. The interaction between the spin and the orbital angular momentum makes that also the spin structure of the deuteron depends on the internal momentum. In the past, one has measured the tensor analyzing power T_{20} in elastic scattering as a third observable, to get a better understanding of the electromagnetic and spin structure of the deuteron. Also our group in Amsterdam has contributed to this effort and a summary of the results from experiments[1-8] and various theoretical predictions[9-12] is shown in Fig. 1.

Another way to get more insight in the deuteron structure is via quasifree scattering. Here, one scatters an electron with momentum k from one of the nucleons inside the deuteron. For a given k, the transferred four-momentum q=(υ,-**q**) is determined by the momentum of the final-state electron (k') and the scattering angle (θ_e) in the scattering plane (see Fig. 2). The final-state proton has a momentum p_f in the reaction plane. The kinematics is determined by the missing momentum **p**$_m$, which is the transferred momentum not carried by the proton, i.e. **p**$_m$ = **q** - **p**$_f$. If one assumes two-body breakup, then one can construct the missing energy E_m = υ + m_d - E_p - E_n, where E_p (E_n) is the energy of the proton (undetected neutron) and m_d the mass of the deuteron. The energy E_m should be equal to the deuteron binding energy. In plane wave impulse approximation (PWIA), the neutron is only a spectator during the scattering process, and p_m = -p. Therefore, assuming PWIA a measurement of the response as function of missing momentum gives direct insight in the momentum distribution inside the deuteron.

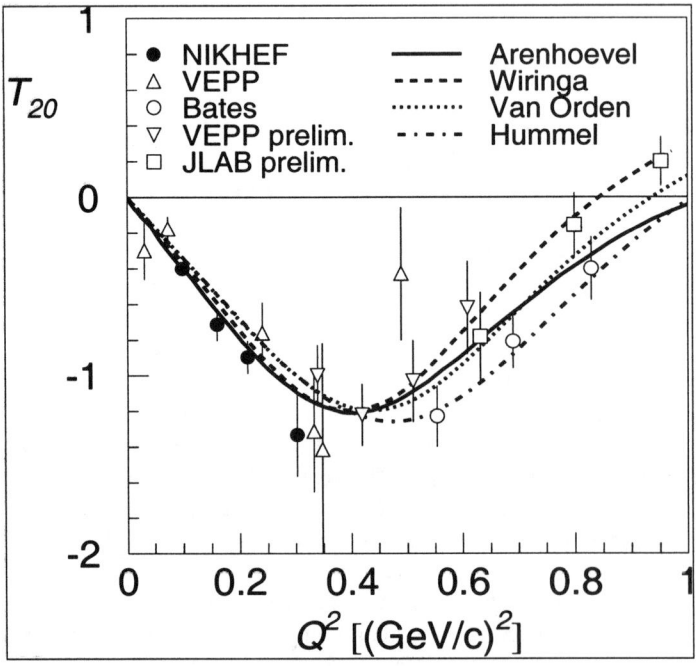

FIGURE 1. *Data and model predictions of the analyzing power T_{20} in elastic electron-deuteron scattering. The data are from NIKHEF [Bouwhuis, 1999; Ferro-Luzzi, 1996], MIT-Bates [Garcon, 1994; Schulze, 1984] and VEPP-3 [Dmitriev, 1985; Gilman, 1990]. Also shown are preliminary results from VEPP-3 [Nikolenko, 1999] and JLAB [Furget, 1998]. The curves are model predictions from non-relativistic models of Arenhoevel et al. [Arenhoevel, 1999c] and Wiringa et al. [Wiringa, 1995] and relativistic models from Van Orden et al, [Orden, 1995] and Hummel and Tjon [Hummel, 1990].*

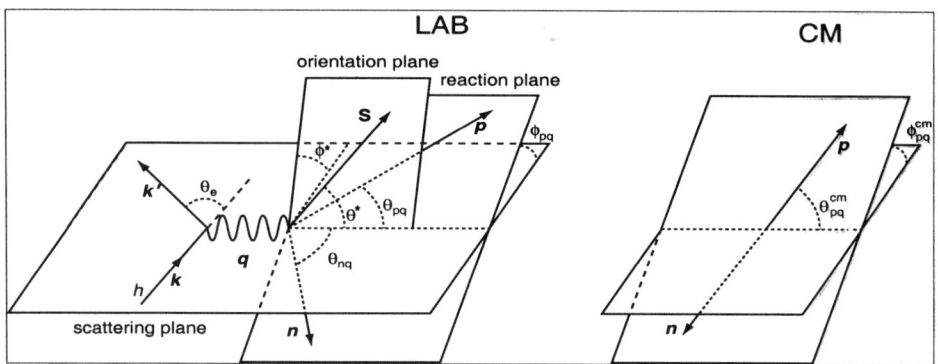

FIGURE 2. *Definition of the most relevant kinematic quantities for spin-dependent quasifree electron scattering from vector polarized deuterium.*

To enhance the sensitivity to the D wave, one can study spin-dependent observables in quasifree scattering. The polarization of a proton (P_z^p) inside a vector polarized deuteron (polarization P_1^d) depends on whether the proton and neutron are in a relative S or D wave. The z-component of the spin of the deuteron is a sum of the z-components of the proton and neutron spin, plus the z-component of the orbital angular momentum (L_z). In an S wave one has that $L_z = 0$, so for a fully vector polarized deuteron ($S_z = 1$), both the proton and neutron spins have to be aligned with the deuteron spin. In a D-wave configuration, also L_z can contribute to S_z, and therefore the proton and neutron spin can also be antialigned. The exact relation between P_1^d and P_z^p is

$$P_z^d = \sqrt{\frac{2}{3}}\, P_1^d (P_S - \frac{1}{2} P_D),$$

where P_S (P_D) are the S-wave (D-wave) components of the ground-state wave function of the deuteron. The square root factor is a convergion factor between different notations for polarization.

In elastic electron-proton scattering with longitudinally polarized electrons and polarized protons, an asymmetry in the cross section exists that depends only on the kinematics of the scattering, and on the electromagnetic form factors of the proton. These form factors are wellknown for $Q^2 \leq 3$ (GeV/c)^2, and it is therefore possible to calculate this asymmetry with high precision. In PWIA, the asymmetry in the proton knock-out reaction from vector polarized deuterium (A_V^{ed}) is the same, if one takes into account the relation given in the equation listed above. The simple PWIA results are modified by reaction-mechanism effects such as meson-exchange currents (MEC), isobar configurations (IC) and final-state interactions (FSI). Here, we describe the results of a measurement of A_V^{ed} performed at NIKHEF (Amsterdam), which uses a stored polarized electron beam and a vector-polarized deuterium target. This measurements gives detailed information about the S- and D-wave contributions as function of missing momentum, and gives information about the contributions of various reaction-mechanism effects. The results have been obtained simultaneously with data from a measurement of the charge form factor of the neutron[13].

The scattering cross section target for the (e,e'p) reaction with both longitudinal polarized electrons and a polarized deuterium, can be written as

$$S = S_0 \{1 + P_1^d A_d^V + P_2^d A_d^T + h(A_e + P_1^d A_{ed}^V + P_2^d A_{ed}^T)\},$$

where S_0 is the unpolarized cross section, h the helicity of the electrons, and P_1^d (P_2^d) the vector (tensor) polarization of the target. The variable A_e is the beam analyzing power, $A^{V/T}_d$ are the vector and tensor analyzing powers and $A^{V/T}_{ed}$ are the vector and tensor spin-correlation parameters. The target analyzing powers and spin-correlation parameters depend on the orientation of the target spin. The polarization direction of the deuteron is defined by the angles θ_d and Φ_d in the frame where the z-axis is along the direction of the three-momentum transfer **q** and the y-axis is defined by the vector product of the incoming and outgoing electron momenta, **k** × **k'**.

The above discussed formalism can also be used to extract information about the electromagnetic response of the neutron. Although the neutron has no net electric charge, it does have a charge distribution. Precise measurements where thermal neutrons from a nuclear reactor are scattered from atomic electrons[14] indicate that the neutron has a positive core surrounded by a region of negative charge. The actual distribution is described by the charge form factor G_E^n, which enters the cross section for elastic electron scattering. It is related to the Fourier transform of the charge distribution and is generally expressed as a function of Q^2, the square of the four-momentum transfer. Data on G_E^n are important for our understanding of the nucleon and are essential for the interpretation of electromagnetic multipoles of nuclei, e.g. the deuteron.

Since a practical target of free neutrons is not available, experimentalists mostly resorted to (quasi)elastic scattering of electrons from unpolarized deuterium to determine this form factor. The shape of G_E^n as function of Q^2 is relatively well known from high precision elastic electron-deuteron scattering[15]. However, in this case the cross section is dominated by scattering from the proton and, moreover, is sensitive to nuclear-structure uncertainties and reaction-mechanism effects. Consequently, the absolute scale of G_E^n from this type of experiments still contains a systematic uncertainty of about 50 %.

At present, there is a worldwide effort underway to measure the neutron charge form factor by scattering polarized electrons from neutrons bound in deuterium and ^{3}He nuclei, where either the target is polarized or the polarization of the ejected neutron is measured. Experiments with external beams have been carried out at Mainz and MIT. In the present paper we describe a measurement performed at NIKHEF (Amsterdam), which uses a stored polarized electron beam and a vector-polarized deuterium target.

By simultaneously detecting protons and neutrons in the same detector, one can construct asymmetry ratios for the two reaction channels ^{2}H$(e,e'p)n$ and ^{2}H$(e,e'n)p$, in this way minimizing systematic uncertainties associated with the deuteron ground-state wave function, absolute beam and target polarizations, and possible dilution by cell-wall background events.

EXPERIMENTAL SET-UP

The experiment was performed with a polarized gas target[16] internal to the AmPS electron storage ring. An atomic beam source (ABS) was used to inject a flux of 3×10^{16} deuterium atoms/s (in two hyperfine states) into the feed tube of a cylindrical storage cell cooled to 75 K. The cell had a diameter of 15 mm and was 60 cm long, resulting in a typical target thickness of 1.0×10^{14} deuterons/cm^2. An electromagnet was used to provide a guide field of 40 mT over the storage cell which oriented the deuteron polarization axis perpendicular to $\mathbf{q}$ in the scattering plane. The vector polarization of the target was varied every 10 seconds. Fig. 3 shows shows part of the ABS. In the opening one sees the Zeeman transition units and permanent sextupole magnets needed to polarized the deuterium atoms.

FIGURE 3. *Layout of the atomic beam source. On the right-hand side the weak-field transition unit is shown. To the left and right of this unit one can see the holders for the various permanent sextuple magnets. The strong-field unit is located on the left-hand side of the picture.*

Polarized electrons were produced by photo-emission from a strained-layer semiconductor cathode, InGaAsP, accelerated to 720 MeV, and stacked in the AmPS storage ring. In this way, beam currents of more than 100 mA with a lifetime in excess of

15 minutes were obtained. Every 5 minutes, the remaining electrons were dumped, and the ring was refilled after reversal of the electron polarization at the source. The polarization of the stored electrons was maintained by setting the spin tune to 0.5 with a strong solenoid field, using the well-known Siberian snake principle[17].

Optimization of the longitudinal polarization at the interaction point was achieved by varying the orientation of the spin axis at the source and by measuring the polarization of the stored electrons with a polarimeter based on spin-dependent Compton backscattering[18].

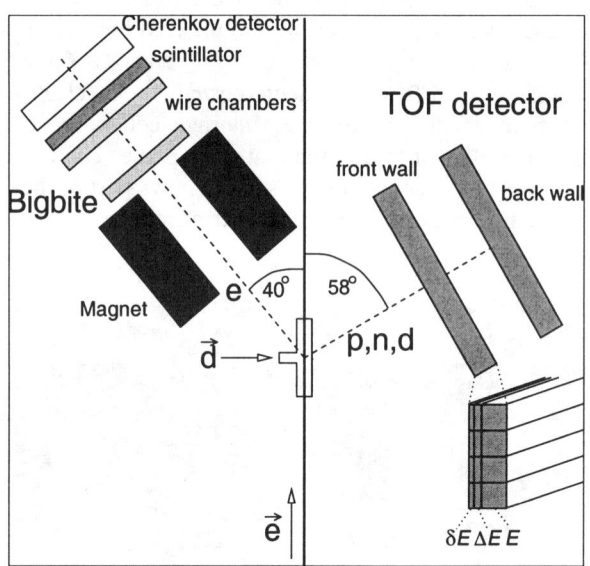

FIGURE 4. *Layout of the detector setup. The electron spectrometer consists of a 1 Tm magnet, two drift chambers of four planes each, a scintillator and a Cherenkov detector. The time of flight system consists of two identical walls of four E-scintillators preceded by two (δE and ΔE) veto scintillators. The second wall was used only for neutron detection, as described in Ref. {Passchier99}.*

Scattered electrons were detected in the large-acceptance magnetic spectrometer Bigbite positioned at a central scattering angle of 40°, with a momentum acceptance from 250 to 720 MeV/c and a solid angle of 96 msr (see Fig. 4). This setting resulted in a central value of $Q^2 = 0.21$ (GeV}/c)^2.

Protons and neutrons were detected in a time-of-flight (TOF) system made of scintillator array, consisting of four 160 cm long, 20 cm high, and 20 cm thick plastic scintillator bars stacked vertically. Each bar was preceded by two (δE and ΔE) plastic scintillators (3 and 10 mm thick, respectively) of equal length and height, used for particle identification. Each of the scintillators was read out at both ends to obtain position information along the bars (resolution of about 4 cm) and good coincidence timing resolution (about 0.5 ns). The TOF detector was positioned at a central angle of 58° and covered a solid angle of about 250 msr. Protons with kinetic energies in excess of 40 MeV were detected with an energy resolution of about 10 %. The $e'N$ trigger was formed by a coincidence between the electron arm trigger and a hit in any one of the TOF bars.

RESULTS

An experimental asymmetry (A_{exp}) can be constructed, via

$$A_{\text{exp}} = \frac{n_+ - n_-}{n_+ + n_-},$$

where $n_\pm$ is the number of events that pass the selection criteria, with hP_1^d either positive or negative, normalized to the integrated luminosity for that state. The contribution of events of electrons scattering from the cell wall has been taken into account by subtracting the normalized rate of cell wall events from $n_\pm$. We have studied the cell wall contribution by measuring with an empty storage cell. The background contribution amounted to 5 % for low missing momentum and increasing to about 40 % for $p_m = 400$ MeV/c. A possible dependence on the target density was investigated by injecting various fluxes of unpolarized hydrogen into the cell and measuring quasi-free nucleon knock-out events. The target density dependence was found to be negligible at ABS operating conditions.

Finite-acceptance effects were taken into account with a Monte Carlo code that interpolated the model predictions between a dense grid of calculations over the full kinematical range and detector acceptance.

Results for the (e,e'p) reaction

Protons are selected by a valid hit in two PMT's of at least one E-bar and a valid hit in both PMT's of one of the preceding ΔE bars. This requirement allowed us to use ΔE - E particle identification in order to discriminate between protons and either deuterons or pions. To select two-body breakup, the electron energy is required to be larger than 450 MeV, and the reconstructed missing energy between –50 MeV and 50 MeV. This requirements resulted in a clean sample of two-body breakup events, with some dilution of cell wall events.

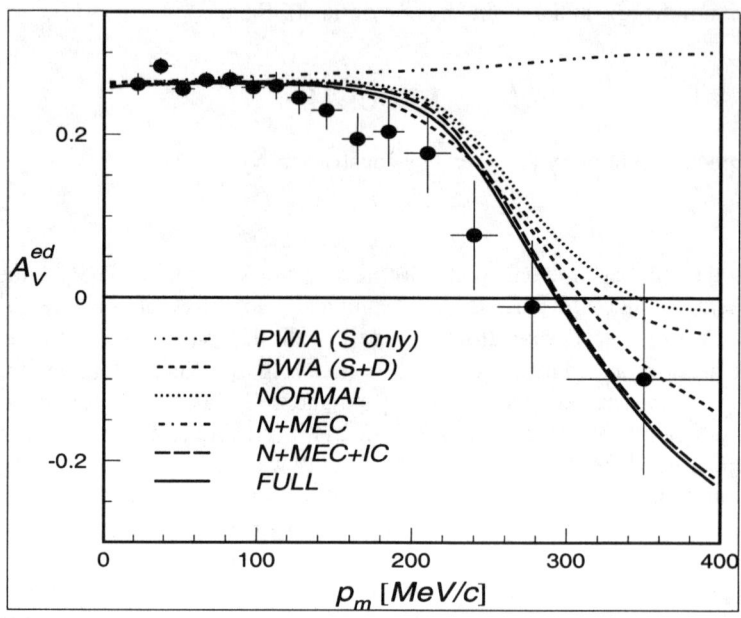

FIGURE 5. *Beam-vector spin correlation parameter as function of missing momentum for the two-body breakup reaction at a four-momentum transfer of 0.21 $(GeV/c)^2$. The dot-dot-dashed curve and dashed curve represent PWIA predictions with and without inclusion of the D-wave, respectively. The other curves are predictions of the model of Arenhoevel et al., for normal (dotted), normal+mec (dashed-dotted), normal+mec+ic (long-dashed) and total (full) calculations, as indicated in Ref. {19}. The predictions are folded over the detector acceptance using Monte Carlo methods.*

This sample has been used (after the background correction, described earlier), to extract $A_V^{ed}(90°, 0°)$. Fig. 5 shows the experimental results, and compares it to the predictions[19] of the model of Arenhoevel et al.

At small missing momentum, the spin correlation function is equal to what is expected for elastic electron-proton scattering, while for increasing momenta, the predictions and the data show that the asymmetry reverses sign, as expected from the increasing D wave in the wave function of the deuteron. It can also be observed, that inclusion of reaction-mechanism corrections, mainly for Isobar Configurations (IC), is required for a better description of the data. We have investigated the dependence of the predictions on the nucleon-nucleon potential for the Bonn, Paris, Argonne and Nijmegen-V_{14} potential. The effect is negligible for $p_m < 200$ MeV/c, and increases to 0.02 for $p_m = 400$ MeV/c, much smaller than the experimental accuracy or the reaction mechanism effects. The data suggests that the model overestimates A_V^{ed} around $p_m = 200$ MeV/c. This could be contributed to an underestimate of the D wave or the effects of IC. Both effects imply that less of the deuteron angular momentum is carried by the spin of the proton and neutron, and more my the orbital angular momentum (the D wave), or virtual Δ particles in the deuteron (IC). This is in agreement with earlier measurements of the deuteron quadrupole moment, unpolarized quasi-elastic electron-deuteron scattering, and measurements of T_{20}.

Results for the (e,e'n) reaction

Neutrons were identified by a valid hit in one E-scintillator or two neighboring E-scintillators (to allow for events that deposit energy in two neighboring E-scintillators) and no hits in the preceding δE and ΔE scintillators, which resulted in an 8- to 12-fold veto requirement. Minimum-ionizing particles and photons were rejected by a cut on the time of flight, resulting in a clean sample of neutrons, with only a small proton contamination. The spin-correlation parameter was obtained from the experimental asymmetry by correcting for the contribution of protons misidentified as neutrons (less than 1 %, as determined from a calibration with hydrogen), and for the product of beam and target polarization, as determined from the (e,e'p) channel.

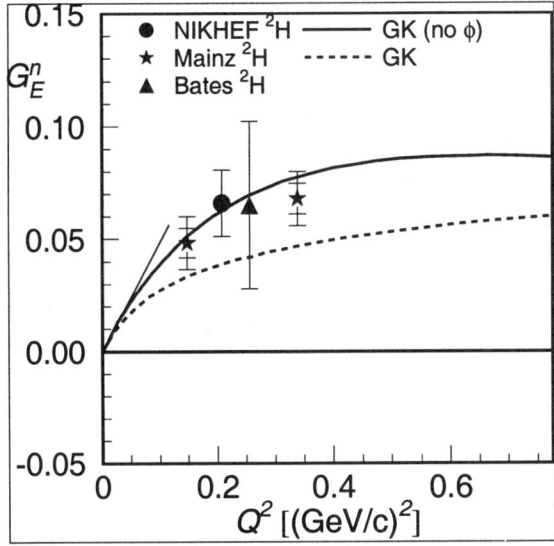

FIGURE 6. Data on the charge form factor of the neutron data from double polarization experiments as function of squared four-momentum transfer. The solid circle shows our result, while the stars correspond to the results obtained at Mainz (see Ref. Herberg, 1999). Both data sets have been obtained on deuterium and have been analyzed with the full model of Arenhoevel et al. The curves represent theoretical predictions according to the VMD model of Gari and Krumpelmann, with and without the inclusion of the φ meson. The line at the origin represents the data on the slope of G_E^n.

The main effect of cell wall events is a reduction of the effective target polarization. Therefore, the effects largely cancel in the asymmetry ratio. We have studied the cell wall contribution by measuring with an empty storage cell. The background contribution to the (e,e'p) and (e,e'n) channels amounted to 5 ± 1 %, stable over the entire run. A possible dependence on the target density was investigated by injecting various fluxes of unpolarized hydrogen into the cell and measuring quasifree nucleon knock-out events. The target density dependence was found to be negligible at ABS operating conditions.

The data are compared to the predictions of the full model of Arenhoevel et al., assuming the dipole parameterization for the magnetic form factor of the neutron and the Paris nucleon-nucleon (NN) potential, folded over the detector acceptance with our Monte Carlo code for various values of G_E^n. Full model calculations are required for a reliable extraction of G_E^n. We extract G_E^n (Q^2=0.21 (GeV/c)^2) = 0.066 ± 0.015 ± 0.004, where the first (second) error indicates the statistical (systematic) uncertainty. The systematic error is mainly due to the uncertainty in the correction for misidentified protons and the orientation of the holding field (thus the contribution of the spin-correlation parameter $A_V^{ed}(0°,0°)$ to our experimental asymmetry).

We have investigated the influence of the NN potential on the calculated spin-correlation parameters using Arenhoevel's full treatment. The results for $A_V^{ed}(90°,0°)$ using the Paris, Bonn, Nijmegen, and Argonne V_{14} NN potential differ by less than 5 % for missing momenta below 200 MeV/c.

In Fig. 6 we compare our experimental result to other data obtained with spin-dependent electron scattering from deuterium[20]. The figure also shows the precize data on the charge radius of the neutron[14]. By comparison to the predictions of the QCD-VM model by Gari and Krumpelmann[21], with and without the inclusion of the coupling of the φ-meson to the nucleon (which these authors identify with the effect of strangeness in the neutron), our datum favors the prediction without strangeness in the neutron included.

SUMMARY

In summary, we presented the first measurement of the sideways spin-correlation parameter $A_{ed}^V(90°,0°)$ in quasifree electron-deuteron scattering for both proton and neutron knockout. High precision data on the spin correlation function for the (e,e'p) reaction allow for a stringent test for any model prediction of polarization observables for quasi-elastic scattering and for nucleon-nucleon potentials. In addition, we obtained asymmetry data for the (e,e'n) reaction, from which we extract the neutron charge form factor at $Q^2 = 0.21$ (GeV/c)^2. When combined with the known value and slope at $Q^2 = 0$ (GeV/c)^2 and the elastic electron-deuteron scattering data from Ref. 15, this result puts strong constraints on G_E^n up to $Q^2 = 0.7$ (GeV/c)^2.

ACKNOWLEDGEMENTS

I would like to thank the the 97-01 collaboration, and the NIKHEF and Vrije Universiteit technical groups for their outstanding support and Prof. H. Arenhoevel for providing the calculations. This work was supported in part by the Stichting voor Fundamenteel

Onderzoek der Materie (FOM), which is financially supported by the Nederlandse Organisatie voor Wetenschappelijk Onderzoek (NOW).

REFERENCES

1. M. Bouwhuis et al., Phys. Rev. Lett. **82**, 3755 (1999).
2. M. Ferro-Luzzi et al., Phys. Rev. Lett. **77**, 2630 (1996).
3. M. Garcon et al., Phys. Rev. **C49**, 2516 (1994).
4. M.E. Schulze et al., Phys. Rev. Lett. **52**, 597 (1984).
5. V.F. Dimitriev et al., Phys. Lett. **B157**, 143 (1985).
6. R. Gilman et al., Phys. Rev. Lett. **65**, 1733 (1990).
7. D. Nikolenko et al., Private communication.
8. C. Furget et al., Acta. Phys. Polon. **B29**, 3301 (1998).
9. H. Arenhoevel, F. Ritz and T. Wilbois, `nucl-th/9910009`.
10. R.B. Wiringa, V.G.J. Stoks and R. Schiavilla, Phys. Rev. **C51**, 38 (1995).
11. J.M.V. Orden, N. Devine and F. Gross, Phys. Rev. Lett. **75**, 4369 (1995).
12. E. Hummel and J.A. Tjon, Phys. Rev. **C42**, 423 (1990).
13. I. Passchier et a.l, Phys. Rev. Lett. **82**, 4988 (1999).
14. S. Kopecki et al., Phys. Rev. Lett. **74**, 2427 (1995); Phys. Rev. **C56**, 2229 (1997).
15. S.Platchkov et al., Nucl. Phys. **A510**, 740 (1990).
16. Z.-L. Zhou et al., Nucl. Instr. Meth. **A378**, 40 (1996).
17. Ya.S. Derbenev and A.M. Kondratenko, Sov. Phys. JETP **37**, 968 (1973); Sov. Phys. Dokl. **20**, 830 (1975).
18. I. Passchier et al., Nucl. Instr. Meth. **A414**, 446 (1998).
19. H. Arenhoevel , W. Leidemann and E.L. Tomusiak, Phys. Rev. **C46**, 455 (1992).
20. C. Herberg et al., Eur. Phys. J. **A5**, 131 (1999).
21. M.F. Gari and W. Krumpelmann, Phys. Lett. **B274**, 159 (1992).

Twenty-Five Years of Progress in the Three-Nucleon Problem

J. L. Friar

Theoretical Division
Los Alamos National Laboratory
Los Alamos, New Mexico, 87545

Abstract. Twenty-five years ago the International Few-Body Conference was held in Quebec City. It became very clear at that meeting that the theoretical situation concerning the ^{3}He and ^{3}H ground states was confused. A lack of computational power prevented converged brute-force solutions of the Faddeev or Schrödinger equations, both for bound and continuum states of the three-nucleon systems. Pushed by experimental programs at Bates and elsewhere and facilitated by the rapid growth of computational power, converged solutions were finally achieved about a decade later. Twenty-five years ago the first three-nucleon force based on chiral-symmetry considerations was produced. Since then this symmetry has been our guiding principle in constructing three-nucleon forces and, more recently, nucleon-nucleon forces. We are finally nearing an understanding of the common ingredients used in constructing both types of forces. I will discuss these and other issues involving the few-nucleon systems and attempt to define the current state-of-the-art.

INTRODUCTION

The purview of my talk is progress that has been made in our understanding of the three-nucleon systems and of the dynamics that underlies that understanding. My emphasis will be on the theoretical side. My reference point in time is 1974, the date when Bates first delivered beam for an experiment. I will survey that progress by referring to two other significant events that occurred in 1974. One of these is personal: I attended the International Few-Body Conference held that year in Quebec City, Canada [1]. The second event is the genesis in that year of three-nucleon forces (3NFs) based on chiral-symmetry considerations [2].

On a personal note it is always a pleasure to return to MIT, where I was a post-doc. Looking back at my work during that period, I find that almost everything dealt with electron scattering, a result of the influence of Bates on the young theorists in the Center for Theoretical Physics. Part of that work involved relativistic corrections to the charge densities of few-nucleon systems, and that motivated my attendance at the Quebec meeting.

There are basically three reasons why three-nucleon physics has become a subfield in its own right. The first is that the trinucleons are rich, nontrivial, and "simple" nuclear systems, and understanding their properties is a minimal criterion for success in this area. The word "simple" in this context means that we are capable of performing the very difficult calculations of three-nucleon properties. Indeed, in recent years we have not only succeeded in performing these calculations, but have achieved an understanding of most of the basic trinucleon properties [3,4].

The second reason is the classic and original goal of the field: using these systems to sort out and refine our understanding of the nuclear force. This is the most important remaining aspect of the problem, which has been greatly aided in recent years by chiral perturbation theory (χPT). Much of our theoretical and experimental attention has been directed at 3NFs, because trinucleon properties show relatively little sensitivity to the details of modern N-N forces. Our remaining problems (though few) are likely due to our lack of understanding of 3NFs [5].

Finally, the lovely techniques used in this field are fun to work with, and this attraction has seduced two generations of theorists. Our efforts have led to the very successful application of few-body methods to heavier systems, which goes far beyond even the dreams of 1974, as shown at this symposium by Vijay Pandharipande.

My strongest impressions of the Quebec meeting are that the field was in a state of confusion. Many calculational techniques were in use, each giving a different answer to the same problem, the ^{3}H bound-state energy. Faddeev methods, hyperspherical expansions, variational bounds, and separable approximations all had their practitioners [1]. There was a 10-20% uncertainty ($\sim$ 1-2 MeV) in the ^{3}H binding energy, implying that most (in retrospect, all) of the calculations were not converged. The situation was similar with respect to scattering calculations. In order to achieve convergence one requires brute-force computational resources on a scale that would not be available for another decade.

NUCLEAR FORCES

The genesis of the computational problem is the spin of the nucleon. Contrary to much folklore, nuclear physics is difficult not because the force is complicated (in shape), but because it is complex (i.e., it has many components). The origin of the problem is the spin and parity of the pion: $J^\pi = 0^-$. The π-nucleon vertex must have a complementary pseudoscalar structure in order to conserve angular momentum and parity, and the dominant form ($\sim 1^+ \cdot 1^-$) is $\vec{\sigma}_N \cdot \vec{q}$, where $\vec{\sigma}_N$ is the nucleon (Pauli) spin and $\vec{q}$ is the pion momentum. This leads immediately to a tensor component of the force (part of the one-pion-exchange potential (OPEP)), which dominates interactions in few-nucleon systems. Indeed, $\langle V_{\text{OPEP}} \rangle$ is roughly 75% of the total potential energy. This spin dependence, together with isospin dependence, accounts for the complexity. Each nucleon has $2 \cdot 2 = 4$ spin-isospin components, implying that there are roughly $(4)^2 = 16$ such components in the

N-N force, which is indeed exemplified by the 18 components of the recent AV18 potential [6]. Dealing with these complexities, in addition to the 3 continuous coordinates specifying the positions of 3 nucleons, is a formidable numerical problem.

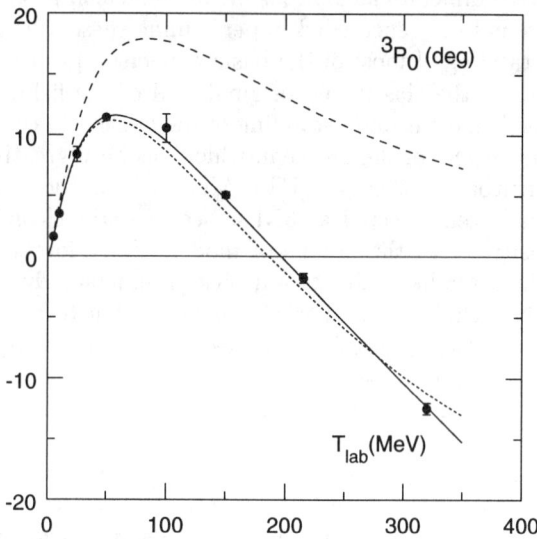

FIGURE 1. $^3P_0$ phase shift calculated with OPEP tail for $r > b$ (dashed line), and with either one (dotted) or 3 (solid) short-range interaction terms added.

The importance of OPEP is illustrated in Fig. (1) from the Nijmegen group [7]. Using a potential that vanishes out to $b = 1.4$ fm and incorporates OPEP plus some two-pion-exchange potential beyond that value leads to the dashed curve. Clearly, the shape of the phase shift is correct. Adding a smooth background contribution from a short-range potential ($r \leq 1.4$ fm) produces the dotted curve, while fine tuning leads to the solid curve. All of the "shape", however, is produced by pion exchange, which is hardly a surprise given that the pion is the lightest of the mesons exchanged between two nucleons.

An obvious question is whether a 1-2 MeV uncertainty is a serious handicap in understanding the physics. Alternatively, if one wishes to probe the nuclear force by examining trinucleon properties, what level of calculational accuracy is a reasonable requirement? The fundamental problem is determining the structure of the N-N force, and this is impossible to achieve using only the N-N scattering data. Imagine that some N-N phase shift is known at all energies and with infinite accuracy (neither assumption is true), and that there is no bound state. Under these idealized conditions a potential $V(r)$ (where r is the separation of the two nucleons) can be deduced that in the Schrödinger equation will reproduce the phase shift. Unfortunately, one can also deduce a $V(r,p)$ (where p is the relative nucleon momentum) that reproduces that phase shift equally well. On-shell (free-nucleon) scattering cannot produce a unique potential. This led to the idea that making the nucleons "off-shell" by placing them in a bound system with a third nucleon might provide enough additional information to fix the potential, since $V(r)$ and $V(r,p)$ defined above will definitely produce different tritons. This is one aspect of what has become known as the "off-shell problem".

We can estimate the uncertainties by noting that the N-N system (with potential V) feels the presence of the third nucleon only through the action of another V and the effect should scale as V^2, which has the wrong dimensions. Another related off-shell problem is that the motion of a pion propagating between nucleons is conventionally specified only by its transferred momentum, $\vec{q}$, while its transferred energy, q_0, is replaced by other variables such as $p^2/2M$. This hints that the effective off-shell interaction scale is set by $\Delta H = V^2/Mc^2$, which is correct [8] in spite of the intuitive derivation. Because V^2 contains terms linking three nucleons together and because of the $1/c^2$, this effect is at the same time a three-body force, an off-shell effect, and a relativistic correction. Using reasonable numbers for the triton we estimate $\langle \Delta H \rangle \sim 0.5\text{-}1.0$ MeV. Thus the previously noted calculational uncertainties ($\sim$ 1-2 MeV) are unacceptably large, and calculational errors $\lesssim 100$ keV (which is approximately 1% of the binding energy) are required in order to investigate the three-nucleon effects discussed above. In addition, 1% absolute experiments are extremely difficult and uncommon. Consequently, 1%-error calculations, known variously as "exact", "complete", or "rigorous", have become the standard of the field. The ability to achieve this has become our field's major success story.

THREE-NUCLEON CALCULATIONS

The types of problems attacked and the period during which success was achieved are shown in Fig. (2) and Table 1 [9]. There are four regions of energy illustrated

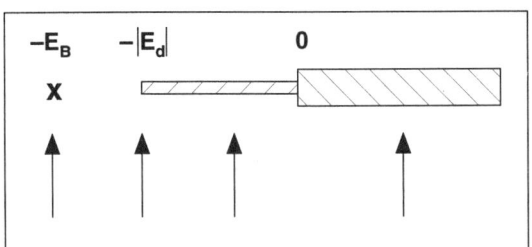

FIGURE 2. Energy spectrum of ^{3}H.

in Fig. (2) (by arrows) that conveniently encompass the three-nucleon problem: (1) the trinucleon bound states (a pole at $-E_B$); (2) zero-energy nucleons scattering from the deuteron; (3) N-d scattering below deuteron-breakup threshold (viz., zero total energy); (4) N-d scattering above breakup threshold. These problems were solved at the 1% level at times indicated in Table 1. The Los Alamos-Iowa group was fortunate enough to have participated in half of the entries (top half) in the table, beginning with the ^{3}H bound state in 1985 [10] and using only N-N forces, then adding a 3NF, and finally solving ^{3}He in 1987 (which includes a p-p Coulomb interaction) [11]. Scattering lengths were calculated a few years later [11]. The bound-state problems are relatively easy, however. Scattering below breakup threshold [12] is nearly an order of magnitude harder than a bound-state problem, and above-breakup scattering is nearly an order of magnitude harder still [13]. Above-breakup p-d scattering is a very recent development [14].

TABLE 1. Complete three-nucleon calculations:
"♦" indicates calculations from mid-late 1980's;
"★" indicates calculations from the early 1990's;
"•" indicates calculations from early-mid 1990's;
"■" indicates very recent calculations.

Type	NN Force	NN + 3NF	Coulomb
$E = -E_B$	♦	♦	♦
$E_{Nd} = 0$	★	★	★
$E < E_{th}$	•	•	•
$E > E_{th}$	♦	•	■

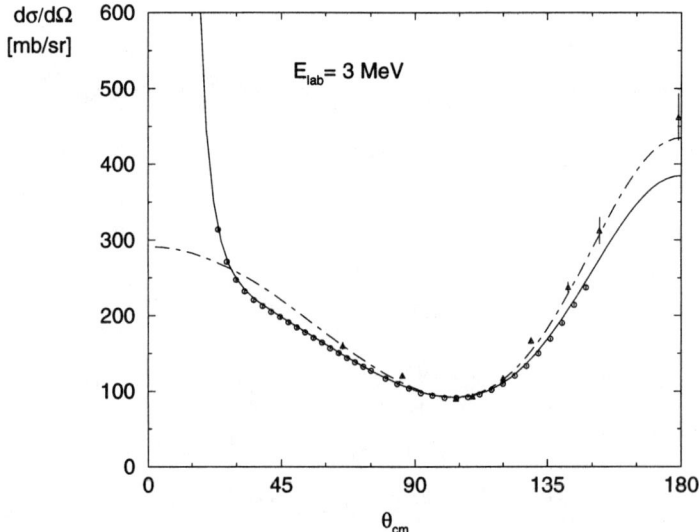

FIGURE 3. N-d scattering at 3 MeV.

A particularly lovely example of this progress is shown in Fig. (3), obtained from the Pisa group [12]. Elastic scattering of 3 MeV nucleons (just below breakup threshold) from deuterons is calculated and compared to data. The solid curve (p-d) agrees superbly well with the dense, accurate data, while sparser n-d data agree well with the (dashed) calculated values. Note the large Coulomb effect at the forward and backward angles. This plot is rather typical of differential cross sections: they are insensitive to the details of the nuclear force and agree very well with data. Most spin observables, such as tensor analyzing powers, also agree well with data.

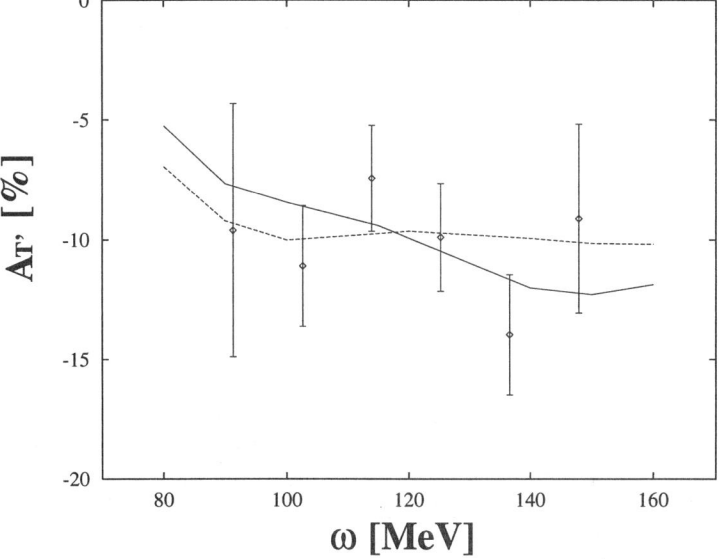

FIGURE 4. The spin-dependent asymmetry $A_{T'}$ in the reaction $^3\vec{\mathrm{He}}(\vec{e},e'n)pp$. The solid curve depicts the full calculation, while the dashed curve lacks final-state interactions.

Figure (4) shows a very recent calculation [15] of an electromagnetic spin observable, $A_{T'}$, in the reaction $^3\vec{\mathrm{He}}(\vec{e},e'n)pp$. The ^{3}He target is polarized along the direction of electron momentum transfer, and the electrons are longitudinally polarized. This spin-dependent asymmetry in a response function is proportional to G_M^n (neutron magnetic form factor) in the most naive description of the reaction. That description is based on the observation that s-waves dominate between the nucleons in ^{3}He. In that case the two protons are required by the Pauli principle to have spins anti-aligned, and the entire spin of the nucleus is carried by the neutron. The protons do contribute to the reaction because the tensor force modifies the simple s-wave picture and the protons' spins will be aligned in D-states, and can contribute to the asymmetry through final-state p-n charge-exchange reactions. The figure illustrates the Bates data [16] compared to two theoretical calculations: the full calculation (solid curve) and a calculation (dashed curve) that neglects all final-state interactions. The latter calculation would be typical of what was available until very recently, which illustrates both the difficulty of the calculations and the progress that has been made.

I would like to summarize this part of my talk as follows:

- We can now accurately calculate three-nucleon properties. Most of these properties, such as differential cross sections and most spin observables (e.g., the tensor analyzing power, T_{22} [4]), agree well with data and depend only weakly on a 3NF. Electromagnetic calculations are very difficult and are the state-of-the-art.

- Spin-isospin degrees of freedom are the biggest impediment to few-nucleon calculations.

- Many different techniques are now successfully employed in performing calculations [3].

- 1% accuracy is needed in order to disentangle the physics.

- The most demanding problems drive the progress, and Bates problems are of this type.

THREE-NUCLEON FORCES

Three-nucleon forces are small, as we argued earlier for a very special case. In fact that argument holds for the whole class of such forces, as we shall see. If they are so small, are they really necessary, or even interesting? The most modern potentials produce ^{3}H bound states that are underbound by up to 1 MeV. This defect can be compensated by the addition of a 3NF. Nevertheless, I do not consider this to be very compelling evidence for three-nucleon forces. Are such forces just "theorists' toys" or is there more compelling experimental evidence?

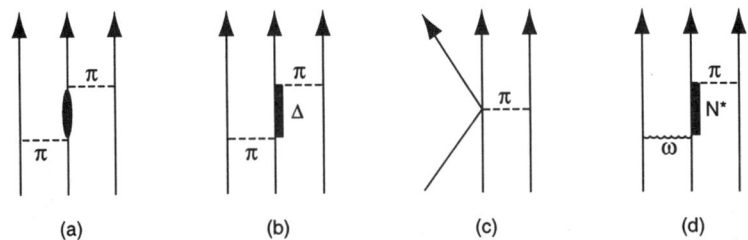

FIGURE 5. Mechanisms that contribute to three-nucleon forces. Two-pion-exchange forces are shown generically in (a), and the important isobar contribution in (b). Chiral perturbation theory predicts a large contribution of the type shown in (c), a specific mechanism of that type being displayed in (d).

In order to answer this question, we must first establish the credentials of the physics underlying the various models of such forces, which are relatively few in number. The longest-range mechanisms are those based on 2π-exchange, and these have been extensively investigated. Figure (5a) illustrates the generic force of this type, while Fig. (5b) shows the single most important ingredient (other ingredients are also important). The history of this field is depicted in Fig. (6), a diagram showing the evolution of these forces, all of which are field-theory based. Time runs vertically and long lines indicate the oldest forces. Near the bottom are the primitive models (PM). The august Fujita-Miyazawa model [17] (FM) is based on Δ-isobars, as is its offshoot the Urbana-Argonne model [18] (UA). To the left are the models based on chiral symmetry, including the Yang model [2] (Y) (the first of this type, published in 1974) and the Tucson-Melbourne model [19] (TM), the oldest such model still in use. The more recent models based on relativistic field theories (RFT) are the Brazil [20] (BR) and RuhrPot [21] (RP) models. Finally, the Texas model [22] (TX) is based on chiral perturbation theory. It is clear from this history that the two key ingredients of 3NFs are:

- adequate phenomenology (such as isobars).
- imposing chiral constraints.

How does one accomplish this?

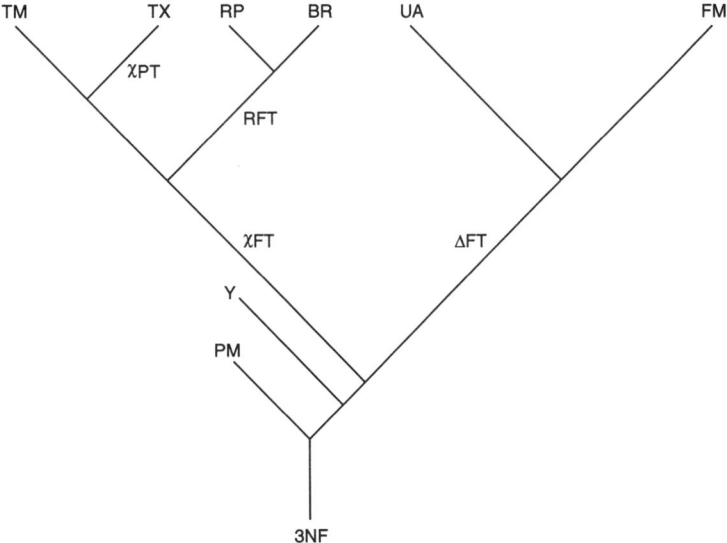

FIGURE 6. Cladogram [23] of 2π-exchange three-nucleon-force models, showing their history with a vertical time line, together with the properties that characterized their development.

It is believed that the theory underlying the strong interactions is QCD. The "natural" degrees of freedom of this theory are quarks and gluons. We aren't required to use these degrees of degrees, however, and traditional nuclear physics uses effective (observable) degrees of freedom: nucleons and pions. One can imagine freezing out all other particles and constructing a theory in this compressed Hilbert space, in the fashion of (Feshbach) [P,Q] reaction theory [24]. Although the resulting operators can be quite complicated, chiral symmetry, that most important ingredient residing in QCD, can be implemented in the new theory. This "QCD in disguise" is better known as chiral perturbation theory, and applies to both particles and nuclei [25].

Only one aspect of that theory is needed here: dimensional power counting [25]. The latter is a kind of (not obvious!) dimensional analysis based on only two QCD internal energy scales. The first scale is f_π, the pion decay constant ($\sim$ 93 MeV), which controls the Goldstone bosons and specifically the pion. The second scale is the energy above which we agree to freeze out all excitations, $\Lambda \sim 1$ GeV, and is the scale appropriate to the QCD bound states, such as the nucleon, ρ and ω resonances, etc. Using these scales, it can be shown [26] that a given term in a Lagrangian should scale as:

$$\mathcal{L}^{(\Delta)} \sim \frac{c}{f_\pi^\beta \Lambda^\Delta} \text{(times various fields)} .$$

Two important properties are that the power Δ (used to classify Lagrangian terms) satisfies $\Delta \geq 0$ (which is a not very obvious chiral-symmetry constraint), while the dimensionless constant c satisfies $|c| \sim 1$, the condition of "naturalness" (an even less obvious constraint). Because freezing out degrees of freedom results in effective interactions with unknown coefficients, the latter condition is the only handle we have on reasonable values for those constants.

This formal scheme can be implemented in nuclei to estimate the size of various contributions to potential energies (among others). An additional nuclear scale is required, the effective momentum or inverse correlation length, which is given by $Q \sim m_\pi c$, where m_π is the pion mass. Then it can be shown that [25]

$$\langle V_\pi \rangle \sim \frac{Q^3}{f_\pi \Lambda} \sim 30 \text{ MeV/pair} ,$$

$$\langle V_{3NF} \rangle \sim \frac{Q^6}{f_\pi^2 \Lambda^3} \sim 1 \text{ MeV/triplet} .$$

The latter relationship can also be written as $\langle V_{3NF} \rangle \sim \langle V_\pi \rangle^2 / \Lambda$, which is equivalent to the expression we developed earlier (since $M \sim \Lambda$) and is also the correct size to explain the ^{3}H binding discrepancy. The use of χPT is finally leading to a consensus on 2π-exchange 3NF terms, and a "standard" model of the 3NF is within reach. All such terms in leading order of χPT have been calculated, although some of them have not yet been implemented.

Several of these terms have been checked by testing the tail of the N-N potential against the set of p-p data. That tail is calculated by using the same Lagrangian

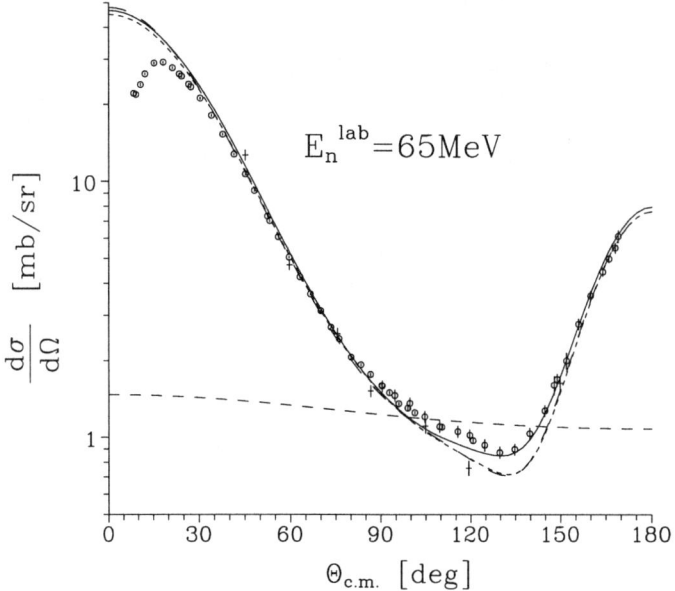

FIGURE 7. Differential cross section for 65 MeV proton-deuteron scattering, showing calculations with N-N forces only (dashed lines), a full calculation that includes the TM 3NF (solid line), and an estimate of the effect of the 3NF alone (long-dashed line).

building blocks that are used to calculate π-N scattering and the 2π-exchange 3NF. Important elements of the 2π-exchange N-N force were verified [27], which validates the corresponding terms in the 3NF.

In addition to the ^{3}H (^{3}He) binding discrepancy, there is one other piece of experimental evidence for a 3NF that is much stronger. The Sagara discrepancy [28] is illustrated in Fig. (7), which shows p-d elastic scattering at 65 MeV. Ignoring the forward direction (where the Coulomb interaction plays a significant role), the agreement is very good between calculations with an N-N force only (dashed lines) and the experimental data except in the diffraction minimum. Adding the TM 3NF produces the solid curve, which is in fairly good agreement with experiment in the minimum. The small 3NF effect is depicted by the long-dashed line, which follows from keeping only those terms linear in the 3NF. This behavior is very reminiscent of Glauber scattering, with a dominant single-scattering contribution falling rapidly with angle until the smaller double-scattering term (which has a reduced slope) becomes significant. This is rather strong evidence for a 3NF, and it persists to higher energies.

Our final topic is the extension of 3NFs beyond 2π-exchange. Chiral perturbation theory predicts that there are two mechanisms that have pion range in one pair of nucleons and short range in a second pair, and they should be comparable in size to the 2π-exchange mechanisms. The generic force in χPT is shown in Fig. (5c), and a particular example (the so-called d_1-term) is illustrated in Fig. (5d). All mechanisms affect the ^{3}H binding energy, so this is a poor test of a *specific* mechanism. A tedious examination of low-energy observables [29] finds that the d_1-mechanism makes a potentially large contribution to the n-d asymmetry, A_y. This observable at 3 MeV is depicted in Fig. (8). The calculation with only N-N forces is the solid line, which is about 30% lower than the data. The long-dashed curve includes the effect of the TM force, which accounts for only about 1/4 of the discrepancy. Adding the d_1-term in the 3NF with a dimensionless coefficient, $c_1 = -1$, produces the short-dashed curve. The size and sign of that coefficient are unknown, and the sign was chosen to move the prediction upward. Although it appears that a choice of $c_1 = -3$ (and quite acceptable in size) would resolve the problem, the algorithms used in our codes failed to converge for such a value, and that final conclusion could not be checked at the time this manuscript was written.

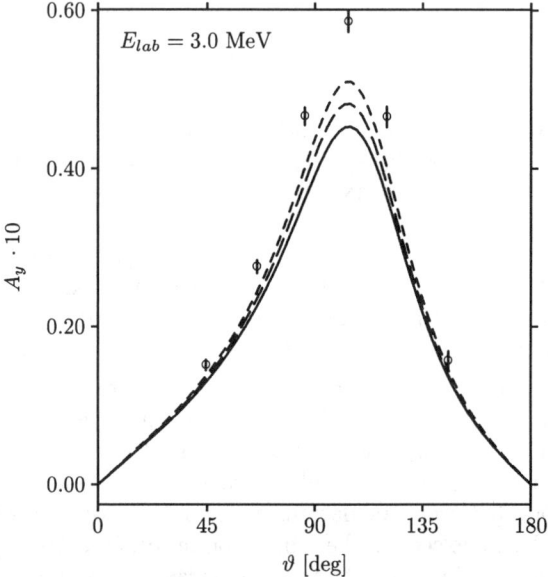

FIGURE 8. The asymmetry, A_y, for 3 MeV neutron-deuteron scattering, calculated using N-N forces only (solid), incorporating the TM force (long-dashed), and further adding a d_1-type force (short-dashed).

Nevertheless, it appears that this mechanism could resolve the low-energy A_y puzzle, which has existed for many years and in many forms, for both p-d and n-d scattering and in electromagnetic reactions [30]. It remains to be seen whether this mechanism is compatible with the $A > 3$ bound states and other data. We summarize this section as follows.

- Most three-nucleon observables are insensitive to 3NFs.

- 3NFs are small in size but appear necessary to reproduce the ^{3}H binding energy, the Sagara discrepancy, and the A_y puzzle.

- Chiral symmetry provides a unified approach to 3NFs; power counting identifies dominant mechanisms.

- The leading-order (dominant) 2π-exchange 3NFs have been calculated; they have large isobar contributions.

- New short-range plus pion-range mechanisms may resolve the low-energy A_y puzzle.

- Although much remains to be investigated, a consensus appears to be developing for the bulk of 3NF terms, and a "standard model" of 3NFs may be possible in the near future.

- The basic building blocks of 3NFs have been recently validated by verifying the corresponding elements in the tail of the N-N potential.

ACKNOWLEDGEMENTS

We thank Alejandro Kievsky of the Univ. of Pisa and Jacek Golak of the Univ. of Cracow for providing figures. Walter Glöckle of Ruhr-Universität Bochum engaged in a helpful correspondence. The work of J.L.F. was performed under the auspices of the United States Department of Energy.

REFERENCES

1. *Few Body Problems in Nuclear and Particle Physics*, Proceedings of the International Conference on Few Body Problems in Nuclear and Particle Physics, edited by R. J. Slobodrian, B. Cujec, and K. Ramavataram, Laval University Press, Québec, 1975.
2. Yang, S. N., *Phys. Rev. C* **10**, 2067 (1974).
3. Carlson, J., and Schiavilla, R., *Rev. Mod. Phys.* **70**, 743 (1998).
4. Glöckle, W., Witała, H., Hüber, D., Kamada, H., and Golak, J., *Phys. Rep.* **274**, 107 (1996).
5. Friar, J. L., Hüber, D., and van Kolck, U., *Phys. Rev. C* **59**, 53 (1999).
6. Wiringa, R. B., Stoks, V. G. J., and Schiavilla, R., *Phys. Rev. C* **51**, 38 (1995).
7. Stoks, V. G. J., Klomp, R. A. M., Rentmeester, M. C. M., and de Swart, J. J., *Phys. Rev. C* **48**, 792 (1993).
8. Friar, J. L., *Phys. Rev. C* **60**, 34002 (1999).

9. Friar, J. L., in *Workshop on Electronuclear Physics with Internal Targets and the BLAST Detector*, edited by R. Alarcon and M. Butler, World-Scientific, Singapore, 1993, p. 230.
10. Chen, C. R., Payne, G. L., Friar, J. L., and Gibson, B. F., *Phys. Rev. C* **31**, 2266 (1985).
11. Friar, J. L., and Payne, G. L., in *Coulomb Interactions in Nuclear and Atomic Few-Body Collisions*, edited by F. S. Levin and D. A. Micha, Plenum Press, New York, 1996, p. 97.
12. Kievsky, A., Rosati, S., Tornow, W., and Viviani, M., *Nucl. Phys.* **A607** 402 (1996); Kievsky, A., Viviani, M., and Rosati, S., *Phys. Rev. C* **52**, R15 (1995).
13. Witała, H., Cornelius, Th., Glöckle, W., *Few-Body Systems* **5**, 89 (1988).
14. Kievsky, A., Rosati, S., and Viviani, M., *Phys. Rev. Lett.* **82**, 3759 (1999).
15. Ishikawa, S., Golak, J., Witała, H., Kamada, H., and Glöckle, W., *Phys. Rev. C* **57**, 39 (1998).
16. Gao, H., *et al*, *Phys. Rev. C* **50**, R546 (1994).
17. Fujita, J.-I., and Miyazawa, H., *Prog. Theor. Phys.* **17**, 360 (1957).
18. Carlson, J., Pandharipande, V. R., and Wiringa, R. B., *Nucl. Phys.* **A401**, 59 (1983).
19. Coon, S. A., Scadron, M. D., McNamee, P. C., Barrett, B. R., Blatt, D. W. E., and McKellar, B. H. J., *Nucl. Phys.* **A317**, 242 (1979).
20. Coelho, H. T., Das, T. K., and Robilotta, M. R., *Phys. Rev. C* **28**, 1812 (1983); Robilotta, M. R., and Coelho, H. T., *Nucl. Phys.* **A460**, 645 (1986).
21. Eden, J. A., and Gari, M. F., *Phys. Rev. C* **53**, 1510 (1996).
22. van Kolck, U., Thesis, University of Texas, (1993); Ordóñez, C., and van Kolck, U., *Phys. Lett.* **B 291**, 459 (1992); van Kolck, U., *Phys. Rev. C* **49**, 2932 (1994).
23. Harvey, P. H., and Pagel, M. D., *The Comparative Method in Evolutionary Biology*, Oxford University Press, Oxford, 1991.
24. Levin, F. S., and Feshbach, H., *Reaction Dynamics*, Gordon and Breach, New York, 1973.
25. Friar, J. L., *Few-Body Systems* **22**, 161 (1997).
26. Manohar, A., and Georgi, H., *Nucl. Phys.* **B234**, 189 (1984); Georgi, H., *Phys. Lett.* **B 298**, 187 (1993).
27. Rentmeester, M. C. M., Timmermans, R. G. E., Friar, J. L., and de Swart, J. J., *Phys. Rev. Lett.* **82**, 4992 (1999).
28. Witała, H., Glöckle, W., Hüber, D., Golak, J., and Kamada, H., *Phys. Rev. Lett.* **81**, 1183 (1998).
29. Hüber, D., Friar, J. L., Nogga, A., Witała, H., and van Kolck, U., *Few-Body Systems* (submitted).
30. Hüber, D., and Friar, J. L., *Phys. Rev. C* **58**, 674 (1998).

Inclusive Scattering from Polarized ^{3}He and Neutron Form Factors

Haiyan Gao*

*Laboratory for Nuclear Science and Department of Physics[1]
Massachusetts Institute of Technology
Cambridge, Massachusetts 02139, U.S.A.

Abstract. Because of the lack of a free neutron target, deuterium targets had been used extensively in studying the neutron structure in the past from unpolarized electron-deuteron scattering experiments. Only recently polarized electron-deuteron scattering measurements have been performed which yield more precise information on the charge form factor of the neutron. The unique spin structure of the ^{3}He ground state and the recent developments in polarized target technologies make polarized ^{3}He targets very effective neutron targets. In this talk, I review the current status of the polarized ^{3}He targets and focus on the quasielastic asymmetry measurement from inclusive $^3\vec{H}e(\vec{e},e')$ process and the neutron form factors. I discuss the results of the MIT-Bates experiment 88-25 and the preliminary results of the recently completed JLab experiment E95-001 in which precision measurements of the neutron magnetic form factor at low Q^2 are aimed.

INTRODUCTION

Electromagnetic form factors are of fundamental importance for an understanding of the underlying structure of nucleons. Knowledge of the distribution of charge, magnetization within the nucleons provides a sensitive test of models based on Quantum Chromodynamics (QCD), as well as a basis for calculations of processes involving the electromagnetic interaction with complex nuclei. The understanding of the nucleon structure in terms of quark and gluon degrees of freedom of QCD will provide a basis to understand more complex strongly interacting matter at the level of quarks and gluons. While the proton form factors are known with excellent precision over a large range of four-momentum transfer Q^2, the corresponding data for the neutron are of inferior quality due to the lack of free neutron targets. Over the past decade, with the advent of improved experimental techniques, the precise determination of both the neutron electric form factor, G_E^n, and the magnetic

---

[1] This work is supported in part by the U.S. Department of Energy under contract number DE-FC02-94ER40818.

form factor, G_M^n, has become a focus of experimental activity. While improving the precision of G_M^n is interesting in itself, it also benefits experiments designed to determine G_E^n, which usually measure the ratio G_E^n/G_M^n. Furthermore, precise data for the nucleon electromagnetic form factors are essential for the analysis of parity violation experiments designed to measure the strangeness content of the nucleon.

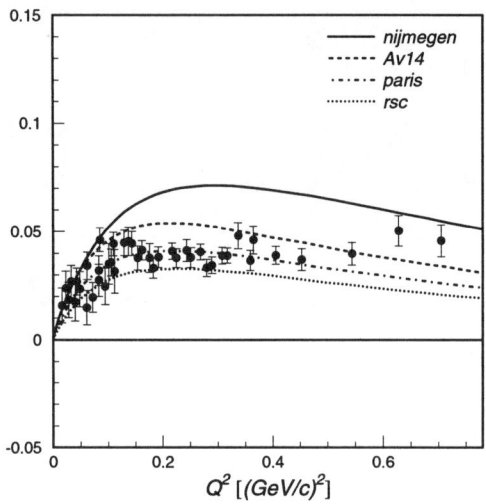

FIGURE 1. The electric form factor of the neutron as a function of four-momentum transfer from Platchkov *et al.* [1]. Data taken from [1].

The most precise information on G_E^n at low Q^2 prior to any polarization experiment is from elastic electron-deuteron scattering experiment by Platchkov *et al.* [1]. However, the extracted G_E^n values are extremely sensitive to the deuteron structure. Fig. 1 shows the G_E^n values extracted with the Paris potential together with the fit of the data (dash-dotted curve). Fits from fitting the G_E^n data extracted with the Nijmegen potential, the Argonne V14 (AV14) and the Reid-Soft Core (RSC) NN potentials are shown as solid, dashed and dotted curves, respectively. The large spread represents the uncertainty of G_E^n due to the deuteron structure, and the absolute scale of G_E^n contains a systematic uncertainty of about 50% from the measurement by Platchkov *et al.* [1].

The development of polarized targets and beams has allowed more complete studies of electromagnetic structure than has been possible with unpolarized reactions. In quasielastic scattering, the spin degrees of freedom introduce new response functions into the differential cross section, thus providing additional information on nuclear structure [2]. Experiments with longitudinally polarized electron beams and recoil neutron polarimeters have been carried out at MIT-Bates [3] and Mainz [4,5] and G_E^n has been extracted from the $d(\vec{e}, e'\vec{n})$ process. Recently, the neutron electric form factor was extracted for the first time [6] from $\vec{d}(\vec{e}, e'n)$ reaction in which a vector polarized deuteron target from an atomic beam source was employed.

Using the polarization degrees of freedom, the proton contribution to the scattering process is suppressed and more precise information on the neutron charge form factor can be extracted. Currently, our best knowledge of G_E^n from these polarization measurements is $\sim 30\%$ for $Q^2 \leq 0.6 (\text{GeV}/c)^2$.

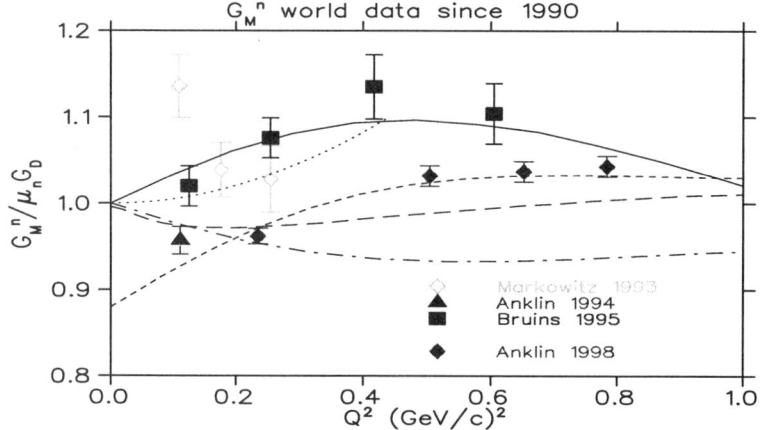

FIGURE 2. The neutron magnetic form factor G_M^n in units of the standard dipole parameterization, $\mu_n G_D$, in the low Q^2 region, as determined in several recent measurements: Markowitz et al. [7] (open diamonds) using $d(e,e'n)$; Anklin et al. [8] (triangle), Bruins et al. [9] (squares), and Anklin et al. [10] (solid diamonds) using the ratio $d(e,e'n)/d(e,e'p)$; and Gao et al. [11] (circle) using $^3\vec{\text{He}}(\vec{e},e')$.

Until recently, most data on G_M^n had been deduced from elastic and quasielastic electron-deuteron scattering. For inclusive measurements, this procedure requires the subtraction of a large proton contribution and suffers from large theoretical uncertainties due to the deuteron model employed and corrections for final-state interactions (FSI) and meson-exchange currents (MEC), limiting the precision of G_M^n to $\sim 20\%$ at low Q^2. The proton subtraction is avoided in coincidence $d(e,e'n)$ experiments [7], and the sensitivity to nuclear structure can be greatly reduced by measuring the cross section ratio $d(e,e'n)/d(e,e'p)$ in quasielastic kinematics. Several recent experiments [8–10] have employed the latter technique to extract G_M^n with uncertainties of <2% in the momentum transfer range $Q^2 = 0.1$ to 0.8 $(\text{GeV}/c)^2$. While this precision is excellent, the results of the experiments [7–10] are not fully consistent (cf. Figure 2). Furthermore, the two most precise data sets [9,10] of G_M^n seem to suggest a very different Q^2 dependence. In addition to the existing data, several theoretical calculations of G_M^n are shown in Figure 2. The solid curve is an improved quark model calculation by Lu, Thomas and Williams [12] with a bag constant of 0.9 fm. The dotted curve is the minimal vector dominance model by Meissner [13]. The short and long dashed curves are the non-relativistic and relativistic quark model calculations by Eich [14] and Schlumpf [15], respectively. The dash-dotted line is a recent calculation by Mergell et al. [16] based on a fit of the proton data using dispersion theoretical arguments. These calculations clearly show very different behavior in the low Q^2 region. Thus, further data are desirable to clarify the situation with respect to the discrepancies among different measurements and theoretical calculations.

An alternative approach to a precision measurement of G_M^n is the inclusive quasielastic $^3\vec{\text{He}}(\vec{e}, e')$ process. In comparison to deuterium experiments, this technique employs a different target and relies on polarization degrees of freedom. It is thus subject to completely different systematics. A first such experiment was performed at the MIT-Bates [17] laboratory, and a result for G_M^n was extracted as shown in Figure 2.

The rest of the paper is organized as follows: Section II contains a discussion on the polarized ^{3}He target followed by Section III on the formalism for the spin-dependent inclusive quasielastic scattering, Section IV and V describe the MIT-Bates experiment 88-25 and JLab experiment E95-001, respectively.

I POLARIZED ^{3}HE TARGETS

Optical pumping technique is widely used to polarize a sample of atoms by transferring angular momentum from a pump light beam, typically a laser beam, to the sample atoms. In the case of ^{3}He, direct optical pumping between its ground state and the first excited state is not possible because of the energy difference involved. Metastability-exchange optical pumping and spin-exchange optical pumping are two indirect optical pumping techniques commonly used. In this section, I will review both techniques.

A Metastability-Exchange Optical Pumping

The metastability-exchange optical pumping technique was developed in the early 1960s at Rice university [18] to polarize ground state ^{3}He or ^{4}He atoms through metastability-exchange collisions with optically pumped ^{3}He or ^{4}He metastable atoms. This method involves optical pumping of 2^3S_1 metastable state atoms, then transferring the polarization to ^{3}He ground state atoms through metastability-exchange collisions, in which the excitation of the electronic cloud is exchanged leaving the ground state polarized after the collision.

Metastability-exchange optical pumping of ^{3}He works as following: metastable 2^3S_1 atoms are produced by an electrodeless weak radio frequency (RF) discharge in a glass cell filled to a pressure of order 1 torr of pure ^{3}He. The ratio of the ground state atoms to the 2^3S_1 atoms is about $10^6 : 1$; the exact number depends on discharge characteristics such as intensity, uniformity and the discharge frequency. The sample is placed in a weak uniform magnetic field which defines the spin direction of the sample. Right-handed or left-handed circularly polarized light (defined by the right-hand rule used in atomic physics) at $\lambda = 1083.4\ nm$ corresponding to the transition of $2^3S_1 \to 2^3P_0$ excites transitions between the 2^3S_1 and 2^3P_0 states with the selection rule $\Delta m = \pm 1$ depending on the helicity of the incident light (+ for the right-handed circularly polarized light and − for the left-handed case). The pumping light excites atoms from the $m_F = -\frac{1}{2}$ and $m_F = -\frac{3}{2}$ sublevels of the metastable state to the 2^3P_0 level which then decay back to all sublevels of 2^3S_1 through spontaneous emission. The result is that atoms from lower magnetic sublevels of the 2^3S_1 level ($m_F = -\frac{1}{2}$, $m_F = -\frac{3}{2}$) are transferred to higher sublevels of the 2^3S_1 level ($m_F = \frac{1}{2}$, $m_F = \frac{3}{2}$), hence the metastable atoms become polarized. In metastable state, hyperfine interaction mixes electronic polarization into nuclear polarization. The polarization of the metastable atoms is then transferred to the ground state through metastability-exchange collisions in which only the excitation of the electronic cloud is exchanged. If the ground state of ^{3}He is polarized, then the nucleus is polarized because the atom is in a $J = 0$ state. This process can be expressed schematically as:

$$^3He + {}^3\vec{He}^* \to {}^3\vec{He} + {}^3He^* \tag{1}$$

where $*$ denotes the 2^3S_1 metastable state and the vector notation indicates that the nucleus is polarized. This optical pumping technique only works for relatively low pressure conditions (0.1 torr to 10 torr). Destruction of metastables at the wall of the container dominates the relaxation at pressures below about 0.1 torr and at high pressures the lifetime of the metastable state atoms limits the optical pumping efficiency. It is also experimentally difficult to maintain a uniform discharge under high pressure conditions. Furthermore, it is necessary to operate optical pumping around room temperature to achieve efficient optical pumping because metastability exchange cross-section, σ_e, which is very temperature dependent for the case of ^{3}He atoms [19] [20]. σ_e decreases roughly two orders of magnitude between 300 K and 4.2 K. To make a relatively dense target for nuclear physics experiments, the compression and low temperature techniques have been used.

People have tried successfully to make a dense nuclear physics target by mechanically compressing the polarized gas. At Toronto, Timsit et al. in the early 1970s [21] constructed a dense target and achieved a density of $0.7 \times 10^{19} cm^{-3}$ with 3% polarization by compressing the gas with liquid mercury. However, the performance of these targets was severely limited by the absence of laser sources for optical pumping. At Mainz Otten et al. have designed and built a new type of dense polarized ^{3}He target using the compression method [22] [23] and have achieved pressures around 1 bar with 38% of polarization. Currently, such a target is operated at a pressure of ~ 6.0 bar with a target polarization of $\sim 32\%$ at an incident electron beam current of $\sim 10 \mu A$ [24].

Low temperature is another approach to take to construct a dense polarized ^{3}He target. A double-cell system consisting of a pumping cell and a target cell is a practical design for a polarized ^{3}He nuclear target. The pumping cell is at room temperature where metastability-exchange optical pumping can be performed efficiently and the target cell is cooled to low temperature where a practical luminosity can be achieved for a nuclear physics experiment. This idea was first explored at Rice University [25]. The ^{3}He nuclei in the target cell become polarized because the polarized ^{3}He atoms diffuse between the pumping cell and the target cell, reaching an equilibrium polarization state. External polarized ^{3}He targets based on this technique were employed in the MIT-Bates experiments [26] [17]. Optical measurement of the atomic polarization in the pumping cell provides a good monitor of the target nuclear polarization during the experiment. The atomic polarization and the nuclear polarization are related because of the hyperfine coupling and this indirect optical measurement of the ^{3}He nuclear polarization was calibrated carefully with the NMR technique at Caltech [27]. The metastability-exchange optical pumping technique has the unique features of pure atomic species and fast pumping rate for constructing an internal polarized ^{3}He target. The internal polarized ^{3}He targets based on this technique have been used successfully at IUCF, DESY and NIKHEF.

B Spin-Exchange Optical Pumping

In spin-exchange optical pumping, circularly polarized resonance light is absorbed by a saturated vapor of alkali-atoms contained in a glass cell. The cell also included a much larger quantity of noble-gas atoms. The spin angular momentum is transferred to the alkali-metal atoms, thereby spin-polarizing the valence electrons of the alkali-metal atoms. Subsequent spin-exchange collisions between the alkali-metal atoms and noble gas atoms transfer some of the electron-spin polarization to the nuclei of the noble gas.

Rubidium has been commonly used in polarized ^{3}He targets based on spin-exchange optical pumping technique. The high vapor pressure of rubidium allows operation at modest temperatures where chemical attack on the glass container is not a problem. The resonance line (794.7 nm corresponding to the transition between $5S_{1/2}$ and $5P_{1/2}$ levels) lies in a spectrum region where intense laser light is available from sources such as dye lasers, Ti:sapphire lasers, and more recently diode laser arrays.

A central feature of the target will be sealed glass target cells, which will contain a ^{3}He pressure of about 10 atmospheres. The cells will have two chambers, an upper chamber in which the spin exchange takes place, and a lower chamber, through which the electron beam will pass. In order to maintain the appropriate number density of alkali-metal (Rb) the upper chamber will be kept at a temperature of 170–200°C using an oven constructed of the high temperature plastic Torlon. A small amount of nitrogen gas is typically used in this type of target to quench the $P_{1/2}$ states, thus reducing the radiation trapping effect in order to reach high optical pumping efficiency. Radiation trapping occurs when the mean free path for the unpolarized photons is much shorter than the dimensions of the pumping vessel. The incident photons can be reemitted, i.e. resonantly scattered, and depolarized. Thus, additional dilution factor comes from the nitrogen present in the target. The dilution factor from the amount of rubidium in the target is typically negligible ($n_{Rb}/n_{He} \sim 10^{-4}$). Although the spin-exchange optical pumping technique is capable of producing a dense target (10 amagat), slow pumping time resulting from small spin-exchange cross section of ^{3}He atom and the Rb atom makes this technique only suitable for external polarized ^{3}He targets. NMR technique is typically used to measure the target polarization for this type of the targets. External polarized ^{3}He targets based on this technique have been used in experiments at JLab, MIT-Bates, SLAC and TRIUMF. Figure 3 shows the schematics of the spin-exchanged polarized ^{3}He target employed in electron scattering experiments in Hall A at Jefferson Laboratory (JLab) recently.

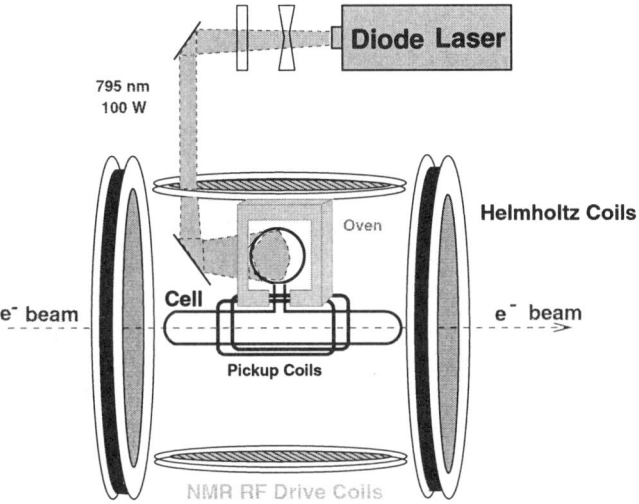

FIGURE 3. The schematics of the JLab Hall A polarized ^{3}He target based on the technique of the spin-exchange optical pumping.

II INCLUSIVE QUASIELASTIC SCATTERING OF POLARIZED ELECTRONS FROM POLARIZED ^{3}HE TARGETS

Polarized ^{3}He is a good candidate for an effective neutron target because its ground state wave function is dominated by the S-state in which the proton spins cancel and the nuclear spin is entirely due to the neutron. Therefore, inelastic scattering of polarized electrons from polarized ^{3}He in the vicinity of the quasielastic peak should be useful for studying the neutron electromagnetic form factors.

The idea of using polarized ^{3}He nuclear target as an effective neutron target was first investigated by Blankleider and Woloshyn in closure approximation [28]. Friar et al. [29] have studied the model dependence in the spin structure of the ^{3}He wave function and its effect on the quasielastic asymmetry. The plane wave impulse approximation (PWIA) calculations performed independently by two groups [30,31] using spin-dependent spectral functions show that the spin-dependent asymmetries are very sensitive to the neutron electric or magnetic form factors at certain kinematics near the top of the quasielastic peak. Recently, Fadeev calculations have been carried out which include the final state interaction (FSI) [32], FSI and meson exchange current (MEC) [33]. These state-of-the-art three-body calculations are very important for extracting the neutron form factors from double polarization electron-^{3}He scattering experiments.

The differential cross section for the process $^3\vec{\text{He}}(\vec{e},e')$ in the scattering plane can be written in terms of four nuclear response functions $R_K(Q^2,\nu)$ [2] as

$$\frac{d^2\sigma}{d\Omega dE'} = \sigma_{Mott}\left[v_L R_L + v_T R_T - h(\cos\theta^* v_{T'} R_{T'} + 2\sin\theta^*\cos\phi^* v_{TL'} R_{TL'})\right], \quad (2)$$

where θ^* and ϕ^* are the polar and azimuthal angles defining the direction of the target spin with respect to the momentum transfer vector $\vec{q}$ the v_K are kinematic factors, ν is the electron energy loss, h is the helicity of the incident electron, and $Q^2 \equiv \vec{q}^2 - \nu^2$. $R_{T'}$ and $R_{TL'}$ are two response functions arising from the polarization degrees of freedom. An experimentally clean signature of these observables is the spin-dependent asymmetry, defined as

$$A = \frac{\sigma_+ - \sigma_-}{\sigma_+ + \sigma_-} = -\frac{\cos\theta^* v_{T'} R_{T'} + 2\sin\theta^*\cos\phi^* v_{TL'} R_{TL'}}{v_T R_T + v_L R_L}, \quad (3)$$

where the subscript $+$ $(-)$ refers to the electron helicity h. By orienting the target spin at $\theta^* = 0°$ or $\theta^* = 90°$, corresponding to the spin direction either along the 3-momentum transfer vector $\vec{q}$ or normal to it, one can select the transverse asymmetry $A_{T'}$ (proportional to $R_{T'}$) or the transverse-longitudinal asymmetry $A_{TL'}$ (proportional to $R_{TL'}$).

In plane wave impulse approximation (PWIA) the cross section for $^3\vec{\text{He}}(\vec{e},e')$ at the center of the quasielastic peak is (roughly) proportional to the sum of the en plus twice the ep elastic cross section. The cross section for polarized electron-nucleon scattering is

$$\frac{d\sigma_{eN}}{d\Omega} = \sigma_{Mott}\frac{E'}{E}\left[v_L(1+\tau)G_E^2 + v_{TL'}2\tau G_M^2 - h\bar{p}_N(\cos\theta^* v_{T'} 2\tau G_M^2 \right.$$
$$\left. -2\sin\theta^*\cos\phi^* v_{TL'}\sqrt{2\tau(1+\tau)}G_M G_E)\right], \quad (4)$$

where $\tau = Q^2/(4M^2)$, and $\bar{p}_N$ is the effective nucleon polarization. As a consequence of the S-state dominance, the neutron in ^{3}He is almost fully polarized,

$\bar{p}_n \approx 1$, at the quasielastic peak while the remaining small components of the ^{3}He ground state, the D state ($\sim 8\%$) and the mixed-symmetry S' state ($\sim 1\%$), give rise to a small net proton polarization of $\bar{p}_p \approx -0.03$ [29]. Combining the above equations, the transverse asymmetry can be written

$$A_{T'} \propto \frac{\sigma_n}{(\sigma_n + 2\sigma_p)} \bar{p}_n (G_M^n)^2 + \frac{2\sigma_p}{(\sigma_n + 2\sigma_p)} \bar{p}_p (G_M^p)^2, \qquad (5)$$

where σ_n and σ_p are the electron-neutron and electron-proton elastic scattering cross sections, respectively. Since G_M^n and G_M^p are comparable in magnitude, but $|\bar{p}_p| \ll |\bar{p}_n|$, $A_{T'}$ is dominated by the neutron contribution and so is essentially proportional to $(G_M^n)^2$. Note that it is mostly the proton contribution that is sensitive to details of the ^{3}He ground state wave function.

The strong sensitivity of $A_{T'}$ to $(G_M^n)^2$ in quasielastic kinematics has been verified in a number of recent calculations [30-33]. The most advanced of these include corrections for FSI [32] and FSI and MEC [33], which are relatively small at the quasielastic peak. One may conclude that $A_{T'}$ depends only weakly on the details of the ^{3}He nuclear ground state and the reaction mechanism. Thus, a precision measurement of $A_{T'}$ is suitable to extract precise information on G_M^n.

While A_{TL} is sensitive to the product of G_E^n and G_M^n, a dominant proton contribution to $A_{TL'}$ greatly reduces the sensitivity to G_E^n at low Q^2 from inclusive $^3\vec{H}e(\vec{e}, e')$ process.

III MIT-BATES EXPERIMENT 88-25

The experiment was performed at the MIT-Bates Linear Accelerator Center in spring 1993 using a 370 MeV longitudinally polarized electron beam. A Wien spin rotator was employed to produce longitudinally polarized electrons at the target. The average beam current during the experiment was 25 μA and the average beam polarization was determined using a Møller apparatus [34] to be 36.5%.

The polarized ^{3}He target used in this experiment was a double-cell system consisting of a glass pumping cell and a copper target cell. The target was polarized by the metastability-exchange optical pumping technique [35]. The target was operated at 13 K during the experiment with a ^{3}He gas pressure of 2.2 torr. The target wall was coated with a thin layer of nitrogen to maintain a sufficiently long relaxation time at low temperature. A holding field of 36 gauss provided by a pair of Helmholtz coils defined the target spin quantization axis. The target spin direction was aligned at an angle of 42.5° to the electron beam. The pumping cell polarization was measured continuously by monitoring the circular polarization of the 668-nm line excited by the ^{3}He discharge. The target polarization was inferred from the polarization of the pumping cell and the time constants of the coupled system. This optical measurement of the ^{3}He nuclear polarization was calibrated by an NMR measurement [27] with an accuracy of ±2%. With 25μA of beam, the target polarization was 38% or greater.

The scattered electrons were detected in the Medium Energy Pion Spectrometer (MEPS) and the One Hundred Inch Spectrometer (OHIPS) configured at an electron scattering angle $\theta = 91.4°$ to the left of the beam and $70.1°$ to the right of the beam, respectively. The MEPS spectrometer central momentum was 250 MeV/c corresponding to $Q^2 = 0.19$ (GeV/c)^2 and $\theta^* = 8.9°$ or $171.1°$ for positive or negative target polarization, respectively. The OHIPS spectrometer had a central momentum of 285 MeV/c, corresponding to the QE kinematic setting for $A_{TL'}$ measurement at a $Q^2 = 0.14$ (GeV/c)^2. While the MEPS detector package consisted of two vertical drift chambers, three planes of trigger hodoscopes, and an Aerogel Čerenkov detector, the OHIPS detector package consisted of two crossed drift chambers, three planes of plastic scintillators, and an isobutane gas Čerenkov detector. The triggers were formed by events for which all three hodoscopes fired in each spectrometer. The Čerenkov detectors were used for pion rejection.

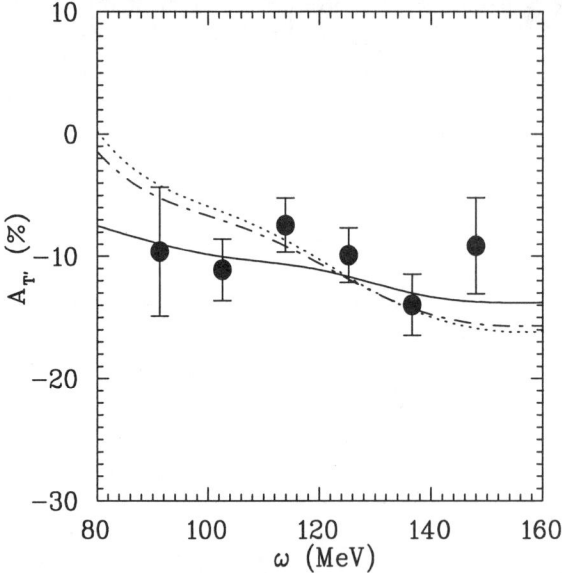

FIGURE 4. Transverse asymmetry $A_{T'}$ from the MIT-Bates experiment 88-25 [17] as a function of electron energy loss ω. The data are shown with statistical uncertainties only. The solid line is the calculation by Ishikawa et al.. The dashed line is the calculation by Salm'e et al. and the dash-dotted line is the calculation by Schulze et al..

Fig. 4 shows the measured ^{3}He inclusive spin-dependent quasielastic transverse asymmetry $A_{T'}$ [17], as a function of the electron energy transfer, ω, together with the two PWIA calculations and the calculation by Ishikawa et al. [32]. The deviation of the result by Ishikawa et al. from those of PWIA calculations [30] [31] is significant away from the quasielastic peak. The agreement between the data on $A_{T'}(\omega)$ and the calculation by Ishikawa et al. is excellent in terms of the magnitude of the asymmetry and also the shape. Unfortunately, because of

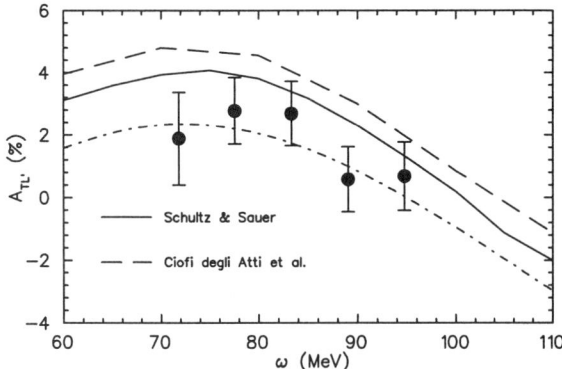

FIGURE 5. Longitudinal-transverse asymmetry $A_{TL'}$ from MIT-Bates experiment [17] as a function of electron energy loss ω. The data are shown with statistical uncertainties only. The solid line is the calculation by Ishikawa et al.. The dashed line is the calculation by Salmé et al. and the dash-dotted line is the calculation by Schulze et al..

the large errors associated with the measured $A_{T'}(\omega)$ as shown in Fig. 4, it is not possible to put constraints on the theoretical calculations of the ^{3}He inclusive spin-dependent quasielastic asymmetry.

Because of the limitation of the statistics of this measurement, the measured quasielastic asymmetry, $A_{T'}(\omega)$, averaged over the experimental ω acceptance was used in extracting $G_M^{n\,2}$ together with the calculation of Ishikawa et al. [32]. The G_M^n value extracted from this experiment at $Q^2 = 0.19$ (GeV/c)^2 is shown (Fig. 3, closed circle) with total uncertainty dominated by the statistical uncertainty.

On the other hand, $A_{TL'}$ from quasielastic $^3\vec{He}(\vec{e},e')$ at low Q^2 is ($Q^2 \leq 0.3\,(GeV/c)^2$) is dominated by the proton contribution largely because of the smallness of $G_E^{n\,2}$ and the small non-S state part of the ^{3}He ground state wave function. Thus, it is questionable to extract information on G_E^n at low Q^2 from $^3\vec{He}(\vec{e},e')$. It is possible to go to higher Q^2 ($Q^2 \geq 0.3(GeV/c)^2$) to extract $G_E^{n\,2}$ with respectable accuracy from quasielastic $^3\vec{He}(\vec{e},e')$ measurement where the proton contribution to $A_{TL'}$ is under better control. Fig. 5 shows $A_{TL'}$ from the MIT-Bates experiment [17] as a function of ω together with the PWIA calculations and the calculation by Ishikawa et al. [32] which included FSI. The PWIA calculations are consistently higher than the measured asymmetry. The calculation with FSI is in better agreement with the data.

To extract precise information on G_M^n from inclusive quasielastic $A_{T'}$ measurement, it is important to measure $A_{T'}$ with high precision across the ^3He quasielastic peak. As away from the quasielastic peak, predictions from different models deviate. Thus, one can constrain theoretical model using high precision data of $A_{T'}$ in the wings of the QE peak. To extract precise information on G_M^n, one then use the measured $A_{T'}$ right on top of the quasielastic peak, this is a procedure much less sensitive to model dependence than the procedure used in extracting G_M^n from the

MIT-Bates experiment 88-25.

IV   JLAB EXPERIMENT E95-001

The experiment was carried out in Hall A at JLab in early 1999 using a longitudinally polarized CW electron beam at energies of 0.778 and 1.727 GeV and 10 μA of beam current. A high pressure ^{3}He gas target was polarized via spin-exchange optical pumping at a density of 2.5×10^{20} nuclei/cm^3. The effective target length seen by the spectrometers was 10 cm/sinθ_e, where θ_e is the electron scattering angle. To facilitate optical pumping, the target contained small admixtures of nitrogen ($\sim 10^{18}$ cm^{-3}) and rubidium ($\sim 10^{14}$ cm^{-3}). Background from the nitrogen was determined in calibration measurements using a reference cell which has the same dimensions as those of the ^{3}He target cell and is corrected for in the analysis; the contribution from rubidium is negligible. The beam and target polarizations were approximately 70% and 30%, respectively. The beam helicity was reversed randomly at a rate of 1 Hz. A total beam charge of approximately 22 C was accumulated, resulting in a total data set of 1.3×10^9 quasielastic events after background subtraction.

Six kinematic points were measured corresponding to $Q^2 = 0.1$ to 0.6 (GeV/c)^2 in steps of 0.1 (GeV/c)^2. To maximize sensitivity to $A_{T'}$, the target spin was oriented at $-62.5°$ with respect to the beam direction, resulting in a contribution to the asymmetry due to $R_{TL'}$ of less than 2% for all kinematic points. The target spin direction was rotated by 180° every 24-48 hours for systematic checks, causing the asymmetry to change sign. The beam helicity definition was also reversed every 24 hours for systematic checks associated with the electron beam helicity. The scattered electrons were observed in the two Hall A High Resolution Spectrometers, HRSe and HRSh. Both spectrometers were configured to detect electrons in single-arm mode using nearly identical detector packages consisting of two dual-plane vertical drift chambers for tracking, two planes of segmented plastic scintillators for trigger formation, and a CO_2 gas Cherenkov detector and Pb-glass total-absorption shower counter for particle identification. Pion background was rejected using the Cherenkov and shower counter information. The spectrometer momentum and angular acceptances were approximately ±4.5% and 5.5 msr, respectively. The level of background from the walls of the glass target was measured at regular intervals with the target cell empty and was less than about 5% of the full target yield. The HRSe was set for quasielastic kinematics while the HRSh detected elastically scattered electrons. The elastic asymmetry can be calculated to better than 2% using the well-known elastic form factors of ^{3}He [36], and so the elastic measurement allows precise monitoring of the product of beam and target polarizations. Standard Møller and NMR polarimetry was also performed and served as a cross-check.

A statistical precision in $A_{T'}$ of $\sim 2.0\%$ was achieved for each Q^2 point in a ±15 MeV bin around the center of the quasielastic peak. Fig. 6 shows the preliminary E95-001 result of $A_{T'}$ with statistical errors as a function of ω at $Q^2 = 0.1$

Preliminary E95-001 $A_{T'}(Q^2)$ Result

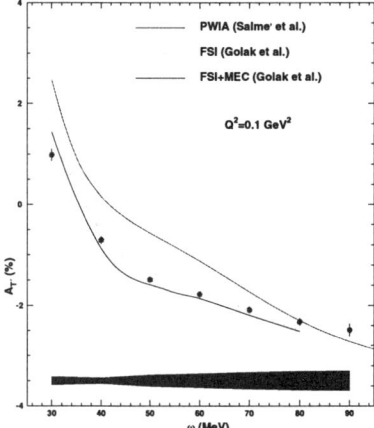

FIGURE 6. The preliminary E95-001 $A_{T'}$ result as a function of ω at $Q^2 = 0.1$ (GeV/c)^2.

(GeV/c)^2 together with the PWIA calculation of Ref. [31] and calculations with FSI and FSI+MEC by Golak *et al.* [33]. The error band shows the preliminary systematic uncertainty of the measurement, which will be reduced to a level comparable to the statistical uncertainty of the measurement after detailed analysis. This precision is better by about a factor of five that that of our previous experiment on ^{3}He [17] at $Q^2 = 0.19$ (GeV/c)^2. Extraction of G_M^n from the data requires the use of theoretical calculations. Currently the Bochum-Krakow group [33] is carrying out extensive calculations of $A_{T'}$ as a function of G_M^n for the kinematics of this experiment which include the FSI and MEC effects. This is the state-of-the-art three-body calculation. In addition, an independent calculation of the ^{3}He quasielastic asymmetry which will include FSI and MEC effects is currently in progress [37]. The Bochum-Krakow calculation will be convoluted with the experimental acceptances, and G_M^n will be determined using a best fit of $A_{T'}(G_M^n)$ to the data in the vicinity of the quasielastic peak. Fig. 7 shows the expected precision for G_M^n from this experiment. The data also allow a detailed analysis of the dependence of $A_{T'}$ on the electron energy transfer ω. The regions away from the quasielastic peak are expected to be sensitive to details of the reaction mechanism. Thus, the shape of $A_{T'}(\nu)$ can be used to constrain calculations that include FSI and MEC corrections. Data analysis is currently in progress and results are expected in late 1999.

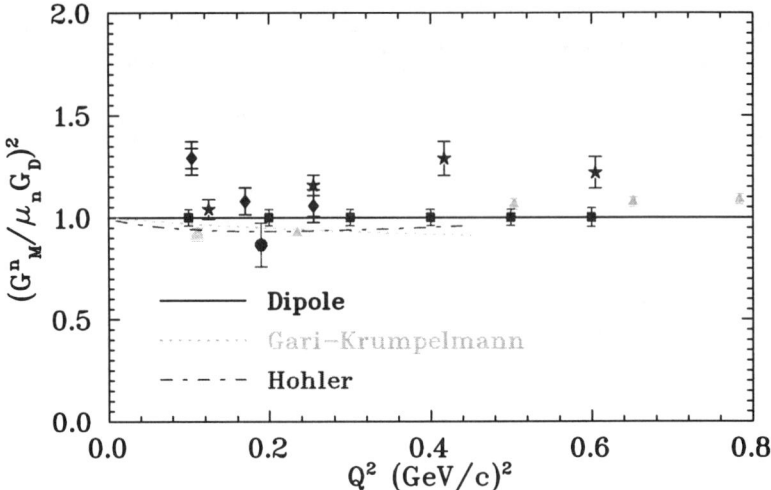

FIGURE 7. The neutron magnetic form factor $G_M^{n\,2}$ in units of the standard dipole parameterization, $(\mu_n G_D)^2$, in the low Q^2 region, as determined in several recent measurements. The projected results from the JLab experiment (E95-001) are shown as solid squares using the $^3\vec{\text{He}}(\vec{e}, e')$ process.

REFERENCES

1. S. Platchkov et al., Nucl. Phys. **A510**, 740 (1990).
2. T.W. Donnelly and A.S. Raskin, Ann. Phys. (N.Y.) **169** (1986) 247.

3. T. Eden *et al.*, Phys. Rev. **C50**, R1749 (1994).
4. M. Ostrick *et al.*, Phys. Rev. Lett. **83**, 276 (1999).
5. C. Herberg *et al.*, Eur. Phys. Jour. **A5**, 131 (1999).
6. I. Passchier *et al.*, Phys. Rev. Lett. **82**, 4988 (1999).
7. P. Markowitz *et al.*, Phys. Rev. **C48** (1993) R5.
8. H. Anklin *et al.*, Phys. Lett. **B336** (1994) 313.
9. E.E.W. Bruins *et al.*, Phys. Rev. Lett. **75** (1995) 21.
10. H. Anklin *et al.*, Phys. Lett. **B428** (1998) 248.
11. H. Gao *et al.*, Phys. Rev. **C50** (1994) R546.
12. D.H. Lu, A.W. Thomas, A.G. Williams, Phys. Rev. **C57**, 2628 (1998).
13. U.-G. Meissner, Phys. Rep. **161**, 213 (1988).
14. E. Eich, Z. Phys. **C45**, 627 (1988).
15. F. Schlumpf, J. Phys. **G20** 237, 1994.
16. P. Mergell, U.-G. Meissner, D. Dreshel, Nucl. Phys. **A 596**, 367 (1996).
17. H. Gao *et al.*, Phys. Rev. **C50**, R546 (1994); H. Gao, Ph.D. thesis, California Institute of Technology (unpublished, 1994); J.-O. Hansen *et al.*, Phys. Rev. Lett. **74**, 654 (1995).
18. F.D. Colegrove, L.D. Shearer, and G.K. Walters, Phys. Rev. Lett. **132**, 2561 (1963).
19. P.J. Nacher and M. Leduc, J. Phys. **46**, 2057 (1985).
20. F.D. Colegrove, L.D. Shearer, and G.K. Walters, Phys. Rev. **135**, A353 (1964).
21. R.S. Timsit *et al.*, Can. J. Phys. **49**, 509 (1971).
22. G. Eckert, W. Heil *et al.*, Nucl. Inst. Meth. **A320**, 53 (1992).
23. M. Meyerhoff *et al.*, Doctoral thesis, Mainz University (1994).
24. R. Surkau *et al.*, Nucl. Instrum. Methods Phys. Res. Sect. **A384**, 444 (1997).
25. H.H. Mc Adams, Phys. Rev. **170**, 276 (1968). **A47**, 468 (1993).
26. C.E. Woodward *et al.*, Phys. Rev. Lett. **65**, 698 (1990).
27. W. Lorenzon, T.R. Gentile, H. Gao, and R.D. McKeown, Phys. Rev.
28. B. Blankleider and R.M. Woloshyn, Phys. Rev. **C29**, 538 (1984).
29. J.L. Friar *et al.*, Phys. Rev. **C42** (1990) 2310.
30. R.-W. Schulze and P.U. Sauer, Phys. Rev. **C48** (1993) 38.
31. C. Ciofi degli Atti, E. Pace and G. Salmè, Phys. Rev. **C51** (1995) 1108; C. Ciofi degli Atti, E. Pace and G. Salmè, in *Proceedings of the 6th Workshop on Perspectives in Nuclear Physics at Intermediate Energies*, ICTP, Trieste May 1993, (World Scientific); C. Ciofi degli Atti, E. Pace and G. Salmè, Phys. Rev. **C51**, 1108 (1995); G. Salmè, private communication.
32. S. Ishikawa *et al.*, Phys. Rev. **C57** (1998) 39; and private communication.
33. J. Golak, private communication.
34. J. Arrington, E.J. Beise, B.W. Filippone, T.G. O'Neill, W.R. Dodge, G.W. Dodson, K.A. Dow, and J.D. Zumbro, Nuclear Instrument and Methods **A311**, 39 (1992).
35. R.G. Milner, R.D. McKeown, and C.E. Woodward, Nucl. Instrum. Methods Phys. Res., Sec. **A274**, 56 (1989); C.E. Jones *et al.*, Phys. Rev. **C47**, 110 (1993).
36. A. Amroun *et al.*, Nucl. Phys. **A579** (1994) 596.
37. G. Salmé, private communication.

Studies of the electric form factor of the neutron

Hartmut Schmieden

Institut für Kernphysik, Universität Mainz, Germany

Abstract. Elastic form factors provide information about the charge and current distribution inside proton and neutron. Recent coincidence experiments improved our knowledge of the neutron form factors significantly. The results from the $^3\vec{\text{He}}(\vec{e},e'n)$ and $D(\vec{e},e'\vec{n})$ double polarization experiments at MAMI indicate that the neutron electric form factor is almost a factor of two larger than previously assumed from $D(e,e')$ elastic scattering.

I INTRODUCTION

The internal structure of the nucleon at very high energies is currently understood in terms of quarks and gluons, and described by perturbative QCD. However, at distances of the size of the nucleon and corresponding momentum transfers of about 1 GeV the strong coupling diverges. Therefore, perturbative methods completely fail to understand the variety of low energy phenomena of the nucleon such as anomalous magnetic moments, polarizabilities or radii.

The elastic form factors parameterize the ability of the composite nucleon to incorporate a momentum transfer, $\vec{q}$, coherently, i.e. without excitation and particle emission. They are related to the distribution of charge and currents and therefore of fundamental importance for the understanding of nucleon structure.

Since R. Hofstadter's pioneering experiments elastic electron scattering has become a unique tool for the investigation of the electromagnetic structure of the nucleon. In the simplest approximation, it is characterized by the exchange of one virtual photon, which transfers the momentum $\vec{q}$ and the energy ω. Electron scattering covers the spacelike region, because the squared four-momentum transfer, $q^2 = \omega^2 - \vec{q}\,^2$, is always negative. It is therefore usually expressed by the positive quantity $Q^2 = -q^2 > 0$. The timelike region, where $Q^2 < 0$, can be accessed through electron-positron annihilation into a pair of proton and anti-proton [1] or neutron and anti-neutron [2].

In elastic electron-nucleon scattering the charge-current density of the nucleon can be written in the form [3]:

$$\bar{N}\Gamma_\mu N = \bar{N}\left[\gamma_\mu F_1(Q^2) + \frac{i\sigma_{\mu\nu}q^\nu}{2m_N}\kappa F_2(Q^2)\right]N. \tag{1}$$

The *Dirac* form factor, $F_1(Q^2)$, modifies the vector current of charge and normal magnetic moment. Additionally, the *Pauli* form factor, $F_2(Q^2)$, parameterizes the effect of the anomalous magnetic moment, κ, as motivated by the Gordon decomposition of the electromagnetic current. Linear combinations of F_1 and F_2 constitute the electric and magnetic *Sachs* form factors [4]

$$G_E^{n,p}(Q^2) = F_1^{n,p}(Q^2) - \tau\kappa_{n,p}F_2^{n,p}(Q^2) \tag{2}$$
$$G_M^{n,p}(Q^2) = F_1^{n,p}(Q^2) + \kappa_{n,p}F_2^{n,p}(Q^2), \tag{3}$$

where $\tau = Q^2/4m_N^2$ is a dimensionless measure of the squared four-momentum transfer in units of the nucleon rest mass, m_N. These form factors correspond in the static limit, $Q^2 \to 0$, to the total charge and magnetic moment of protons and neutrons:

$$G_E^{n,p}(Q^2 \to 0) = 0, 1 \tag{4}$$
$$G_M^{n,p}(Q^2 \to 0) = -1.91, 2.79 \tag{5}$$

In the particular reference frame with vanishing energy transfer, the *Breit* frame, G_E and G_M can be interpreted as the Fourier transforms of the corresponding distributions of charge and magnetism. The interpretation of G_E^n in terms of the rest frame charge distribution is recently discussed again [5].

With these Sachs form factors the cross section for elastic electron-nucleon scattering can be written in the famous *Rosenbluth* form,

$$\frac{d\sigma}{d\Omega} = \left(\frac{d\sigma}{d\Omega}\right)_{\text{Mott}} \cdot \left(\frac{G_E^2 + \tau G_M^2}{1+\tau} + 2\tau G_M^2 \tan^2\frac{\vartheta_e}{2}\right), \tag{6}$$

where $\left(\frac{d\sigma}{d\Omega}\right)_{\text{Mott}}$ is the Mott cross section for electron scattering off a pointlike spin-$\frac{1}{2}$ object and ϑ_e denotes the electron scattering angle. Due to their different angular weights in Eq.6, G_E and G_M can be experimentally separated at constant Q^2.

This method of Rosenbluth-separation has been used in elastic electron-proton scattering [6,7] to determine G_E^p and G_M^p up to $Q^2 \simeq 9\,(\text{GeV}/c)^2$. At higher Q^2 up to $30\,(\text{GeV}/c)^2$, G_M^p dominates the cross section and has thus been determined directly [8]. The result is that G_E^p approximately follows the so-called dipole form,

$$G_E^p \simeq G_D = \left(1 + \frac{Q^2}{0.71\,(\text{GeV}/c)^2}\right)^{-2} \quad \text{and}$$
$$G_M^p \simeq \mu_p G_D \tag{7}$$

scales with its magnetic moment. However, recent results [9] from measurements of the ratio of G_E^p/G_M^p in double polarized elastic scattering, $p(\vec{e},e'\vec{p})$, seem to confirm the substantial deviation from the scaling behaviour at $Q^2 \geq 1.5\,(\text{GeV/c})^2$, which had already been indicated by older low statistics measurements [10,11].

Generally, the measurement of the neutron form factors raises more difficulties, because there is no free neutron target available which is suited for electron scattering experiments. The best possible approximation of free electron-neutron scattering is quasifree scattering off the lightest nuclei, usually the deuteron. In single arm experiments the dominating proton contribution has to be subtracted, with corresponding large uncertainties [12]. A Rosenbluth-separation has nevertheless been achieved [13] up to $Q^2 = 4\,(\text{GeV/c})^2$. The neutron magnetic form factor exhibits the scaling behaviour, too,

$$G_M^n \simeq \mu_n G_D. \tag{8}$$

Coincidence experiments allow the explicit tagging of electron-neutron scattering [11,14–17]. The influence of binding effects can be minimized through the simultaneous measurement of the $D(e,e'p)$ reaction.

The situation concerning the neutron electric form factor, G_E^n, is most unfavourable. Due to the zero charge of the neutron it must vanish in the static limit

$$G_E^n(Q^2 \to 0) = 0. \tag{9}$$

The smallness of $(G_E^n)^2$ compared to $\tau(G_M^n)^2$ makes a Rosenbluth decomposition according to Eq.6 very difficult. Due to the corresponding large errors in the small quantity the extracted values for G_E^n are compatible with zero [11,13]. Therefore, in the momentum transfer range $Q^2 < 1\,(\text{GeV/c})^2$ the most precise data came from *elastic* electron-deuteron scattering [18,19], where the structure function $A(Q^2)$ depends on the isoscalar form factor $(G_E^p + G_E^n)^2$ and thus provides a higher sensitivity to G_E^n through its interference with the large G_E^p. However, the necessary unfolding of the deuteron wavefunction introduces a substantial model dependence in the extracted neutron electric form factor.

II G_E^N DOUBLE POLARIZATION EXPERIMENTS

Double polarization observables in quasifree electron-deuteron scattering with longitudinally polarized electrons offer high sensitivity to G_E^n, due to an interference with the large G_M^n, combined with negligible dependence on the deuteron wavefunction [20]. For the ideal case of free electron-neutron scattering, $n(\vec{e},e'\vec{n})$, Arnold, Carlson and Gross [21] obtained for the components of the recoil polarization

$$P_x = -P_e \frac{\sqrt{2\tau\epsilon(1-\epsilon)}G_E^n G_M^n}{\epsilon(G_E^n)^2 + \tau(G_M^n)^2} \tag{10}$$

$$P_y = 0 \tag{11}$$

$$P_z = P_e \frac{\tau\sqrt{1-\epsilon^2}(G_M^n)^2}{\epsilon(G_E^n)^2 + \tau(G_M^n)^2}, \tag{12}$$

where $\hat{x}$ is in the electron scattering plane perpendicular to the direction of the momentum transfer, $\hat{y}$ is normal to the scattering plane, and $\hat{z}$ points into the direction of the momentum transfer, $\vec{q}$ (compare Fig.2). $\epsilon = (1 + \frac{2|\vec{q}|^2}{Q^2} \tan^2 \frac{\vartheta_e}{2})^{-1}$ is the photon polarization parameter and P_e denotes the longitudinal polarization of the electron beam.

In the completely equivalent scattering of longitudinally polarized electrons off a polarized neutron target, $\vec{n}(\vec{e}, e'n)$, the cross section asymmetry with regard to reversal of the electron beam polarization is given by

$$A = -P_e \frac{\sqrt{2\tau\epsilon(1-\epsilon)} G_E^n G_M^n \cdot \tilde{P}_x + \tau\sqrt{1-\epsilon^2}(G_M^n)^2 \cdot \tilde{P}_z}{\epsilon(G_E^n)^2 + \tau(G_M^n)^2}, \quad (13)$$

where now $\tilde{P}_{x,z}$ are the components of the initial state neutron polarization. The polarized target neutrons can be provided by polarized ^{3}He [22–24], where the neutron carries approximately 87 % of the polarization of the nucleus [25]. In the asymmetry ratio $\frac{A_x(\tilde{P}_z=0)}{A_z(\tilde{P}_x=0)}$ the absolute degree of the target polarization cancels out [26].

A Polarized target experiments

First, pioneering experiments at Bates aimed at the extraction of G_E^n from the inclusive quasielastic reaction $^3\vec{\text{He}}(\vec{e}, e')$. Two different types of targets were used which were based on metastability exchange optical pumping and Rubidium exchange optical pumping. Though the principle feasability of such kind of experiments was successfully demonstrated, none of the targets allowed satisfactory statistical accuracy due to limitations both in density and average polarization.

The magnetic moment of ^{3}He within 10 % agrees with the free neutron one. Therefore the proton contribution in the measured asymmetries (Eq.13) first was expected to be small [25]. Later calculations however showed that the remaining impact of the protons on the measured asymmetries is large enough to prohibit a reliable extraction of G_E^n [27].

This problem can be overcome, if the occurence of e-n scattering is explicitly tagged through the detection of the outgoing neutron in coincidence with the scattered electron. Such an exclusive $^3\vec{\text{He}}(\vec{e}, e'n)$ experiment was performed for the first time by the A3-collaboration at MAMI [26] at a squared four-momentum transfer of $Q^2 = 0.31 \,(\text{GeV/c})^2$. With the detector-setup described in the following section the statistics was improved later on [28]. The most recent experiment at $Q^2 = 0.67 \,(\text{GeV/c})^2$ used the 3-spectrometer setup [29] of the A1-collaboration at MAMI [30].

In this experiment the target gas was polarized by metastable optical pumping and subsequently compressed to 6 bars. The glass target cell was then transported in a weak holding field to the experimental place. Under beam conditions an average polarization of 32 % was achieved. The target thickness was $24\,\text{mg}/\text{cm}^2$. Thinned down to $25\,\mu\text{m}$, the beam windows of the glass cell could not stand the 6 bar pressure. Therefore *un*polarized ^{3}He gas of the same pressure was contained between the glass windows and $10\,\mu\text{m}$ thick havar foils, resulting in an empty-target thickness of $27\,\text{mg}/\text{cm}^2$.

The relaxation time of approximately one day required twice a day the replacement of the target cell by a freshly polarized one. Quasielastic measuremets were performed with target spin aligned perpendicular and parallel to the momentum transfer direction in order to access both the transverse asymmetry, A_x, and the longitudinal asymmetry, A_z. This enabled a measurement of the ratio A_x/A_z, which is directly proportional to G_E^n/G_M^n but independent of both the absolute degrees of beam and target polarization, P_e and P_T, respectively. Furthermore, the product $P_e \cdot P_T$ was monitored through the *elastic* measurement $^3\text{He}(e,e')$ in spectrometer B of the 3-spectrometer setup.

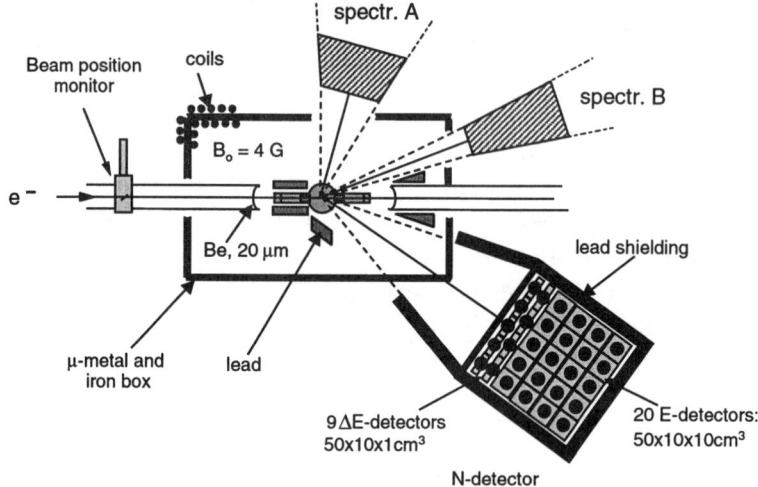

FIGURE 1. Target area of the $^3\vec{\text{He}}(\vec{e},e'n)$ experiment at MAMI. The cell with polarized gas is magnetically shielded against the spectrometer's fringe fields.

As shown in Fig.1, the quasielastically scattered electrons were detected in spectrometer A, the neutrons in coincidence in a dedicated neutron detector provided by the University of Basel. It consisted of four layers of five plastic scintillators of dimensions $50 \times 10 \times 10\,\text{cm}^3$, which were equipped with photomultipliers at both ends. Charged particles could be rejected by means of two layers of 1 cm thick ΔE-counters. The neutron detector was shielded with 2 cm of lead against direct target sight in order to reduce the charged background.

The setup of the JLab Hall C $\vec{D}(\vec{e},e'n)$ experiment in principle is similar [31]. Here the scattered electrons are detected in the HMS spectrometer in coincidence with neutrons in a segmented plastic scintillator. At the required luminosity of

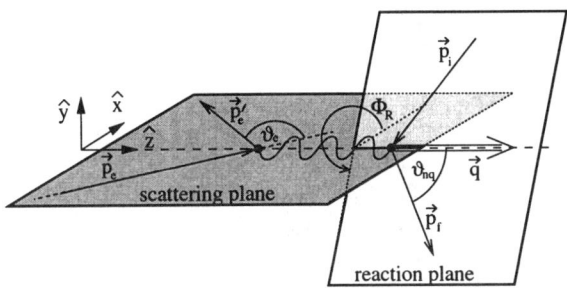

FIGURE 2. Kinematics of the quasifree $n(\vec{e}, e'\vec{n})$ reaction with definition of the relevant coordinate frames.

$1 \cdot 10^{35}$ cm^{-2}s^{-1} a 40% polarization of the ND$_3$ target is achieved by the technique of dynamic nuclear polarization. The Deuteron nuclei are polarized by microwave irradiation at temperatures around 1 K in a strong magnetic field of 5 T. In the measurement of A_x this field deflects the incoming electron beam by as much as 4°. This has to be compensated by a meagnetic chicane in order to guarantee horizontal beam at the center of the target. First data have been taken in the Q^2 range between 0.5 and 2 (GeV/c)^2. The analysis is in progress.

In contrast to MAMI and JLab, the NIKHEF $\vec{D}(\vec{e}, e'n)$ experiment was performed with a vector polarized internal gas target at the AmPS electron storage ring [32]. The scattered electrons were detected in coincidence simultaneously with protons and neutrons. Thus the asymmetry ratio between the $\vec{D}(\vec{e}, e'n)$ and $\vec{D}(\vec{e}, e'p)$ reactions could be determined. For a detailed description the reader is referred to J. van den Brand's contribution [33].

B The $D(\vec{e}, e'\vec{n})$ recoil polarization experiment at MAMI

As in the case of polarized targets a pioneering recoil polarization experiment was performed at MIT-Bates [34]. Electrons and neutrons from the $D(\vec{e}, e'\vec{n})$ reaction were detected in coincidence according to the kinematics depicted in Fig. 2 and the transverse neutron polarization, P_x, was measured. However, due to the low duty cycle of the Bates linac only modest statistical accuracy could be achieved. Furthermore, the external absolute calibration of the neutron polarimeter's effective analyzing power remained unsatisfactory.

The full potential of the recoil polarization method was exploited for the first

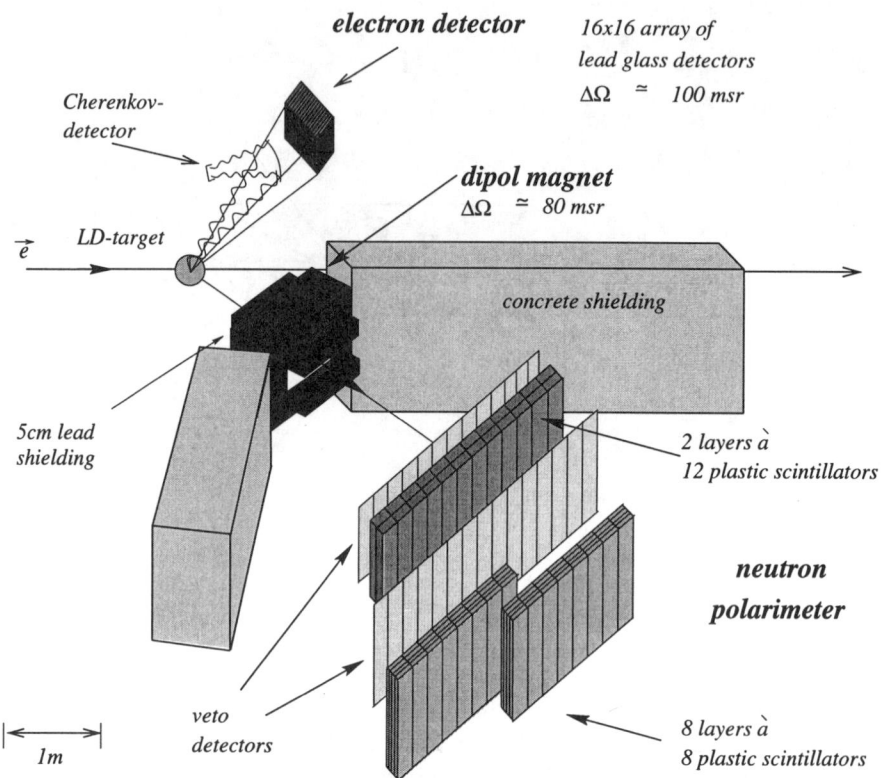

FIGURE 3. Setup of the $D(\vec{e}, e'\vec{n})$ experiment at MAMI

time at MAMI. Fig.3 shows the experimental setup of this experiment. The longitudinally polarized electron beam ($I \simeq 2.5\,\mu\text{A}$, $P_e \simeq 75\,\%$) hit a 5 cm long liquid deuterium target and the scattered electrons were detected in a 256 element lead glass array, which covered a solid angle of $\Delta\Omega \simeq 100\,\text{msr}$. The energy resolution of $\delta E/E \simeq 25\,\%$ was sufficient to suppress pion production events. Only the electron angles, which were measured with an accuracy of $\delta\vartheta, \delta\phi \simeq 3.5\,\text{mrad}$ entered the event reconstruction, which became kinematically complete through the measurement of the neutrons time-of-flight and hit position in the front plane of the neutron detector.

The neutron polarization can be analyzed in the detection process itself [35]. This required a second neutron detection in one of the rear detector planes, which yielded the polar and azimuthal angles, Θ'_n and Φ'_n, of the analyzing scattering in the front wall. With the number of events $N^\pm(\Phi'_n)$ for $\pm$ helicity states of the electron beam the azimuthal asymmetry, $A(\Phi'_n)$, was determined through the ratio

$$\frac{1 - A(\Phi'_n)}{1 + A(\Phi'_n)} = \sqrt{\frac{N^+(\Phi'_n) \cdot N^-(\Phi'_n + \pi)}{N^-(\Phi'_n) \cdot N^+(\Phi'_n + \pi)}}, \tag{14}$$

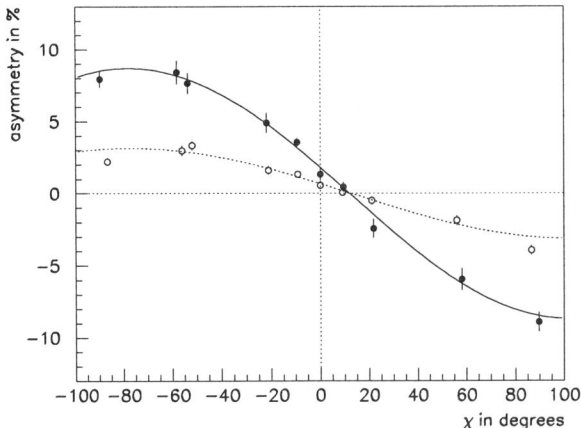

FIGURE 4. Transverse asymmetries as a function of the spin precession angle. The zero crossing is not affected by the different effective analyzing powers obtained by different conditions in the offline analysis (full and open points).

which is insensitive to variations of detector efficiency and luminosity. The extraction of P_x from $A(\Phi'_n) = \epsilon_{\text{eff}} \cdot P_x \cdot \sin \Phi'_n$ requires the calibration of the effective analyzing power, ϵ_{eff}, of the polarimeter. This, however, varies strongly with the event composition as determined by hardware conditions during data taking and software cuts applied in the offline analysis.

For the first time, the problem of calibration of the effective analyzing power has been avoided by controlled precession of the neutron spins in the field of a dipole magnet in front of the polarimeter [36] (see Fig.3). After precession by the angle χ the transverse neutron polarization behind the magnet, $P_\perp$, is a superposition of x and z components, and likewise is the measured asymmetry:

$$A_\perp = A_x \cos \chi - A_z \sin \chi. \tag{15}$$

In the particular case of the zero crossing, $A_\perp(\chi_0) = 0$, one immediately gets the relation

$$\tan \chi_0 = \frac{A_x}{A_z} = \frac{\epsilon_{\text{eff}} \cdot P_e \cdot \sqrt{2\tau\epsilon(1-\epsilon)}\, G_E^n \cdot G_M^n}{\epsilon_{\text{eff}} \cdot P_e \cdot \tau\sqrt{1-\epsilon^2}\, (G_M^n)^2}. \tag{16}$$

Obviously, this ratio is independent of both the degree of electron beam polarization, P_e, and the polarimeter's effective analyzing power. It therefore directly

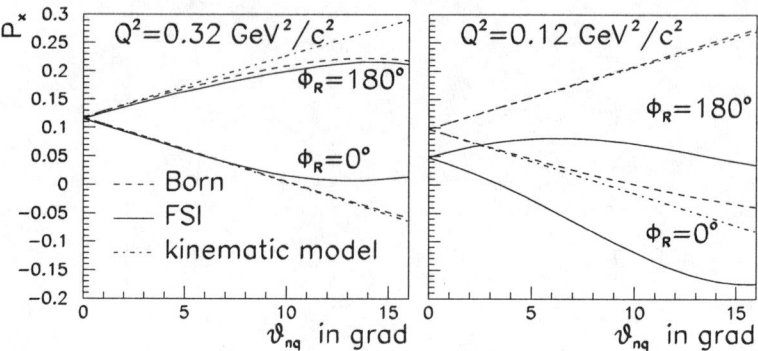

FIGURE 5. The polarization component P_x, calculated for the event population of the $D(\vec{e},e'\vec{n})$ experiment at MAMI, as a function of the angle between the outgoing neutron and the momentum transfer direction, ϑ_{np}, without final state interaction (Born) and with final state interaction (FSI). Left and right parts are for $Q^2 = 0.32\,(\text{GeV}/c)^2$ and $Q^2 = 0.12\,(\text{GeV}/c)^2$, respectively. The kinematical model takes into account only the proper rotation of quantization axes between neutron rest frame and laboratory frame required by fermi motion.

yields G_E^n/G_M^n. Different effective analyzing powers do change the magnitude of the transverse asymmetry, $A_\perp$, but not the zero crossing angle, χ_0 [36]. This is demonstrated in Fig. 4, where $A_\perp$ is plotted for two different cut conditions in the offline analysis (full and open points) as a function of the spin precession angle, χ. The magnitude of the asymmetry is affected but the zero crossing remains unchanged.

C Final state interaction

Data have been taken around $Q^2 = 0.32\,(\text{GeV}/c)^2$ and $Q^2 = 0.12\,(\text{GeV}/c)^2$. According to calculations of H. Arenhövel [37] the effect of final state interaction on P_x – which is dominated by charge exchange of the outgoing nucleons – is almost negligible at the higher momentum transfer. However, due to the small relative energy in the n-p final state, it becomes important at the small Q^2. This is illustrated in Fig.5. P_x, as calculated for the event population of the present experiment [38], is plotted as a function of the angle ϑ_{nq} of the outgoing neutron with respect to $\vec{q}$ for the two extreme situations $\Phi_R = 0$ (left of $\vec{q}$) and $\Phi_R = 180°$ (right of $\vec{q}$). For $\vartheta_{nq} \leq 10°$ the general behaviour is well approximated by purely kinematical effects due to fermi motion and can therefore be understood within a kinematical model [38] (dashed-dotted curves). The effect of final state interaction is exhibited in the difference between the dashed lines (Born calculation, i.e. without FSI) and the full curves, which include FSI. As expected, this difference

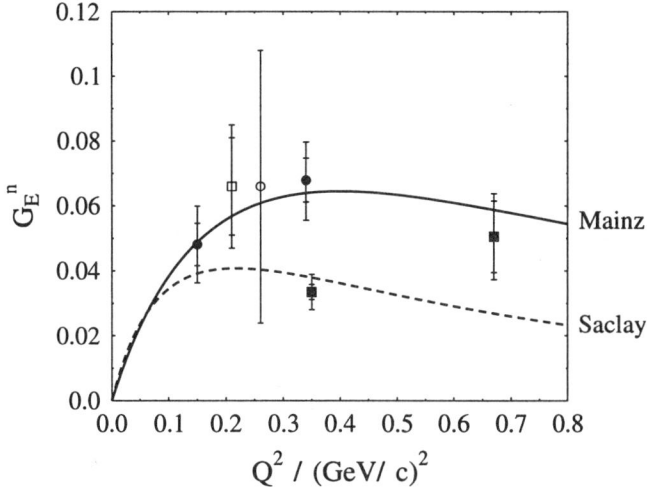

FIGURE 6. Results for G_E^n from the $D(\vec{e}, e'\vec{n})$ experiment at MAMI (full circles) [36,38] along with the $^3\vec{\text{He}}(\vec{e}, e'n)$ results from MAMI (full squares) [28,30], the $D(\vec{e}, e'\vec{n})$ result from Bates (open circle) [34], and the $\vec{D}(\vec{e}, e'n)$ one from NIKHEF (open square) [32]. Except the Bates data point the deuterium results are FSI-corrected, whereas the $^3\vec{\text{He}}$ results are not. The curves are discussed in the text.

is very small at $Q^2 = 0.32\,(\text{GeV}/c)^2$ (Fig.5, left), but it is roughly 50 % at $Q^2 = 0.12\,(\text{GeV}/c)^2$ (Fig.5, right). Despite its size, the required correction of the G_E^n/G_M^n ratio extracted from P_x has only small uncertainties, because it is insensitive to the choice of N-N potential and G_E^n parameterization [37]. Therefore, even at $Q^2 = 0.12\,(\text{GeV}/c)^2$ a reliable extraction of G_E^n/G_M^n is possible. The longitudinal neutron polarization, P_z, remains almost unaffected by FSI. The neutron electric form factor, G_E^n, has been extracted from the ratio G_E^n/G_M^n relying on the dipole values for G_M^n (Eq.8).

III RESULTS

Fig. 6 gives a summary of the recent double polarization measurements with statistical (inner) and systematical (outer) errors. The recoil polarization experiments are depicted by circles, full for the MAMI results [36,38] and open for the Bates one [34]. The open square indicates the NIKHEF internal target measurement [32]. The full squares represent the MAMI results for $^3\vec{\text{He}}(\vec{e}, e'n)$. At $Q^2 = 0.67\,(\text{GeV}/c)^2$ [30] FSI is expected to be negligible due to the large kinetic energy of the ejected neutron. However, first, still incomplete Fadeev calculations indicate a substantial correction of the $^3\vec{\text{He}}(\vec{e},e'n)$ data point at $Q^2 = 0.36\,(\text{GeV}/c)^2$ (full square) [28] towards larger G_E^n. At the present status of calculation where no meson exchange currents are yet included the central value of the extracted G_E^n is shifted to 0.052 [39]. Without this data point, a new one parameter fit to the recent MAMI double polarization results yielded

$$G_E^n = -\frac{\mu_n \tau}{1+\eta\tau} \cdot G_D \tag{17}$$

with $\eta = 3.4$ [38]. Such a fit reproduces within the errors the neutron's measured rms charge radius [40], which is related to the slope of $G_E^n(Q^2 \to 0)$. It lies almost a factor of two above the so far favoured result from elastic $D(e,e')$ scattering, where the Paris potential has been used for the unfolding of the wave function contribution [19].

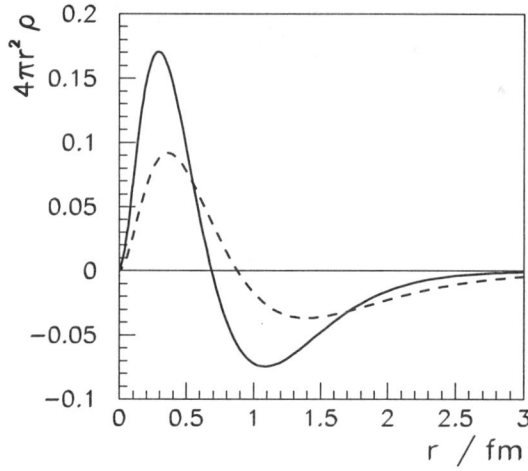

FIGURE 7. Charge density of the neutron in the Breit frame as obtained by Fourier transformation of the Saclay (dashed curve) and Mainz (full curve) G_E^n parameterizations.

This causes a significant difference in the respective Fourier transforms, which correspond to the neutron charge distribution in the Breit frame. The full and broken curves in Fig.7 correspond to the Mainz and Saclay parameterizations of G_E^n, respectively.

IV SUMMARY AND OUTLOOK

Elastic form factors contain information about the distribution of charge and current in the nucleon ground state. Precise proton data are available up to $Q^2 = 30\,(\text{GeV}/c)^2$. The neutron magnetic form factor, G_M^n, could be extracted up to $Q^2 = 4\,(\text{GeV}/c)^2$ from single arm $D(e,e')$ quasielastic scattering. Below $Q^2 =$

1 $(GeV/c)^2$, recent coincidence experiments allowed a precise determination of G_M^n from the ratio $R = \frac{\sigma(e,e'n)}{\sigma(e,e'p)}$ of quasifree electron scattering cross sections off the deuteron. Despite their individual statistical errors of only 2 %, two datasets from ELSA and MAMI differ by as much as 10 - 15 %. A recent reanalysis of a Bates experiment [41] with inclusion of FSI effects tends to prefer the MAMI dataset [42]. This is important as a normalization for the recent double polarization experiments $^3\vec{\text{He}}(\vec{e},e'n)$, $\vec{D}(\vec{e},e'n)$ and $D(\vec{e},e'\vec{n})$ at various laboratories, where the neutron electric form factor is extracted from the ratio G_E^n/G_M^n of electric to magnetic neutron form factors.

In the $D(\vec{e},e'\vec{n})$ experiment at MAMI the neutron polarimeter was, for the first time, supplemented by a spin precessing dipole magnet. This avoided the problem of its analyzing power calibration. The influence of final state interaction has been quantitatively evaluated for the $D(\vec{e},e'\vec{n})$ reaction. At the lowest Q^2 a significant FSI correction is required for the extracted G_E^n. After this correction the data are consistently described by a new one parameter fit which lies almost a factor of two above the so far favoured result from elastic electron deuteron scattering.

At MAMI and JLab there are approved proposals for both polarized target measurements [31,43] and recoil polarimetry [44,45]. These experiments will further exploit the potential of double polarized quasielastic scattering for the measurement of G_E^n over an extended Q^2 range in the near future.

V ACKNOWLEDGEMENTS

The MAMI experiments have been performed within the framework of the collaborations A1 and A3 with contributions of institutes of the universities of Basel, Bonn, Glasgow, Mainz, and Tübingen. H. Arenhövel supplied all the necessary calculations for the FSI correction. The spin precession magnet for the $D(\vec{e},e'\vec{n})$ experiment was provided by the Physikalisches Institut of the university of Bonn. Financial support came from the Deutsche Forschungsgemeinschaft (SFB 201).

REFERENCES

1. T.A. Armstrong et al., *Phys. Rev. Lett.* **70**, 1212 (1993), and
   G. Bardin et al., *Nucl. Phys.* B **411**, 3 (1994) and references therein
2. A. Antonelli et al., *Nucl. Phys.* B **517**, 3 (1998)
3. J.D. Bjørken and S.D. Drell, Relativistic quantum mechanics, McGraw-Hill 1964
4. R.G. Sachs, *Phys. Rev.* **126**, 2256 (1962)
5. N. Isgur, *Phys. Rev. Lett.* **83**, 272 (1999)
6. P. Bosted et al., *Phys. Rev. Lett.* **68**, 3841 (1992)
7. A.F. Sill et al., *Phys. Rev.* D **48**, 29 (1993)
8. R.G. Arnold et al., *Phys. Rev. Lett.* **57**, 174 (1986)
9. M.K. Jones et al., nucl-ex/9910005 and submitted to *Phys. Rev. Lett.*
10. Ch. Berger et al., *Phys. Lett.* B **35**, 87 (1971)

11. W. Bartel et al., *Nucl. Phys.* B **58**, 429 (1973)
12. for a compilation see I. Sick, Proc. 6th Miniconference on electron scattering, p. 193, NIKHEF 1989
13. A. Lung et al., *Phys. Rev. Lett.* **70**, 718 (1993)
14. P. Markowitz et al., *Phys. Rev.* C **48**, R5 (1993)
15. H. Anklin et al., *Phys. Lett.* B **336**, 313 (1994)
16. E. Bruins et al., *Phys. Rev. Lett.* **75**, 21 (1995)
17. H. Anklin et al., *Phys. Lett.* B **428**, 248 (1998)
18. S. Galster et al., *Nucl. Phys.* B **32**, 221 (1971)
19. S. Platchkov et al., *Nucl. Phys.* A **510**, 740 (1990)
20. H. Arenhövel, *Phys. Lett.* B **199**, 13 (1987) and *Z. Phys.* A **331**, 509 (1988)
21. R.G. Arnold et al., *Phys. Rev.* C **23**, 363 (1981)
22. G. Eckert et al., *Nucl. Instr. Meth.* A **320**, 53 (1992) and W. Heil et al., *Phys. Lett.* A **201**, 337 (1995)
23. C.E. Woodward et al., *Phys. Rev. Lett.* **65**, 698 (1990)
24. T.E. Chupp et al., *Phys. Rev.* C **45**, 915 (1992)
25. J. Friar, *Phys. Rev.* C **42**, 2310 (1990)
26. M. Meyerhoff et al., *Phys. Lett.* B **327**, 201 (1994)
27. R. Schulze and P. Sauer, *Phys. Rev.* C **48**, 38 (1993)
28. J. Becker et al., accepted by *Eur. Phys. J. A*
29. K.I. Blomqvist et al., *Nucl. Instr. Meth.* A **403**, 263 (1998)
30. D. Rohe et al., *Phys. Rev. Lett.* **83**, 4257 (1999)
31. D. Day et al., JLab proposal E-93-026 (1993)
32. I. Passchier et al., *Phys. Rev. Lett.* **82**, 4988 (1999)
33. J.F.J. van den Brand, these proceedings
34. T. Eden et al., *Phys. Rev.* C **50**, R1749 (1994)
35. T.N. Taddeucci et al., *Nucl. Instr. Meth.* A **241**, 448 (1985)
36. M. Ostrick et al., *Phys. Rev. Lett.* **83**, 276 (1999)
37. H. Arenhövel, priv. comm., 1998
38. C. Herberg et al., *Eur. Phys. J.* A **5**, 131 (1999)
39. W. Glöckle, priv. comm., and G. Ziemer, doctoral thesis, Bochum (in preparation)
40. S. Kopecki et al., *Phys. Rev.* C **56**, 2229 (1997)
41. H. Gao et al., *Phys. Rev.* C **50**, R546 (1994)
42. H. Gao, these proceedings and priv. comm.
43. W. Heil et al., approved MAMI proposal A1/4-95 (1995)
44. R. Madey and S. Kowalski et al., approved JLab proposal E93-038 (1993)
45. U. Müller and H. Schmieden et al., approved MAMI proposal A1/2-99 (1999)

Essential Differences between the Structure of Nuclei and Nucleons

J.W. Negele

Center for Theoretical Physics[1]
Laboratory for Nuclear Science and Department of Physics
Massachusetts Institute of Technology, Cambridge MA 02139

Abstract. In celebration of 25 years of experiments at the Bates Laboratory, this talk highlights some of the differences between the structure of nuclei and nucleons that have been revealed by precision electromagnetic probes.

I INTRODUCTION

The last quarter century has produced dramatic progress in our understanding of nuclear and subnuclear structure, driven and tested to a very large extent by precision electromagnetic probes. The remarkable feature that has emerged is how fundamentally different the structure of a nucleon is from that of that of a nucleus. Furthermore, since the sizes of nucleons and nuclei differ by at most a single order of magnitude from charge radii of 0.8 fm for the proton to less than 8 fm for a heavy nucleus, the Bates accelerator has been able to play an important role in exploring both systems. Hence, it is appropriate for a theorist to celebrate the contributions at Bates during the last quarter century by highlighting some of the interesting differences in structure of nucleons and nuclei that have been revealed by electromagnetic probes.

II MANY-BODY PHYSICS OF NUCLEAR STRUCTURE

In thinking about nuclear structure, it is useful to first recall our understanding of atomic physics. Although the interactions between electrons and nuclei are generated by the exchange of photons, one does not need to solve QED with photons as explicit dynamical degrees of freedom to understand the essential features of atomic physics. Rather, the exchanged bosons generating the interaction may be subsumed into the static Coulomb potential yielding a many-fermion problem with Coulomb forces. This many-body problem in turn can be approximated quite well in the mean-field approximation.

---

[1] This work is supported in part by funds provided by the US Department of Energy (DOE) under cooperative research agreement #DF-FC02-94ER40818.

One observes quantitative agreement between the radial charge distribution calculated in self-consistent mean field theory and measured in elastic electron scattering, and the case of the Argon atom shown in Ref. [1] is a good example.

In considering nuclear structure, one can again ask the question of whether the mesons whose exchange gives rise to the long-range components of the nucleon-nucleon interaction are essential degrees of freedom. As in the case of atomic physics, it turns out that, to a very good approximation, they may be subsumed into a nucleon-nucleon potential. By virtue of the very strong repulsive and attractive components of the potential, the nuclear wave function has strong short-range correlations that make the many-body physics both difficult and interesting. In the end, using techniques related to what is now known as effective field theory, one can derive an effective interaction [2,3] that incorporates the virtual excitations induced by strong short-range interactions and use this density-dependent effective interaction at tree level in mean-field theory to understand the essential features of nuclear structure.

To make contact with the subsequent discussion of hadron structure, before turning to experimental tests of this mean-field theory, it is useful to comment on its relation to path integrals [4]. Path integrals provide a very fruitful language to formulate many-body physics and field theory. At the stationary-phase level, the many-nucleon path integral yields mean-field theory. However, this path integral can also be evaluated exactly by Monte Carlo sampling techniques, providing the possibility of definitive checks of many-body approximations. Essentially exact Monte Carlo calculations have now been performed in light nuclei, delineating the detailed role of three-body force and of corrections to the mean field. One can thereby see explicitly how similar the essential features of exact solutions are to those of mean-field theory when the effective interaction is defined to have the correct nuclear matter binding energy and density. Thus it is fair to say that in the case of nuclear structure, the essential physics of the full path integral is already revealed at the mean-field level.

Historically, elastic and inelastic electron scattering played a crucial role in confirming this theory of nuclear structure. The left-hand portion of Figure 1, for example, shows precision elastic scattering cross sections for ^{208}Pb measured over eleven orders of magnitude at Saclay [5] compared with the excellent fit by the mean-field theory density predicted prior to the experiment [3]. High precision electron scattering data like these extending to sufficiently high momentum transfer may be combined with muonic x-ray transitions to determine the radial charge distributions of spherical nuclei to high precision, as shown in the right-hand portion of Figure 1. Here one observes systematic agreement between mean-field theory and experiment [4].

A more subtle test of our understanding of nuclear structure is the delicate balance of macroscopic and shell effects responsible for the transition between spherical and deformed nuclei. The left-hand portion of Figure 2 shows the highly deformed shape of ^{238}U predicted in mean-field theory [6]. The only way to test this shape experimentally is to do precision inelastic scattering experiments between the closely spaced states in the ground state rotational band, where the transition form factor to the J^+ rotational state essentially measures the J-th multipole of the deformed shape. Here, the pioneering energy loss spectrometer at Bates played a crucial role in providing the first precision

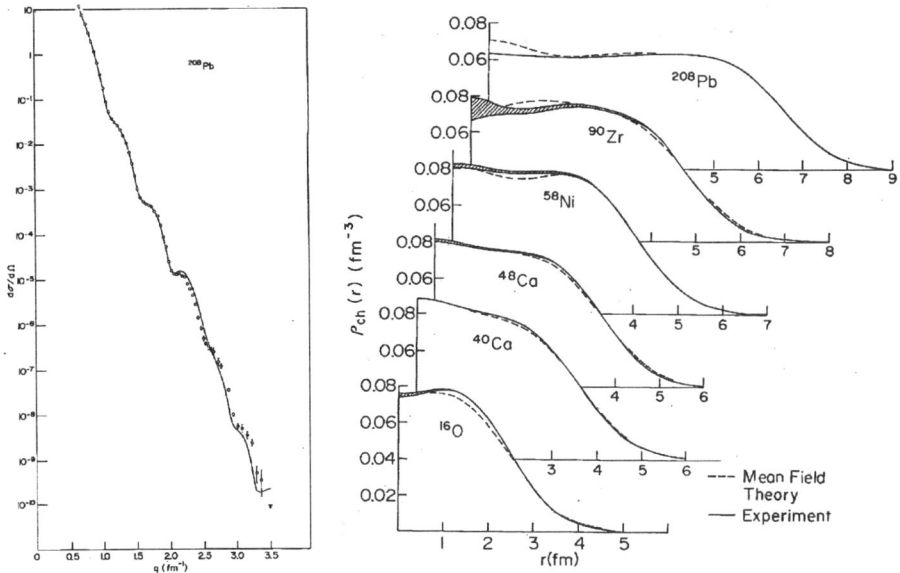

FIGURE 1. Cross sections for elastic electron scattering from ^{208}Pb at 502 MeV [5] compared with mean-field theory predictions [3] (*left-hand plot*) and comparison of mean-field theory with experiment for six spherical nuclei [4] (*right-hand plot*).

measurements of inelastic scattering [7] in the ground state band of ^{238}U shown in the right-hand portion of Figure 2. The detailed agreement with mean-field theory shown in this figure and other cases was one of the key elements in the overall confirmation that mean-field theory described the essential observable features of nuclear structure.

III MANY-BODY PHYSICS OF NUCLEON STRUCTURE

Turning now to nucleon structure, we find the situation is totally different from atomic or nuclear structure. Since the photon is neutral, photons do not interact with each other and the dominant contribution to the QED interaction between two electrons is given by single-photon exchange, which goes to zero at large distances. In contrast, in QCD, gluons carry a color change and hence interact with each other. At large distance, arbitrarily many gluons contribute to the interaction between two quarks, thereby producing confinement and rendering the theory unavoidably nonperturbative. The only known way to solve the resulting theory is by numerical evaluation of the QCD path integral on a space-time lattice. Although one alternative is to simply throw the problem on a computer and only look at the final answer, our goal is to use the numerical evaluation of the path integral as a tool to identify the degrees of freedom and processes that dominate the physics.

One of the great advantages of the path integral formulation of quantum mechanics and field theory is the possibility of identifying non-perturbatively the stationary configura-

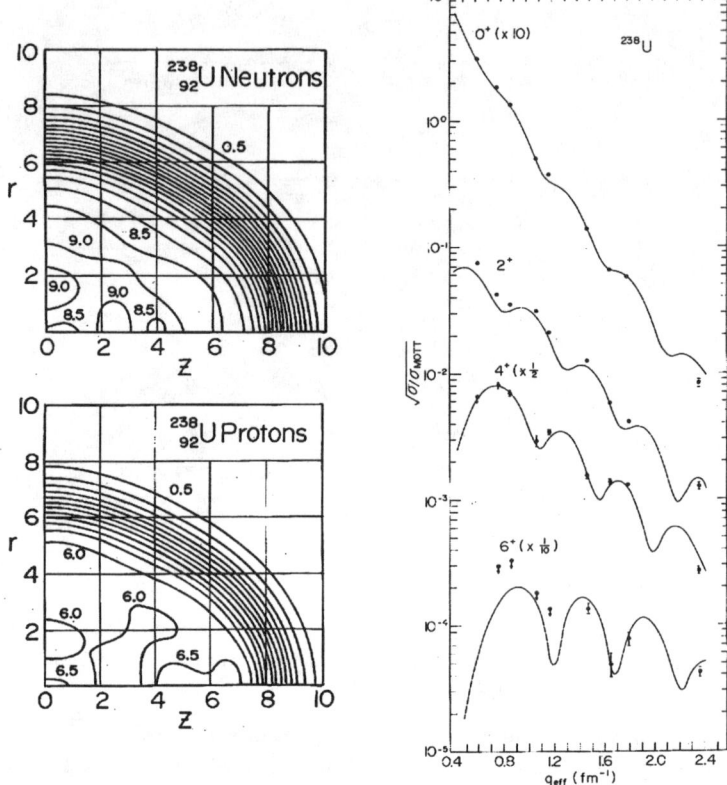

FIGURE 2. Density contours for ^{238}U calculated in mean-field theory [6] (*left graphs*) and comparison of the predicted scattering in the ground state rotational band with experiment [7] (*right-hand plot*).

tions that dominate the action and thereby identify and understand the essential physics of complex systems with many degrees of freedom. Thus, the discovery of instantons in 1975 [8] gave rise to great excitement and optimism that they were the key to understanding QCD. Indeed, in contrast to other many body systems in which the quanta exchanged between interacting fermions can be subsumed into a potential, it appeared that QCD was fundamentally different, with topological excitations of the gluon field dominating the physics and being responsible for a host of novel and important effects including the θ vacuum, the axial anomaly, fermion zero modes, the mass of the η', and the chiral condensate. However, despite nearly a quarter of a century of theoretical effort, it has not been possible to proceed analytically beyond the dilute instanton gas approximation [9]. In the intervening years, the instanton liquid model [10–12] provided a successful phenomenology and qualitative physical understanding, but a quantitative exploration of the role of instantons in nonperturbative QCD has had to wait until lattice QCD became sufficiently sophisticated and sufficient resources could be devoted to the study of instanton physics.

Our basic strategy will be to reverse the usual analytical process of calculating stationary configurations and approximately summing the fluctuations around them. Rather, we will use Monte Carlo sampling of the path integral for QCD on a lattice to identify typical paths contributing to the action and then work backwards to identify the smooth classical solutions about which these paths are fluctuating. The physical picture that arises corresponds closely to the physical arguments and instanton models of Shuryak and Diakonov [10–12] in which the zero modes associated with instantons produce localized quark states, and quark propagation proceeds primarily by hopping between these states.

A Aspects of continuum instanton physics

To put lattice investigations in context, it is useful to recall relevant aspects of continuum instanton physics. Working in Euclidean time, we evaluate a path integral of the form $\int \mathcal{D}[A] e^{\int d^4 x \, S[A]}$. Hence, as in statistical mechanics, the weight of a configuration depends not only on its energy, but also on its entropy – the number of ways it can be realized. In addition, tunneling solutions arise as periodic classical solutions in an inverted potential. Hence, we may expect BPST instantons [8] which connect degenerate minima of differing winding number with the self-dual gauge potential $A_\mu^a(x) = \frac{2\eta_{a\mu\nu} x_\nu}{x^2 + \rho^2}$ having scale-invariant action $S = \frac{1}{4} \int d^4 x \, F_{\mu\nu}^a F_{\mu\nu}^a = 8\pi^2 g^{-2} \equiv S_0$ to have a significant presence in the vacuum due to the high entropy associated with translation, size, and color orientation. The action and topological charge density are localized around the center as $\pm F_{\mu\nu}^a \tilde{F}_{\mu\nu}^a = F_{\mu\nu}^a F_{\mu\nu}^a = \frac{192\rho^4}{(x^2+\rho^2)^4}$. The tunneling rate is $dn_I \sim (\frac{8\pi^2}{g^2})^{2N_c} \frac{d\rho}{\rho^5} d^4 x^{(I)} \exp\{-\frac{8\pi^2}{g^2(\Lambda^{-1})}\} (\Lambda \rho)^{\frac{11N_c}{3}}$, where the prefactor and the running of the coupling constant in the last factor produce a distribution of instantons $\sim \rho^6$ for SU(3). Physically, we expect this distribution to be cut off at large ρ by interactions between instantons and by fluctuations when the amplitude of a sufficiently large instanton becomes small relative to quantum fluctuations. It is the difficulty in treating these infrared effects that has stymied analytic progress.

From the axial anomaly, $\partial_\mu \sum_f \bar{\psi} \gamma_\mu \gamma_5 \psi = 2m \sum_f \bar{\psi} \gamma_5 \psi + \frac{N_f}{16\pi^2} F_{\mu\nu}^a \tilde{F}_{\mu\nu}^a$, the topological charge satisfies the index theorem and, for periodic systems, may be expressed in terms of fermion eigenfunctions $Q = \frac{g^2}{32\pi^2} \int F \tilde{F} = n_L - n_R = m \sum_\lambda \frac{\int \psi_\lambda^\dagger (x) \gamma_5 \psi_\lambda(x)}{m + i\lambda}$, where n_L and n_R denote the number of fermion zero modes. For an isolated instanton, the zero mode is $\psi_0(x) = \frac{\rho \gamma \cdot \hat{x}(1+\gamma_5)}{2\pi(x^2+\rho^2)^{3/2}} \phi$. In the limit of light quarks, the Greens function for N_f quarks reduces to the product of zero modes $\prod_f \det[\not{D} + m] \bar{\psi}_f(x) \psi_f(y) \xrightarrow[m \to 0]{} \prod_f \psi_0(x) \psi_0^\dagger(y)$ and gives rise to the 't Hooft interaction. Thus, light quarks propagate by zero modes which in turn arise from instantons. Based on large N arguments, the Veneziano-Witten formula [13,14] relates the η' mass to the topological susceptibility in the pure gluon sector $\chi \equiv \int \frac{d^4 x}{V} \langle Q(x) Q(0) \rangle = \frac{f_\pi^2}{2N_f}(m_\eta^2 + m_{\eta'}^2 - 2m_K^2)$ yielding the expectation that $\chi = (180 \, MeV)^4$.

The instanton liquid model [10–12] provides an economical phenomenology of instanton mediated quark propagation in the QCD vacuum. The integral over all gluon fields that one evaluates in lattice QCD using an ensemble of configurations sampling the

213

action is replaced by an ensemble of instantons and antiinstantons of size $\rho \sim 1/3$ fm and density $n \sim 1$ fm^4 randomly distributed in space and color orientation, where the values of ρ and n are determined from the physical gluon and chiral condensates. One may think of the 't Hooft interaction as a vertex that absorbs left-handed particles of each flavor and creates corresponding right-handed particles, and *vice versa* for antiinstantons. Mesons then propagate in the QCD vacuum by the hopping of quark-antiquark pairs between these vertices, and the qualitative features of the channel dependence arises naturally with the pion channel being strongly attractive, the scalar channel repulsive, and the interaction in the rho channel very weak. The chiral condensate arises naturally in this picture by the fact that the zero modes for isolated instantons mix in the instanton liquid giving rise to a finite density of states at low virtuality.

B Lattice QCD

A QCD observable is evaluated by defining quark and gluon variables on the sites and links of a space-time lattice, writing a Euclidean path integral of the generic form [15]

$$\langle T e^{-B\widehat{H}} \widehat{\overline{\psi}} \widehat{\overline{\psi}} \widehat{\psi} \widehat{\psi} \rangle = Z^{-1} \int \mathcal{D}(U) \mathcal{D}(\overline{\psi} \psi) e^{-\overline{\psi} M(U) \psi - S(U)} \overline{\psi} \overline{\psi} \psi \psi$$

$$= Z^{-1} \int \mathcal{D}(U) e^{\ln \det M(U) - S(U)} M^{-1}(U) M^{-1}(U)$$

and evaluating the final integral over gluon link variables U using the Monte Carlo method. The link variable is $U = e^{iag A_\mu(x)}$, the Wilson gluon action is $S(U) = \frac{2n}{g^2} \sum_\square (1 - \frac{1}{N} \text{Re Tr } U_\square)$ where $U_\square$ denotes the product of link variables around a single plaquette, and $M(U)$ denotes the discrete Wilson approximation to the inverse propagator $M(U) \rightarrow m + \slashed{\partial} + ig\slashed{A}$.

Vacuum correlation functions for space-like separated hadron currents calculated in lattice QCD display the qualitative behavior expected from the 't Hooft interaction and agree semi-quantitatively with the instanton liquid model. As emphasized in ref [16], correlation functions of the form $R(x) \equiv \langle 0|T J_\mu(x) J_\mu(0)|0\rangle$ characterize the spatial and channel dependence of the interaction between quarks and antiquarks and thus supplement hadron bound state properties like phase shifts supplement deuteron properties in characterizing the nuclear interaction. The ratio $R(x)/R_0(x)$ of the interacting to free correlator has been calculated in quenched QCD for the following meson and baryon currents [17], $J = \overline{u}\gamma_\mu d, \overline{u}\gamma_\mu \gamma_5 d, \overline{u}\gamma_5 d, \overline{u}d, \epsilon_{abc}[c^a C\gamma_\mu u^b]\gamma_\mu \gamma_5 d^c$, and $\epsilon_{abc}[u^a C\gamma_\mu u^b]u^c$. Results are consistent with dispersion analysis of e^+-e^- and other data in relevant channels, and agree in detail with the channel dependence expected from the 't Hooft interaction and the instanton liquid model. Note for subsequent reference, that all correlation functions may be decomposed into a continuum contribution concentrated near the origin and a resonance contribution arising from the lowest bound state or resonance which dominates in the region of 1 fm. Quenched calculations at $\beta = 6.2$ [18] corroborate the original $\beta = 5.7$ results.

C Extraction of the instanton content of lattice gluon configurations

The Feynman path integral for a quantum mechanical problem with degenerate minima is dominated by paths that fluctuate around stationary solutions to the classical Euclidean action connecting these minima. In the case of the double well potential, a typical Feynman path is composed of segments fluctuating around the left and right minima joined by segments crossing the barrier. If one had such a trajectory as an initial condition, one could find the nearest stationary solution to the classical action numerically by using an iterative local relaxation algorithm. In this method, which has come to be known as cooling, one sequentially minimizes the action locally as a function of the coordinate on each time slice and iteratively approaches a stationary solution. In the case of the double well, the trajectory approaches straight lines in the two minima joined by kinks and anti-kinks crossing the barrier and the structure of the trajectory can be characterized by the number and positions of the kinks and anti-kinks.

In QCD, the corresponding classical stationary solutions to the Euclidean action for the gauge field connecting degenerate minima of the vacuum are instantons, and the analogous cooling technique [19,20] reveals the instantons corresponding to each gauge field configuration.

The results of using 25 cooling steps as a filter to extract the instanton content of a typical gluon configuration are shown in Fig. 1 of Ref. [21] using the Wilson action on a $16^3 \times 24$ lattice at $6/g^2 = 5.7$. As one can see, there is no recognizable structure before cooling. Large, short wavelength fluctuations of the order of the lattice spacing dominate both the action and topological charge density. After 25 cooling steps, three instantons and two anti-instantons can be identified clearly. The action density peaks are completely correlated in position and shape with the topological charge density peaks for instantons and with the topological charge density valleys for anti-instantons. Note that both the action and topological charge densities are reduced by more than two orders of magnitude so that the fluctuations removed by cooling are several orders of magnitude larger than the topological excitations that are retained.

Several extensive studies of the instanton content of the SU(3) vacuum have been performed recently and are reviewed in ref [22]. The results consistently indicate that the average size of an instanton when extrapolated to the uncooled vacuum is $0.39 \pm .05$ fm, in agreement with the liquid instanton model, and that the topological susceptibility is $\chi \sim 180$ MeV, in agreement with the Veneziano-Witten formula.

One dramatic indication of the role of instantons in light hadrons is to compare observables calculated using all gluon contributions with those obtained using only the instantons remaining after cooling. Note that there are truly dramatic differences in the gluon content before and after cooling. Not only has the action density decreased by two orders of magnitude, but also the string tension has decreased to 27% of its original value and the Coulombic and magnetic hyperfine components of the quark-quark potential are essentially zero. Hence, for example, the energies and wave functions of charmed and B mesons would be drastically changed.

As shown in Fig. 3, however, when the coupling constant, or equivalently, the lattice spacing, and quark mass are set by the nucleon and pion masses in the usual way, the prop-

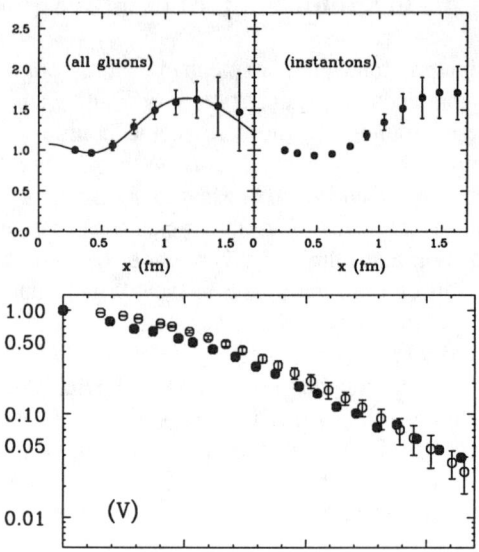

FIGURE 3. Comparison of rho observables calculated with all gluon configurations and only instantons. The upper left-hand plot shows the vacuum correlator in the rho channel calculated with all gluons and the upper right-hand plot shows the analogous result with only instantons. The lower plot shows the ground state density-density correlation function for the rho with all gluons (*solid circles*) and with only instantons (*open circles*).

erties of the rho meson are virtually unchanged. The vacuum correlation function in the rho (vector) channel and the spatial distribution of the quarks in the rho ground state, given by the ground state density-density correlation function [23] $\langle \rho | \bar{q}\gamma_0 q(x) \bar{q}\gamma_0 q(0) | \rho \rangle$, are statistically indistinguishable before and after cooling. Also, as shown in Ref. [21], the rho mass is unchanged within its 10% statistical error. In addition, the pseudoscalar, nucleon, and delta vacuum correlation functions and nucleon and pion density-density correlation functions are also qualitatively unchanged after cooling, except for the removal of the small Coulomb induced cusp at the origin of the pion. Similarly, the axial charge matrix elements specifying the spin content of the nucleon, $\langle \vec{P}\vec{S} | (\bar{q}\gamma^i i\gamma_5 q) | \vec{P}\vec{S} \rangle = 2S^i \Delta q$, are quite similar when calculated with all gluons and only instantons [24].

D Quark zero modes and their contributions to light hadrons

In the continuum limit, the Dirac operator for Wilson fermions,

$$D\psi_x = \psi_x - \kappa \sum_\mu \Big[(r - \gamma_\mu) u_{x,\mu} \psi_{x+\mu} + (r + \gamma_\mu) u^\dagger_{x-\mu,\mu} \psi_{x-\mu} \Big]$$

approaches the familiar continuum result $\frac{1}{m}[m + i(\slashed{\partial} + g\slashed{A})]\psi$.

In the free case, the continuum spectrum is $\frac{1}{m}[m+i|\vec{p}|]$ and the Wilson lattice operator approximates this spectrum in the physical regime and pushes the unphysical fermion modes to very large (real) masses. In the presence of an instanton of size ρ at $x = 0$, it is shown in Ref. [25] that the lattice operator produces a mode with a real eigenvalue which approaches the continuum result $\psi_0(x)_{s,\alpha} = u_{s,\alpha} \frac{\sqrt{2}}{\pi} \frac{\rho}{(x^2+\rho^2)^{3/2}}$ and whose mixing with other modes goes to zero as the lattice volume goes to infinity. In addition, instanton-anti-instanton pairs that interact sufficiently form complex conjugate pairs of eigenvalues that move slightly off the real axis. Thus, by observing the Dirac spectrum for a lattice gluon configuration containing a collection of instantons and anti-instantons, it is possible to identify zero modes directly in the spectrum.

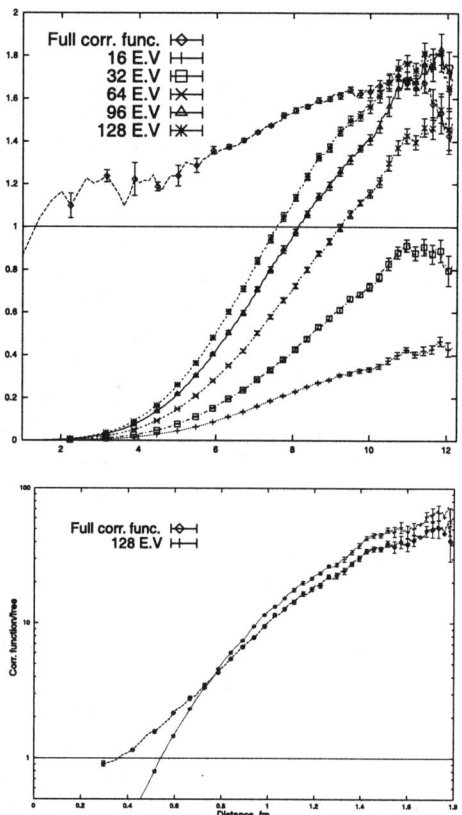

FIGURE 4. Contributions of low Dirac eigenmodes to the vector (*upper graph*) and pseudoscalar (*lower graph*) vacuum correlation functions. The upper graph shows the contributions of 16, 32, 64, 96, and 128 eigenmodes compared with the full correlation function for an unquenched configuration and the lower graph compares 128 eigenmodes with the full correlation function for a quenched configuration.

Ref. [25] shows the lowest 64 complex eigenvalues of the Dirac operator on a 16^4 unquenched gluon configuration for $6/g^2 = 5.5$ and $\kappa = 0.16$, both before and after cooling (where 100 relaxation steps with a parallel algorithm are comparable to 25 cooling steps). The cooled plot has just the structure we expect with a number of isolated instantons with modes on the real axis and pairs of interacting instantons slightly off the real axis. However, even though the uncooled case also contains fluctuations several orders of magnitude larger than the instantons, it shows the same structure of isolated instantons and interacting pairs, clearly revealing the zero modes in the original, uncooled vacuum.

The Wilson–Dirac operator has the property that $D = \gamma_5 D^\dagger \gamma_5$, which implies that $\langle \psi_j | \gamma_5 | \psi_i \rangle = 0$ unless $\lambda_i = \lambda_j^*$ and we may write the spectral representation of the propagator $\langle x | D^{-1} | y \rangle = \sum_i \frac{\langle x | \psi_i \rangle \langle \psi_{\bar{i}} | \gamma_5 | y \rangle}{\langle \psi_{\bar{i}} | \gamma_5 | \psi_i \rangle \lambda_i}$ where $\lambda_i = \lambda_{\bar{i}}^*$. A clear indication of the role of zero modes in light hadron observables is the degree to which truncation of the expansion to the zero mode zone reproduces the result with the complete propagator.

Fig. 4 shows the result of truncating the vacuum correlation functions for the vector and pseudoscalar channels to include only low eigenmodes [25]. On a 16^4 lattice, the full propagator contains 786,432 modes. The top plot of Fig. 4 shows the result of including the lowest 16, 32, 64, 96, and finally 128 modes. Note that the first 64 modes reproduce most of the strength in the rho resonance peak, and by the time we include the first 128 modes, all the strength is accounted for. Similarly, the lower plot in Fig. 4 shows that the lowest 128 modes also account for the analogous pion contribution to the pseudoscalar vacuum correlation function. Similarly, most of the strength of the disconnected graph contribution to the η' correlation function, which should be particularly sensitive to instantons, is already provided by the lowest 32 eigenmodes, and the fermionic definition of the topological charge is nearly saturated by the lowest 8 modes [26].

Finally, it is interesting to ask whether the lattice zero mode eigenfunctions are localized on instantons. This was studied by plotting the quark density distribution for individual eigenmodes in the x-z plane for all values of y and t, and comparing with analogous plots of the action density. As expected, for a cooled configuration the eigenmodes correspond to linear combinations of localized zero modes at each of the instantons. What is truly remarkable, however, is that the eigenfunctions of the uncooled configurations also exhibit localized peaks at locations at which instantons are identified by cooling. Thus, in spite of the fluctuations several orders of magnitude larger than the instanton fields themselves, the light quarks essentially average out these fluctuations and produce localized peaks at the topological excitations. When one analyzes a number of eigenfunctions, one finds that all the instantons remaining after cooling correspond to localized quark fermion peaks in some eigenfunctions. However, some fermion peaks are present for the initial gluon configurations that do not correspond to instantons that survive cooling. These presumably correspond to instanton–antiinstanton pairs that were annihilated during cooling.

E Understanding the strange quark content of the nucleon

The zero-mode dominance displayed in Fig. 4 has an interesting connection with the recent SAMPLE experiments at Bates discussed at this symposium. Utilizing the fact that the parity-violating neutral weak current couples to a different linear combination of up, down, and strange quarks than the electromagnetic current, pioneering parity-violating experiments at Bates have recently succeeded in measuring the contribution to the nuclear magnetic moment by strange quarks. Hence, it is of great interest to calculate this contribution by strange quarks from first principles on the lattice. This process requires the evaluation of notoriously difficult disconnected diagrams, in which a nucleon source and sink are connected by three light valence quarks and a heavy strange quark loop must be integrated over the entire spacetime volume. Using the discovery that only a few hundred eigenmodes associated with instantons are required to calculate quark propagation to a good approximation, it is now feasible to calculate the disconnected diagram contribution from the zero-mode zone exactly, and only use stochastic estimates for the difference between the zero-mode approximant and the full propagator. This ability to calculate disconnected diagrams is a good example of the pragmatic, as well as conceptual, value of having understood the role of instantons in the structure of the nucleon.

IV SUMMARY

I hope this brief survey has provided some appreciation for the dramatic differences between the structure of nuclei, where the exchanged mesons can be subsumed into a static potential, and nucleons, where topological excitations of the gluon field are essential degrees of freedom.

REFERENCES

1. L.S. Bartell and L.O. Brockway, *Phys. Rev.* **90**, 833 (1953).
2. J.W. Negele, *Phys. Rev. C* **1**, 1260 (1970).
3. J.W. Negele and D. Vautherin, *Phys. Rev. C* **5**, 1472 (1972); *Phys. Rev. C* **11**, 1031 (1975).
4. J.W. Negele, *Rev. Mod. Phys.* **54**, 913 (1982).
5. B. Frois et al., *Phys. Rev. Lett.* **38**, 152 (1977).
6. J.W. Negele and G.A. Rinker, *Phys. Rev. C* **15**, 1499 (1977).
7. C. Creswell et al., MIT Preprint (1981).
8. A.A. Belavin, A.M. Polyakov, A.P. Schwartz, and Y.S. Tyupkin, *Phys. Lett. B* **59**, 85 (1975).
9. C.G. Callan, R. Dashen, and D. J. Gross, *Phys. Rev. D* **17**, 2717 (1978).
10. E.V. Shuryak *Nucl. Phys. B* **198**, 83 (1982).
11. D.I. Diakonov and V.Y. Petrov, *Nucl. Phys. B* **245**, 259 (1984); **272**, 457 (1986).
12. T. Schäffer and E.V. Shuryak, *Rev. Mod. Phys.* **70**, 323 (1998), and references therein. therein.
13. E. Witten, *Nucl. Phys. B* **156**, 269 (1979).
14. G. Veneziano, *Nucl. Phys. B* **159**, 213 (1979). *Nucl. Phys. B* **169**, 103 (1980).
15. J. W. Negele, Varenna lectures, hep-lat/9804017.
16. E.V. Shuryak, *Rev. Mod. Phys.* **65**, 1 (1993).

17. M.-C. Chu, J.M. Grandy, S. Huang, and J.W. Negele, *Phys. Rev. D* **48**, 3340 (1993).
18. S.J. Hands, P.W. Stephenson, A. McKerrell, *Phys. Rev. D* **51**, 6394 (1995).
19. B. Berg, *Phys. Lett. B* **104**, 475 (1981).
20. M. Teper, *Nucl. Phys. B (Proc.Suppl.)* **20**, 159 (1991).
21. M.-C. Chu, J.M. Grandy, S. Huang, and J.W. Negele, *Phys. Rev. D* **49**, 6039 (1994).
22. J. W. Negele, hep-lat/9810053.
23. M.-C. Chu, M.Lissia, and J.W. Negele, *Nucl. Phys. B* **360**, 31 (1991); M. Lissia, M.-C. Chu, J.W. Negele, and J.M. Grandy, *Nucl. Phys. A* **555**, 272 (1993).
24. D. Dolgov, R.C. Brower, J.W. Negele, and A. Pochinhsky, hep-lat/9809132.
25. T.L. Ivanenko and J.W. Negele, *Nucl. Phys. B (Proc.Suppl.)* **63**, 504 (1998); T.L. Ivanenko, MIT Ph.D. thesis, 1997; J.W. Negele, hep-lat/9804017.
26. L. Venkataraman and G. Kilcup, hep-lat/9711006.

IV. HADRONIC STRUCTURE

Chair: E.C. Booth

Session IV.......

Nucleon Electromagnetic Form Factors

Kees de Jager

Jefferson Laboratory, Newport News, VA 23606, USA

Abstract. A review of data on the nucleon electromagnetic form factors in the space-like region is presented. Recent results from experiments using polarized beams and polarized targets or nucleon recoil polarimeters have yielded a significant improvement on the precision of the data obtained with the traditional Rosenbluth separation. Future plans for extended measurements are outlined.

INTRODUCTION

The nucleon electromagnetic form factors (EMFF) are of fundamental importance for an understanding of their internal structure. These EMFF, which in the Breit frame can be simply related to the spatial distribution of the nucleon charge and magnetization densities, are measured through elastic electron-nucleon scattering.

In Plane Wave Born Approximation (PWBA) the cross section can be expressed in terms of the so-called Sachs form factors G_E and G_M

$$\frac{d\sigma}{d\Omega} = \frac{d\sigma}{d\Omega_M} f_{rec}^{-1} \left\{ \frac{G_E^2 + \tau G_M^2}{1+\tau} + 2\tau G_M^2 \tan^2(\theta_e/2) \right\} \tag{1}$$

$$f_{rec} = 1 + \frac{2E_e}{m_N}\sin^2(\theta_e/2) \qquad \tau = \frac{Q^2}{4m_N^2}$$

where Q is the four-momentum transfer, σ_M the Mott cross section for scattering off a point-like particle, m_N the mass of the nucleon, θ_e the electron scattering angle and E_e the electron energy. This equation shows that G_E and G_M can be determined separately by measuring at fixed Q^2 over a range of (θ_e, E_e) combinations. This procedure is called the Rosenbluth separation.[1]

The Sachs form factors can be identified with the Fourier transform of the nucleon charge and magnetization density distributions, such that the slope at $Q^2 \to 0$ of the EMFF is related to the charge and magnetization radius. There has been considerable debate over the interpretation of the neutron charge radius.[2,3] The charge radius can be split into two components, one of which (the so-called Foldy[4] term) is related to the magnetic moment, not to the rest-frame charge distribution. However, Isgur[5] and Bawin and Coon[6] have shown that this Foldy term is exactly canceled by a contribution from the Dirac form factor, so that the charge radius is indeed determined solely by the rest-frame charge distribution.

Up until the beginning of the previous decade all available proton EMFF data had been collected using the Rosenbluth separation. This experimental procedure requires an accurate knowledge of the electron energy and the total luminosity. In addition, since the contribution to the elastic cross section from the magnetic form factor is weighted with Q^2, data on G_E^p suffer from increasing systematic uncertainties at higher Q^2-values. Data for the neutron resulted mainly from quasi-elastic scattering off the deuteron, because a free neutron target is not available in nature. This additional constraint caused large uncertainties, especially on the data for G_E^n.

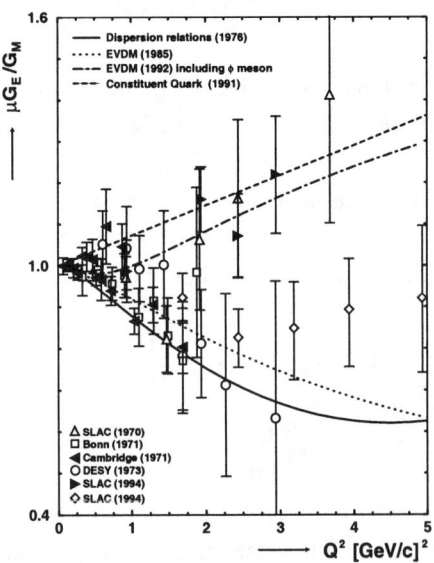

FIGURE 1. The ratio $\mu_p G_E^p / G_M^p$ as a function of Q^2, determined with the Rosenbluth separation technique. Data symbols are explained in ref. 48. Theory: full[9], dotted[11], dashed[16] and dot-dashed.[12]

These restrictions are clearly presented in the review paper by Bosted et al.[7] The then available world data set was compared to the so-called dipole parametrization, which corresponds to exponentially decreasing radial charge and magnetization densities:

$$G_E^p = G_D \qquad G_E^n = 0$$
$$G_M^p = \mu_p G_D \qquad G_M^n = \mu_n G_D \qquad (2)$$
$$G_D = \left(1 + \frac{Q^2}{Q_o^2}\right)^{-2} \quad \text{with } Q \text{ in } GeV/c \text{ and } Q_0 = 0.84 \; GeV/c$$

Accurate data were available for G_M^p up to Q^2-values of over 20 (GeV/c)^2, whereas for G_E^n no significant deviation from zero was measured.[8] For all four EMFF the available data agreed with the dipole parametrization to within 20 %. However, the limitation of the Rosenbluth separation is evident from fig. 1, which shows all available data on G_E^p. Different data sets deviate from each other by up to 50 % at higher Q^2-values, way beyond the already sizeable estimate for the experimental uncertainty.

THEORY

A frequently used framework[9] to describe the EMFF is that of Vector Meson Dominance (VMD), in which one assumes that the virtual photon – after having become a quark-antiquark pair - couples to the nucleon as a vector meson. The EMFF can then be expressed in terms of coupling strengths between the virtual photon and the vector meson and between the vector meson and the nucleon, summing over all possible vector mesons. In some cases additional terms are included to account for the effect of unknown or lesser known mesons.

A common restriction of the VMD models is that they do not predict a correct behaviour of the EMFF at high Q^2-values. The quark-dimensional scaling framework[10] predicts that only valence quark states contribute at sufficiently high Q^2-values. Under these conditions the EMFF Q^2-dependence is determined simply by the number of gluon propagators, causing the Dirac and Pauli form factors to be proportional to Q^{-4} and Q^{-6}, respectively, whereas any VMD-model will predict a Q^{-2} behaviour at large Q^2-values. Gari and Krümpelmann have constructed a hybrid (EVMD) model which combines the low Q^2-behaviour of the VMD model with the asymptotic behaviour predicted by pQCD. In their first paper[11] they consider only coupling to the ρ and ω mesons, whereas later[12] the φ meson was also included.

VMD models form a subset of models using dispersion relations, which relate form factors to spectral functions. These spectral functions can also be thought of as a superposition of vector meson poles, but include contributions from n-particle production continua. This framework allows then a model-independent fit[13] to all available EMFF data in the space- and the time-like region.

Many attempts have been made to enlarge the domain of applicability of pQCD calculations to moderate Q^2-values. Kroll et al.[15] have generalized the hard-scattering scheme by assuming nucleons to consist of quarks and diquarks. The diquarks are used to approximate the effects of correlations in the nucleon wave function. This model is equivalent to the hard-scattering formalism of pQCD in the limit $Q^2 \to \infty$. Chung and Coester[16] have developed a relativistic constituent quark model with effective quark masses and a confinement scale as free parameters.

Lu et al.[17] have recently expanded the cloudy bag model, whereby the nucleon is described as a bag containing three quarks, but including an elementary pion field coupled to them, in such a way that chiral symmetry is restored. Finally, recent developments[18] within the Skewed Parton Distribution formalism indicate a relation between the EMFF behaviour at larger Q^2-values and the nucleon spin.

NUCLEON FORM FACTORS

Over 20 years ago Akhiezer and Rekalo[19] showed that the accuracy of EMFF measurements could be increased significantly by scattering polarized electrons off a polarized target (or by equivalently measuring the polarization of the recoiling nucleon). In the early nineties a series of measurements[20-25] at the MIT-Bates facility showed the feasibility of that measurement principle.

Neutron Magnetic Form Factor

Significant progress has been made in measurements of G_M^n at low Q^2-values by measuring the ratio of quasi-elastic neutron and proton knock-out from a deuterium target. This method is practically insensitive to nuclear binding effects and to fluctuations in the luminosity and detector acceptance. The basic set-up used in all such measurements was very similar: the electron was detected in a magnetic spectrometer with coincident neutron/proton detection in a large scintillator array. The main technical difficulty in such a ratio measurement is the absolute determination of the neutron detection efficiency. For the measurements at Bates[25] and ELSA[26] the efficiency was measured in situ using the $D(\gamma,p)n$ or $p(\gamma,\pi^+)$ reaction with a bremsstrahlung radiator up stream of the experimental target. The hadron detectors used in the experiments at NIKHEF[27] and Mainz[28] were calibrated at the PSI neutron beam using the kinematically complete $p(n,p)n$ reaction.

Figure 2 shows the results of those four experiments. The Mainz G_M^n data are 8-10 % lower than the ELSA ones, despite the quoted uncertainty of appr. 2 %. This discrepancy would require a 16-20% error in the detector efficiency. The contribution from electroproduction in the ELSA set-up, caused by the electron contamination in the bremsstrahlung beam, which could result in a loss of events due to the three-body kinematics in electroproduction, has been extensively investigated.[29] Thus far, the detection inefficiency due to electroproduction has been established at less than 5 %, clearly much smaller than required to explain the discrepancy in the data.

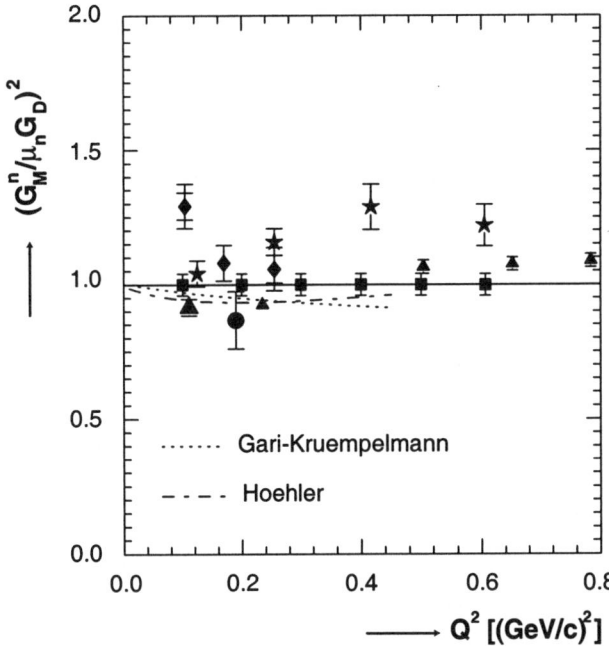

FIGURE 2. The square of the ratio of G_M^n to $\mu_n G_D$ as a function of Q^2, compared to predictions by Gari and Krümpelmann[11] and Höhler[9]. The expected precision of JLab experiment E95-001[30] is indicated by the solid squares. Data: diamonds[25], stars[26], circle[24], large triangle[27], triangles.[28]

Recently, inclusive quasi-elastic scattering of polarized electrons off a polarized ^{3}He target was measured[30] in Hall A at JLab in a Q^2-range from 0.1 to 0.6 (GeV/c)^2. This experiment will provide an independent accurate measurement of G_M^n in a Q^2-range overlapping with that of the ELSA and Mainz data. Measurements of G_M^n at Q^2-

values up to 5 $(GeV/c)^2$ are expected in the near future from a JLab experiment that will measure the neutron/proton quasi-elastic cross-section ratio using the CLAS detector.[31]

Neutron Electric Form Factor

Since a free neutron target is not available, one has to use neutrons bound in nuclei to study the neutron EMFF. The most precise data on G_E^n prior to any spin-dependent experiment were obtained from the elastic electron-deuteron scattering experiment by Platchkov et al.[32] The deuteron elastic form factor contains a term of the form $G_E^n G_E^p$. However, in order to extract G_E^n from the data, one has to calculate the deuteron wave function, which requires a choice of the nucleon-nucleon potential. Figure 3 shows the G_E^n values extracted from the Platchkov data with the Paris potential, while the grey band indicates the range of G_E^n values extracted with the Nijmegen, AV14 and RSC potentials. Clearly, the choice of NN-potentials results in a systematic uncertainty of appr. 50 % in G_E^n. One should realize that all modern NN-potentials yield consistent results for a large variety of two- and three-nucleon observables. Thus, one might expect that a reevaluation of the Platchkov data using modern high-precision NN-potentials and a consistent treatment of exchange currents will yield a reduced potential dependence.

Significant advances have been made in the last decade in the development of electron beams with high polarization and intensity and of reliable polarized targets. This progress has been used in a series of new spin-dependent measurements of G_E^n, which utilizes the fact that the ratio of the beam-target asymmetry with the target polarization perpendicular and parallel to the momentum transfer is directly proportional to the ratio of the electric and magnetic form factors:

$$\frac{G_E^n}{G_M^n} = \frac{A_\perp}{A_{//}} \sqrt{\tau + \tau(1+\tau)\tan^2(\theta_e/2)} \qquad (3)$$

A similar relation can be derived for the reaction $^2H(\vec{e},e'\vec{n})$ when one measures the polarization of the recoiling neutron directly and after having precessed the neutron spin over 90° with a dipole magnet. Figure 3 shows the results of the pioneering experiments of that technique, performed at Bates, using the reactions $^2H(\vec{e},e'\vec{n})$ [20] and $^3\vec{He}(\vec{e},e')$ [21-23] and at Mainz, with the $^3\vec{He}(\vec{e},e'n)$ reaction.[33] These results have not been corrected for rescattering or nuclear medium effects.

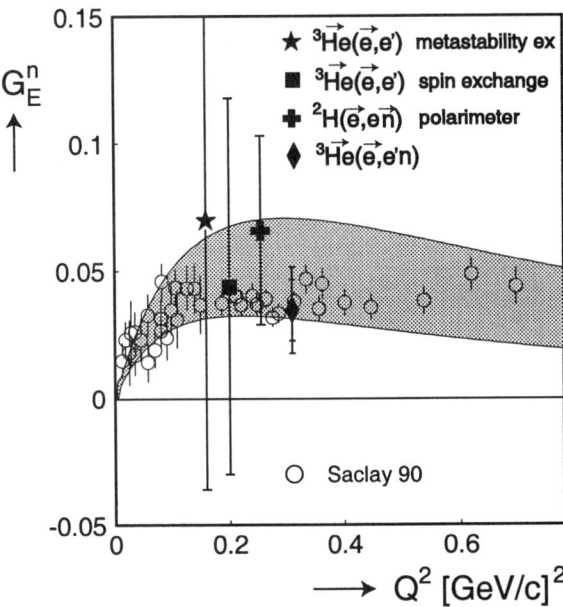

FIGURE 3. Older (star[21,22], square[23], cross[20] and diamond[33]) results for G_E^n as a function of Q^2. The open circles depict the results of Platchkov et al.[32] for the Paris potential, the shaded area the systematic uncertainty due to the choice of NN-potential.

Figure 4 shows the most recent results, obtained through the reaction channels $^2\vec{H}(\vec{e},e'n)$[34], $^2H(\vec{e},e'\vec{n})$[35,36] and $^3\vec{He}(\vec{e},e'n)$[37,38]. At low Q^2-values corrections for nuclear medium and rescattering effects can be sizeable: 65 % for deuterium at 0.15 (GeV/c)^2 and 50 % for ^{3}He[38] at 0.35 (GeV/c)^2. These corrections are expected to decrease significantly with increasing Q, although no reliable results are at present available for ^{3}He above 0.5 (GeV/c)^2. Thus, there are now data from a variety of reaction channels available in a Q^2-range up to 0.6 (GeV/c)^2 with an overall accuracy of appr. 20 %, which are in mutual agreement. However, neither the VMD[11] nor the dispersion relation[14] calculations agree with the data. Only the Galster parametrization[40] which uses a modified version of the dipole form factor, is able to describe the data adequately. A more detailed discussion of these recent results is given by Schmieden.[41] Also shown in fig. 4 are the results expected in the near future, from the $^3\vec{He}(\vec{e},e'n)$ channel at NIKHEF[42] and from the $^2\vec{H}(\vec{e},e'n)$[43] and $^2H(\vec{e},e'\vec{n})$[44] channels at JLab. Finally, in fig. 5 are shown the results expected with the BLAST detector[45] with both the $^2\vec{H}(\vec{e},e'n)$ and the $^3\vec{He}(\vec{e},e'n)$ reaction channels.

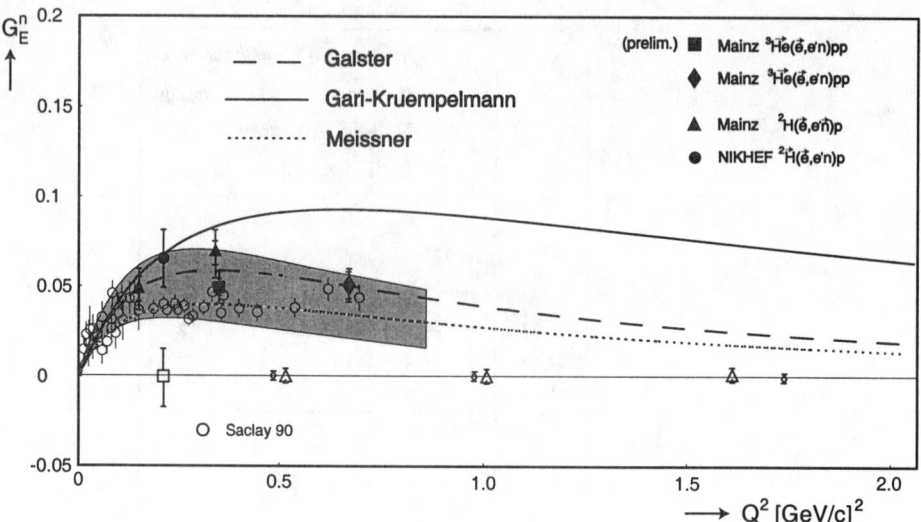

FIGURE 4. Recent (circle[34], triangle[35,36], square[37,39] and diamond[38]) and future (open square[42], open diamonds[44] and open triangles[43]) results for G_E^n as a function of Q^2, compared to three theoretical calculations (full[11], dashed[14] and dotted[40]).

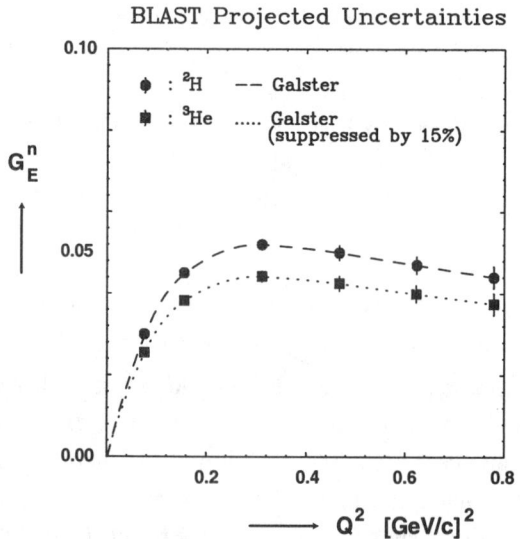

FIGURE 5. Predicted accuracy of G_E^n data to be obtained with the BLAST detector.

Proton Electric Form Factor

Arnold et al.[46] have shown that the systematic error in a measurement of G_E^p, inherent to the Rosenbluth separation, can be significantly reduced by scattering longitudinally polarized electrons off a hydrogen target and measuring the ratio of the transverse to longitudinal polarization of the recoiling proton.

$$\frac{G_E^p}{G_M^p} = -\frac{P_t}{P_l}\frac{(E_e + E_{e'})}{2m_p}\tan(\theta_e/2) \qquad (4)$$

This ratio of the two polarization components can be measured in a focal plane polarimeter, while neither the beam polarization nor the polarimeter analyzing power need be known. This method was first used by Milbrath et al.[47] at MIT-Bates to measure the ratio G_E^p / G_M^p at low Q^2.

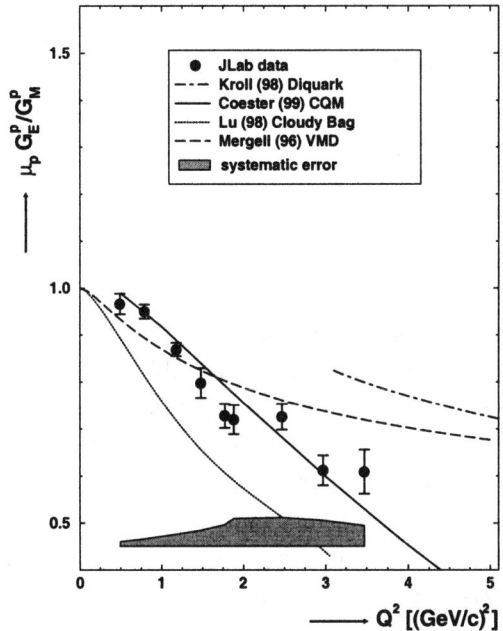

FIGURE 6. The JLab data[48] for the ratio $\mu_p G_E^p / G_M^p$ as a function of Q^2, compared to recent theoretical predictions (full[16], dotted[17], dashed[13] and dot-dashed[15]). The shaded area denotes the size of the systematic error in the data.

Recently a similar experiment[48] was performed in Hall A at JLab. Longitudinally polarized electrons with energies between 0.9 and 4.1 GeV were scattered in a 15 cm long liquid hydrogen target. For the four highest Q^2-values the beam conditions were 39 % polarization at currents up to 115 μA, while at the lower Q^2-values a 60 % polarization was obtained at currents up to 15 μA. Elastic *ep* events were selected by detecting electrons and protons in coincidence in the two identical HRS spectrometers. The polarization of the recoiling proton was determined with a Focal Plane

Polarimeter (FPP) in the hadron HRS, consisting of two pairs of straw chambers with a carbon analyzer in between. Instrumental asymmetries are cancelled by taking the difference of the azimuthal distributions of the protons scattered in the analyzer for positive and negative beam helicity. A Fourier analysis of this difference then yields the transverse and normal polarization components at the FPP. The data were analyzed in bins of each of the target coordinates. No dependence on any of these variables was observed.

The results for the ratio G_E^p / G_M^p are shown in fig. 6. The most striking feature of the data is the sharp decline as Q^2 increases. Since it is known that G_M^p closely follows the dipole parametrization, it follows that G_E^p falls more rapidly with Q^2 than the dipole form factor G_D. A comparison with fig. 1 confirms the expected improvement in accuracy of such a spin-dependent measurement over the Rosenbluth separation. None of the theoretical models shown in fig. 6 is able to adequately describe the new data. An extension[49] of this experiment to a Q^2-value of 5.6 (GeV/c)^2 has been scheduled for the fall of 2000.

CONCLUSIONS

Recent advances in polarized electron sources, polarized nucleon targets and nucleon recoil polarimeters have made it possible to accurately measure the spin-dependent elastic electron-nucleon cross section. New data on nucleon electromagnetic form factors with an unprecedented precision have (and will continue to) become available in an ever increasing Q^2-domain. These data will form tight constraints on models of nucleon structure and will hopefully incite new theoretical efforts. In addition they will significantly improve the accuracy of the extraction of strange form factors from parity-violating experiments.[50]

ACKNOWLEDGMENTS

The author expresses his gratitude to Ulf Meissner and Mark Jones for fruitful discussions and for receiving their results prior to publication. This work was supported in part by the U.S. Department of Energy.

REFERENCES

1. Rosenbluth, M.N., *Phys. Rev.* **79**, 615-619 (1950).
2. Alexandrov, Yu.A., *Neutron News* **5**, 1-20-22 (1994).
3. Byrne, J., *Neutron News* **5**, 4-15-17 (1994).
4. Foldy, L.L., *Rev. Mod. Phys.* **30**, 471 (1958).
5. Isgur, N., *Phys. Rev. Lett.* **83**, 272-275 (1999).
6. Bawin, M., and Coon, S.A., *Phys. Rev. C* **60**, 025207-1-3 (1999).
7. Bosted, P., et al., *Phys. Rev. C* **51**, 409-411 (1995).
8. Lung, A.F., et al., *Phys. Rev. Lett.* **70**, 718-721 (1993).
9. Höhler, G., et al., *Nucl. Phys. B* **114**, 505-534 (1976).
10. Brodsky, S.J., and Farrar, G., *Phys. Rev. D* **11**, 1309-1330 (1975).
11. Gari, M.F., and Krümpelmann, W., *Z. Phys. A* **322**, 689 (1985).
12. Gari, M.F., and Krümpelmann, W., *Phys. Lett. B* **274**, 159-162 (1992).
13. Mergell, P., et al., *Nucl. Phys. A* **596**, 367-396 (1996).
14. Meissner, U.-G, eprint: hep-ph/9907323; private communication (1999).
15. Kroll, P., Schürmann, M., and Schweiger, W., *Z. Phys. A* **338**, 339-348 (1991); private communication (1998).
16. Chung, P.L., and Coester, F., *Phys. Rev. D* **44**, 229-241 (1991); private communication (1999).
17. Lu, D.H., Thomas, A.W., and Williams, A.G., *Phys. Rev. C* **57**, 2628-2637 (1998).
18. Ji, X., *Phys. Rev. D* **55**, 7114-7125 (1997); *Phys. Rev. Lett.* **78**, 610-613 (1997).
19. Akhiezer, A.I., and Rekalo, M.P., *Sov. J. Part. Nucl.* **3**, 277 (1974).
20. Eden, T., et al., *Phys. Rev. C* **50**, R1749-R1753 (1994).
21. Woodward, C.E., et al., *Phys. Rev. Lett.* **65**, 698-700 (1990).
22. Jones-Woodward, C.E., et al., *Phys. Rev. C* **44**, R571-R574 (1991).
23. Thompson, A.K., et al., *Phys. Rev. Lett.* **68**, 2901-2904 (1992).
24. Gao, H., et al., *Phys. Rev. C* **50**, R546-R549 (1994).
25. Markowitz, P., et al, *Phys. Rev. C* **48**, R5-R9 (1993).
26. Bruins, E.E.W., et al., *Phys. Rev. Lett.* **75**, 21-24 (1995).
27. Anklin, H., et al., *Phys. Lett. B* **336**, 313-318 (1994).
28. Anklin, H., et al., *Phys. Lett. B* **428**, 248-253 (1998).
29. Schoch, B., private communication (1999).
30. Gao, H., and Hansen, O., JLab experiment E95-001.
31. Brooks, W., and Vineyard, M.F., JLab experiment E94-017.
32. Platchkov, S., et al., *Nucl. Phys. A* **510**, 740-758 (1990).
33. Meyerhoff, M., et al., *Phys. Lett. B* **327**, 201-207 (1994).
34. Passchier, I., et al., *Phys. Rev. Lett.* **82**, 4988-4991 (1999).
35. Herberg, C., et al., *Eur. Phys. Jour. A* **5**, 131-135 (1999).
36. Ostrick, M., et al., *Phys. Rev. Lett.* **83**, 276-279 (1999).
37. Becker, J., et al., *Eur. Phys. Jour. A* **6**, 329-344 (1999).
38. Rohe, D., et al., *Phys. Rev. Lett.* **83**, 4257-4260 (1999).
39. Glöckle, W., private communication (1999).
40. Galster, S., et al., *Nucl. Phys. B* **32**, 221 (1971).
41. Schmieden, H., these proceedings.
42. Brand, J.F.J. van den, and Ferro-Luzzi, M., NIKHEF experiment 94-05.
43. Day, D., and Mitchell, J., JLab experiment E93-026.

44. Anderson, B.D., Kowalski, S., and Madey, R., JLab experiment E93-038.
45. Bates Large Acceptance Spectrometer Toroid, http://mitbates.mit.edu/blast.
46. Arnold, R., Carlson, C., and Gross, F., *Phys. Rev. C* **23**, 363-374 (1981).
47. Milbrath, B., et al., *Phys. Rev. Lett.* **80**, 452-455 (1998); erratum, *Phys. Rev. Lett.* **82**, 2221 (1999).
48. Jones, M.K., et al., *Phys. Rev. Lett.* **84**, to be published (2000).
49. Perdrisat, C., et al., JLab experiment E99-007.
50. Souder, P., these proceedings.

Nucleon Structure Studied Through VCS and the $N \to \Delta$ Transition

Costas N. Papanicolas

Department of Physics, University of Athens, Athens, Greece
and
Institute of Accelerating Systems and Applications, P.O. Box 17214, Athens 10024, Greece

Abstract. In the last decade the detailed understanding of the structure of hadrons and in particular that of the nucleon has emerged as one of the key issues in intermediate energy electromagnetic investigations. Fundamental questions concerning the shape, the polarizability, the origin of the spin of the nucleon are currently being explored both theoretically and experimentally. We present here an overview of the ongoing investigations of: a) nucleon deformation through electro-excitation to the first nucleon resonance, the $\Delta^+(1232)$, and quantified through the quadrupole amplitudes in the $\gamma^* N \to \Delta$ transition and b) the generalized polarizabilities of the nucleon through Virtual Compton Scattering (VCS). Recent results and planned experiments at intermediate energy facilities world wide, with particular emphasis at the experimental programs at MIT/Bates are reviewed.

INTRODUCTION

As we celebrate the anniversary of 25 years of beam on experiment at Bates, a remarkable transformation has taken place in its scientific program during this period. Investigation of the structure of the nucleon, dominates the on-going and planned scientific program. Twenty five years ago, even fifteen years ago, not a single experiment was pursuing a similar goal. This does not reflect a peculiarity of Bates; a similar transformation has taken place in every intermediate energy laboratory world wide. This transformation reflects the new awareness that has come about, that our understanding of hadrons is very primitive, and cannot explain or account for some obvious and far reaching questions: what is the origin of the nucleon spin, in view of the finding that only a fraction of it comes from the intrinsic quark spins; the distribution of charge in the neutron, a neutral object made up of charged constituents. Even more profound questions concerning the shape of the nucleon and theoretical claims that it is oblate, could not be confirmed or denied. It may appear surprising that at this time such central questions, concerning the most fundamental constituents of matter, are not yet resolved. The underlying common difficulty concerns our lack of understanding of the subtle many body manifestations of a complex and theoretically intractable system, that of quarks and gluons strongly interacting at low energies. The sophistication of the nuclear community in dealing with complex many body systems, as it was demonstrated in the seventies and eighties in the study of the nuclear many body problem, is now being engaged in elucidating the subtleties of the quark-gluon system.

Many presentations in this remarkable meeting convey the excitement and the impressive progress that is being made in addressing the issues mentioned above both theoretically and experimentally. I will try to summarize and present an outlook for two big programs being pursued world wide and which figure prominently in the Bates program: a)the study of the $N \to \Delta$ transition and b) the study of the nucleon through Virtual Compton Scattering (VCS). Bates is playing a leading role in both programs, introducing novel techniques and novel instrumentation. Given the opportunity of this talk and given the pioneering role of Bates in the development of these programs, my brief presentation needs to be Bates-centric.

THE ISSUE OF NUCLEON DEFORMATION

The possibility of nucleon deformation was first raised by Glashow 20 years ago [1] and has since remained an important open question. Because the static quadrupole moment of the nucleon vanishes identically, on account of its $J = 1/2$ nature, experimental and theoretical investigations have focused on the search for (transition) quadrupole strength in the $N \to \Delta$ transition.

The physical origin of resonant quadrupole strength is however interpreted in different terms in the various nucleon models that predict such an effect. For instance, in "QCD-inspired" constituent quark models, it arises from intra-quark effective color-magnetic tensor forces [2,3], a situation analogous to that of the intra-nucleon tensor interaction in the deuteron which leads to its deformation while in chiral bag models [9] [8] where a meson cloud is allowed, most of the deformation can be attributed to the pressure exerted by the pions which couple preferentially along the axis of spin of the nucleon.

Spin-parity selection rules in the $N(J^\pi = 1/2^+) \to \Delta(J^\pi = 3/2^+)$ transition allow magnetic dipole ($M1$) and electric quadrupole ($E2$) or Coulomb quadrupole ($C2$) multipoles. In pion production, amplitudes are denoted by $M^I_{l\pm}$, $E^I_{l\pm}$, and $S^I_{l\pm}$, thus indicating their character (magnetic, electric, or scalar), their isospin (I), and their total orbital angular momentum ($J = l \pm \frac{1}{2}$). Thus, the resonant photon multipoles $M1$, $E2$, and $C2$ correspond to the pion multipoles $M^{3/2}_{1+}$, $E^{3/2}_{1+}$, and $S^{3/2}_{1+}$, respectively. The Electric- and Scalar(Coulomb)-to-Magnetic-Ratios of amplitudes are defined as $EMR = R_{EM} = \Re e(E_{1+}/M_{1+})$ and $CMR = R_{SM} = \Re e(S_{1+}/M_{1+})$ respectively. In the spherical quark model of the nucleon, the $N \to \Delta$ excitation is a pure $M1$ transition. Early electroproduction experiments [4–7] indeed found the $M1$ amplitude to dominate. However, more refined models and analysis [14,8,15] find values of R_{SM} in the range of -1% to -4%, at momentum transfer square $Q^2 \approx 0.1$ (GeV/c)^2.

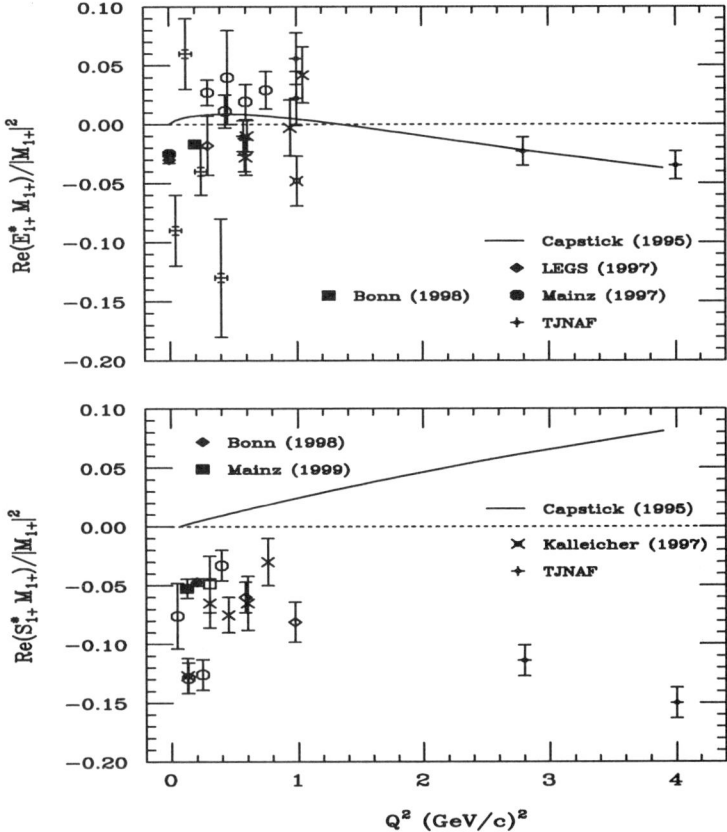

FIGURE 1. The available R_{EM} and R_{CM} values. Filled symbols denote results from recent experiments. In most cases the error shown is purely statistical.

The interpretation of the experimentally determined R_{EM} and R_{SM} is severely complicated by the presence of processes which are coherent with the resonant excitation of the $\Delta(1232)$. These processes (such as the pion pole, Born terms, tails of higher resonances, off-shell effects and pion loops) give rise to additional non-resonant quadrupole amplitudes which further complicate the isolation of the resonant quadrupole amplitudes. These interfering processes, termed "background contributions", which obviously have nothing to do with experimental backgrounds, need to be constrained in order to isolate the resonant contributions to R_{EM} and R_{SM} which could then be interpreted as signals for "deformation".

While real photons are used to extract R_{EM} at $Q^2 = 0$, R_{SM} and the Q^2 evolution of R_{EM} can be investigated only through electro-excitation. Recent precision measurements with polarized tagged photons have resulted in an $R_{EM}^{3/2} = \Re(E_{1+}^{3/2}/M_{1+}^{3/2})$ at resonance of $(-3.0 \pm 0.3)\%$ [10] and $(-2.5 \pm 0.3)\%$ [11], in good agreement with theoretical calculations. The situation is quite different for electron scattering investigations and the resulting R_{EM} and R_{SM} determination. Early experiments of the late sixties or early seventies conducted at Q^2 up to 1 $(GeV/c)^2$ have yielded an R_{SM} of around -7%, but with large statistical and systematic errors [4-7] essentially providing minimal information on the deformation issue. The new generation of high precision data are just beginning to emerge and promise to be of adequate accuracy so as to provide guidance to the models of the nucleon that have emerged.

It has become standard practice to report the findings of these experiments in terms of R_{EM} and R_{SM}, although these quantities are invariably extracted with substantial model error which is rarely quoted. This practice, reminiscent of the days when transition radii and $B(E\lambda)$ values were quoted instead of form factors and transition densities, has already led to confusion and superficial contradictions. We present in Fig. 1 the current status of the these standard benchmarks as a function of Q^2, bearing this caveat in mind.

EXPERIMENTAL METHODS AND EQUIPMENT

As mentioned earlier, the panoply of the field in terms of highly exclusive coincident experiments is being used to access the small quadrupole amplitudes of interest. Experiments involving all possible decay channels (π^0, π^+, and γ) in plane and out-of-plane geometry, focal plane polarimetry experiments where the polarization of the recoiling proton in a $H(e,e'\vec{p})\pi^0$ reaction is detected and finally experiments with polarized targets comprise the main reactions used.

The experimental situation as it currently stands is depicted graphically in Fig. 2 where each major experimental program is depicted by a row giving the range of momentum transfer that is to be investigated. The shaded symbols or bars indicate experiments that have taken data, while the empty symbols denote planned (approved) experiments. It is seen that based on this chart, the low Q^2 is expected to be reasonably well explored in the next few years. The Q^2 evolution of R_{EM} and R_{SM} are clearly of interest, as a crucial test of nucleon models is their ability to predict their magnitude. The first measurement at moderate Q^2 attempting to establish the growth of the absolute values R_{EM} and R_{SM} as a function of Q^2, predicted by all models, was reported by Frolov et al [12] from measurements performed at Jefferson.

The difficult issues that concern the investigation of the $N \to \Delta$ transition are well understood and have been identified early on [13]. The experimentalists are faced with the challenge of addressing the following issues:

- The amplitudes of interest contribute only to a tiny fraction of the reaction cross section.

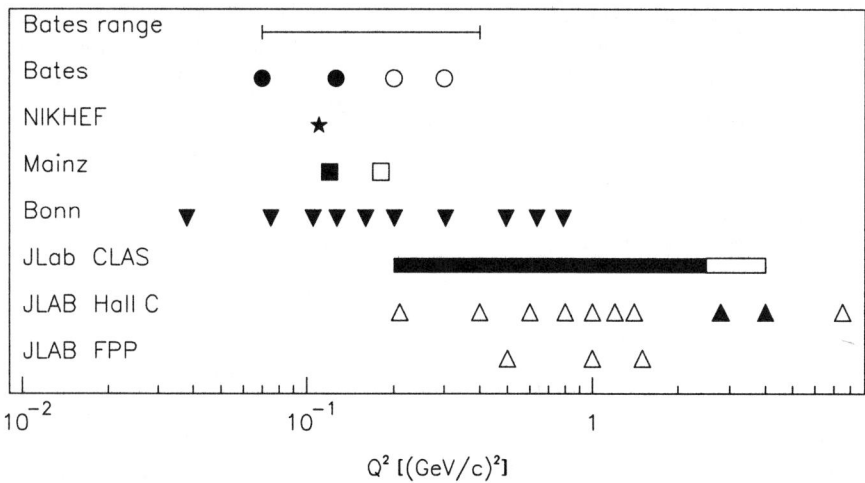

FIGURE 2. A chart of the active programs investigating the $N \to \Delta$ transition. Filled symbols and bars indicate experiments that have taken data.

- Coherent processes ("background terms") are involved that make the isolation of the physics of interest in a model independent fashion impossible.

The first issue raised leads to the conclusion that the most sensitive responses, the ones which will "carry the signal", will be interference responses in which the weak quadrupole amplitudes will manifest themselves through interference with the dominant dipole amplitude. The interference of the $C2$ amplitude with the $M1$ will obviously lead to Longitudinal - Transverse (LT) type responses, while the interference of the $E2$ amplitude with the $M1$ will lead to Transverse - Transverse (TT) type responses. A second obvious conclusion is that because of the "smallness" of the "signal", but not of the cross section, the measurements are extremely demanding in the control of systematic error. This invariably leads to the conclusion that in addition to requiring the ultimate in instrumentation, techniques involving simultaneous measurements and whenever possible helicity and polarization asymmetry measurements need to be employed.

The second issue, concerning the isolation of "background terms", can be addressed through a detailed and exhaustive investigation of the responses and through redundant measurements. This means the detailed mapping in terms of W, Q^2 and angular dependence of the responses and the simultaneous measurement of as many of them as practically possible. We have stressed in the past, the very important role that the imaginary responses could play in elucidating this issue [13] [21]. Since they arise as a result of the interference between amplitudes of different phase (such as those between the resonant and background terms) they are particularly useful in helping constrain the background and therefore isolate the resonant contribution.

The measurement of the π^0 and π^+ channels offers the possibility of isospin decomposition. If only proton targets are utilized, then two of the three independent isospin amplitudes can be determined, the resonant $I = 3/2$ amplitude $(A_{1+}^{3/2})$ and the non-resonant $I = 1/2$ amplitude $(A_{1+}^{1/2})$ which is the sum of isoscalar and isovector terms. Measurements for the $p\pi^0$ and $n\pi^+$ channels give a different weighting of the resonance and "background" amplitudes [18].

$$A_{1+}^{3/2} = \sqrt{\frac{1}{3}} A_{1+}(\gamma p \to \pi^+ n) + \sqrt{\frac{2}{3}} A_{1+}(\gamma p \to \pi^0 p) \tag{1}$$

$$A_{1+}^{1/2} = \sqrt{\frac{2}{3}} A_{1+}(\gamma p \to \pi^+ n) - \sqrt{\frac{1}{3}} A_{1+}(\gamma p \to \pi^0 p) \tag{2}$$

Finally the study of the γ channel offers an entirely different reaction channel for accessing the same physics. It therefore provides yet another independent cross check on the results.

Interference amplitudes manifest themselves in coincidence experiments. Each particular approach (e.g. using polarized targets, focal plane polarimetry etc) offers access to different interference functions, often complementary to each other. For instance, the information obtained from FPP measurements is complementary but not identical to that obtained form polarized target measurements. The simplest case that shows the possibilities discussed above is the case of coincident electron scattering with polarized beam from unpolarized target, without detection of the polarization state of the decay product.

For the case of polarized beam and out-of-plane detection (Fig. 1) the cross section for the $A(\vec{e}, e'x)B$ reaction is [21]:

$$d\sigma = d\sigma_{\text{Mott}}(v_L R_L + v_T R_T + v_{LT} R_{LT} \cos\phi_{xq} + v_{TT} R_{TT} \cos 2\phi_{xq} \\ + hv'_{LT} R'_{LT} \sin\phi_{xq}) \tag{3}$$

where ϕ_{xq} is the azimuthal reaction angle for the detected particle, $v_{\alpha\beta}$ are kinematic factors related to the lepton tensor and h is the beam helicity. The longitudinal-transverse (R_{LT}) and transverse-transverse (R_{TT}) structure functions contain the interference terms $\Re e(S_{1+}^* M_{1+})$ and $\Re e(E_{1+}^* M_{1+})$ respectively. They are thus sensitive to the small quadrupole multipoles by amplifying their contribution through their interference with the the large and reasonably well-understood M_{1+}. The so-called 'fifth' structure function (R'_{LT}) is the imaginary analog to R_{LT}. It contains the term $\Im m(S_{1+}^* M_{1+})$ and is therefore sensitive to the relative phase of the resonant amplitudes. Because for an isolated resonance this observable would be zero, it plays a key role in the separation of resonant from competing channels in the study of nucleon resonances. It can be only measured if the incident electrons are longitudinally polarized and out-of-plane detection is implemented.

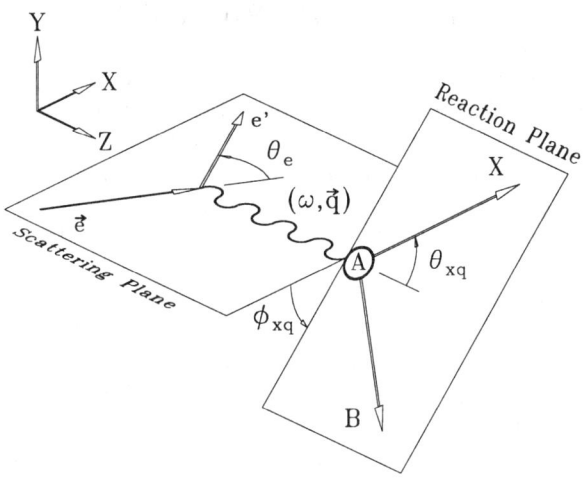

FIGURE 3. Kinematic definitions for the $A(\vec{e}, e'x)B$ reaction.

The isolation of all five responses of Eq. 3 is possible by placing detectors at optimally chosen positions, using ϕ_{xq} as a lever arm. The need of out-of-plane detection is obvious. If, in addition, the measurements are performed simultaneously with multiple detectors, systematic errors can be substantially reduced, as the structure functions are derived - up to a normalization factor - from cross section ratios. It is often useful to form the asymmetries A_{LT}, A_{TT} and A'_{LT} which are related to the corresponding structure functions LT, TT and LT'. The helicity asymmetry can be measured precisely by rapid revearsal of the beam polarization:

$$A_h = A'_{LT} = \frac{d\sigma(+h) - d\sigma(-h)}{d\sigma(+h) + d\sigma(-h)}. \qquad (4)$$

Different laboratories have refined different experimental techniques and developed "niches' in their programs. We see this most clearly in how each laboratory has decided to pursue the $N \to \Delta$ transition. At Bates out-of-plane detection has been chosen form the very beginning as the technique of choice. However, the availability of Focal Plane Polarimeter (FPP) gave the for the first ever measurement of a $H(e, e'\vec{p})\pi^0$ reaction and the determination of the R^n_{LT} response [17]. In the future, with the availability of polarized internal targets and the BLAST facility, Bates plans to pursue this reaction using the $\vec{H}(\vec{e}, e')$ reaction. Already such a first measurement was performed during the closing days of NIKHEF, using the internal target facility there.

THE BATES PROGRAM

The experimental program at Bates for next few years is entirely relying on exploiting the unique features of the unique Out-Of-Plane Spectrometer (OOPS) instrumentation. It is worth pointing out that OOPS has been constructed with the $N \to \Delta$ serving as the guiding experiment for defining its kinematic constraints and tolerances. The approved program [22–24] is geared towards mapping all five responses over the dynamical variables (W, Q^2, θ_{pq}) for all possible Δ decay channels, i.e. $p\pi^0, n\pi^+$ and $p\gamma$.

The out-of-plane capability at Bates-MIT is realized with the OOPS facility. Four identical spectrometers [25,26] have been constructed, so that they can be positioned in a cluster symmetrically around the momentum transfer axis in a × or a + configuration. The OOPSs are 16-ton 850 MeV/c magnetic spectrometers of momentum resolution of $\Delta P/P = 0.5\%$ and a large flat bite of $\pm 10\%$. They can be placed with position and orientation accuracies of better than 1 mm and 1 mr respectively. The focal plane instrumentation of an OOPS spectrometer consists of three horizontal drift chambers for track reconstruction and three scintillators for triggering.

The first measurements [27] at Bates on the $N \to \Delta$ were conducted with beams of 1% duty factor and at energies of 719 and 799 MeV. The differential cross section for proton detection along the momentum transfer direction is shown in Fig.4. For such 'parallel' kinematics, R_{LT} and R_{TT} vanish, and the cross section $(d\sigma_\parallel)$ is dominated by R_T, which contains the term $|M_{1+}|^2$. The data are shown as a function of invariant mass W and exhibit a distinct resonant shape peaking at 1202.0 ± 1.2 MeV. The data are compared with predictions of the model of Sato and Lee, which is an extension of their photo-production model [8], of Mehrotra and Wright [31], and of Drechsel et al. [15]. All calculations predict the position of the maximum correctly, although they differ in their detailed shape and especially in magnitude. The 'deformed' (non-zero quadrupole $\gamma^* N \to \Delta$ form factors) model of Sato and Lee [8] is the one closest to these data.

Fig. 5 shows R_{LT} plotted vs. the proton angle to the momentum transfer axis θ_{pq} in the CM frame. Clearly, a consistent description of the small amplitudes which build the longitudinal-transverse interference response is not given by any of the available theoretical models.

Based on these data we could proceed to extract under certain (drastic) model assumptions the R_{EM} and R_{SM} values, as has been the practice in the field, a procedure clearly unwarranted in view of the previous comments. Assuming vanishing contributions form the background, except in the M_{1+}, M_{1+} and S_{1+} channels, and either zero or finite R_L, constrained through Siegerts prescription, we could arrive at the values shown in table 1. Alternatively one can adjust the amplitudes within a given theoretical framework which provides a consistent prescription of the background. Such a procedure was followed by Frolov [12]. We explored this model extraction using the models of RPI [16] and MAID [15]. As it is evident from the table the disagreement among the values extracted using the standard empir-

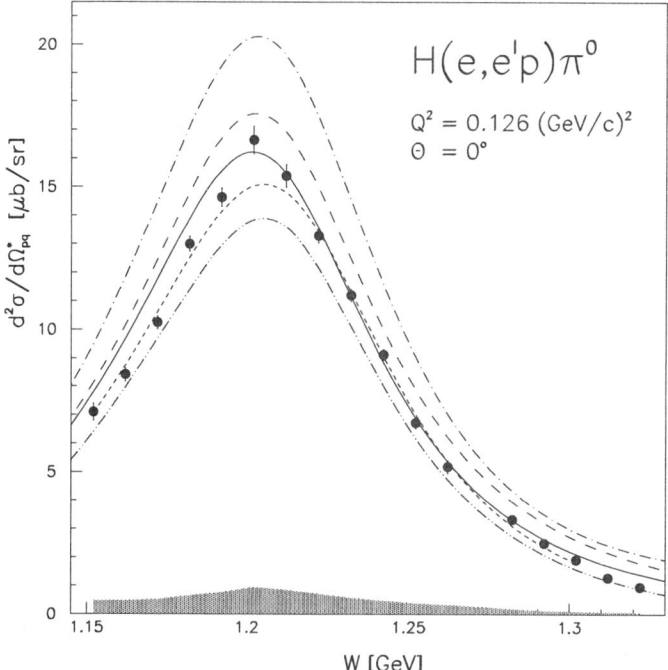

FIGURE 4. The CM cross section in parallel kinematics $d\sigma_\parallel$ [27,30]. The solid curve is the "deformed" and the dashed curve the "non-deformed" prediction of Ref. [8]. The dot-dashed and dot-dot-dashed curves are the calculation of Refs. [31] and [15] respectively. The short-dashed curve ia a calculation of Ref. [16] The shaded band shows the value of the systematic error.

ical approach or the two available models, vary considerably demonstrating the need of quantifying the model error, before further progress can be achieved. The shortcomings of this empirical extraction become obvious when compared with the results of another experiment (at the same Q^2) [20], where cross section was measured at a CM-backward angles. The same analysis with the same assumptions yields inconsistent results yielding a value of $R_{SM} = -13\%$. The disagreement is a strong indication that "background" contributions to the resonant and non-resonant multipoles cannot be ignored in the interpretation of these data for the extraction of R_{SM}.

Using FPP [17], a separate measurement of the polarization of the recoiling proton at $W = 1232$ MeV was performed. With an unpolarized electron beam and target, the final state proton polarization in parallel kinematics has only one component (P_n) normal to the scattering plane, which is $P_n = v_{LT} R^n_{LT}/d\sigma_\parallel$. The

	Assumptions	M1 (10^{-3}GeV$^{-1/2}$)	EMR (%)	CMR (%)	χ^2
Empirical (1+ multipoles only)	$R_L = 0$, EMR=0	284 ± 3	0	-7.6 ± 0.3	0.3
	R_L finite, EMR=0	278 ± 3	0	-7.9 ± 0.4	0.2
	R_L finite, Siegert EMR	309 ± 3	-3.3 ± 0.1	-6.5 ± 0.2	0.7
Theory	RPI	287 ± 5	$+0.9 \pm 0.9$	-9.1 ± 0.8	2.2
	MAID 2000 (preliminary)	—	-2.5 ± 0.2	-6.1 ± 0.2	1.3

TABLE 1. $N \rightarrow \Delta$ amplitudes extracted from the present data at W=1.232 GeV. The errors in the empirical analyses are statistical only.

response function R_{LT}^n is similar to R'_{LT}, in that to leading order it is proportional to the term $\Im m(S^*_{1+}M_{1+})$ which would be zero in the case of a pure resonance. The large measured value [17] of $P_n = -0.397 \pm 0.055 \pm 0.009$ provides another indication of strong "background" contributions in the Δ electro-excitation. The P_n result is in good agreement with that of Drechsel et al. [15], but about a factor of 2 larger than the prediction of the model of Sato and Lee [8,17]. A far more complete FPP measurement [19] has been recently completed at Mainz, performed with polarized beam. This allows the extraction of three responses. In addition to the P_n measurement, the complete information allows for an extraction of R_{sm}

For the first ever out-of-plane measurements of the $\Delta \rightarrow p\pi^0$ channel, two OOPS spectrometers were positioned at $\phi_{pq} = 45, 135$ deg. In all but one measurement, longitudinal polarization of $\approx 37\%$ was delivered, produced by circularly polarized laser light hitting a GaAs crystal.

The measurements were made at angles considerably larger than in the in-plane case. This gives a lever arm in the θ_{pq} angle which is very useful in distinguishing among theoretical models, as clearly shown in Fig. 5.

The first measurement [28,29] of R'_{LT} (Fifth Response) in the $p\pi^0$ channel is shown in Fig. 6. The measured point lies are several standard deviations away from zero, thus providing yet another indication of non-zero "background". As in the case of P_n measurement, the MAID results [15] describe the data well while the Sato Lee calculations [8] underpredict the effect on resonance. The consistent result from the two imaginary responces measured is rewarding.

The motivation to study the $\Delta \rightarrow n\pi^+$ channel [23] is to add isospin discrimi-

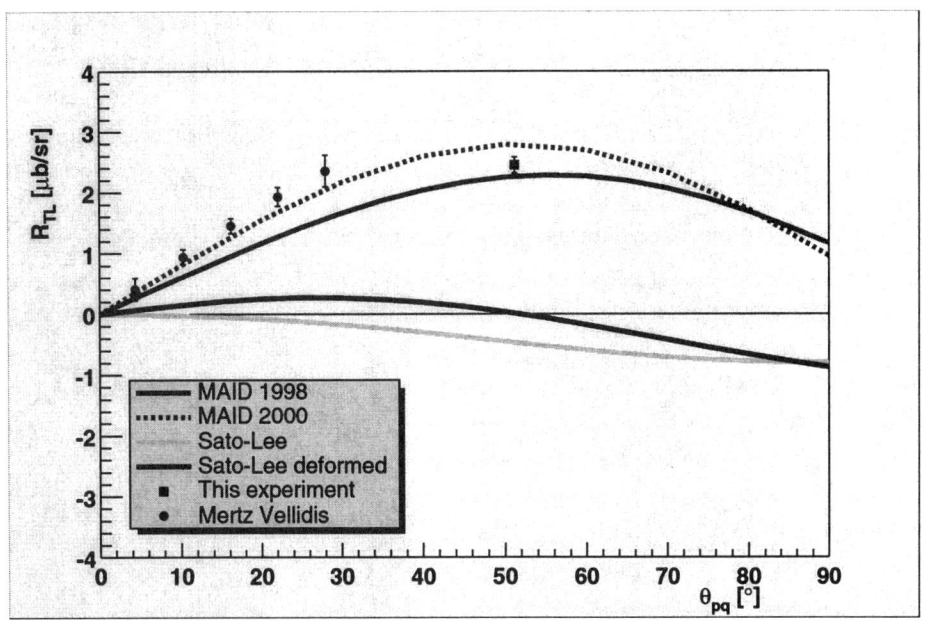

FIGURE 5. The longitudinal-transverse response, as a function of the CM proton polar angle [27,28,30]. The dense dotted and dotted curves are the "deformed" and "non-deformed" cases of Ref. [14]. The notation of the other curves is explained in Fig. 4. The hatched band is the projection of the result of Ref. [20], as explained in the text. The shaded bands show the values of the systematic error.

nation to the determination of resonant and "background" amplitudes. The Bates measurements program for the $H(\vec{e},e'\pi^+)n$ reaction parallels the one of the $p\pi^0$ channel. So far, one measurement has been completed at $Q^2 = 0.126(GeV/c)^2$ The LT and LT' responses and asymmetries were measured on the resonance, at the same momentum transfer as in the previously described experiments on the $p\pi^0$ channel. A preliminary data analysis indicates that A'_{LT} is much larger than in the π^0 case.

A second generation of measurements will soon be possible at Bates due to the availability of high quality continuous-wave beams and to the full 4-spectrometer OOPS apparatus. For the first time, the simultaneous measurement of five responses at identical kinematics will be possible over a large range of kinematics and with superior control of the systematic uncertainties. New measurements will extend to cover larger regions of W, Q^2 and θ_{pq} angle. Also the R_{TT} response and asymmetry will be accessed, opening the study of the Q^2 evolution of R_{EM}, on which little is known. Finally, the availability of beams with polarization higher than 60% will greatly reduce the statistical error of A'_{LT}, which is much larger than

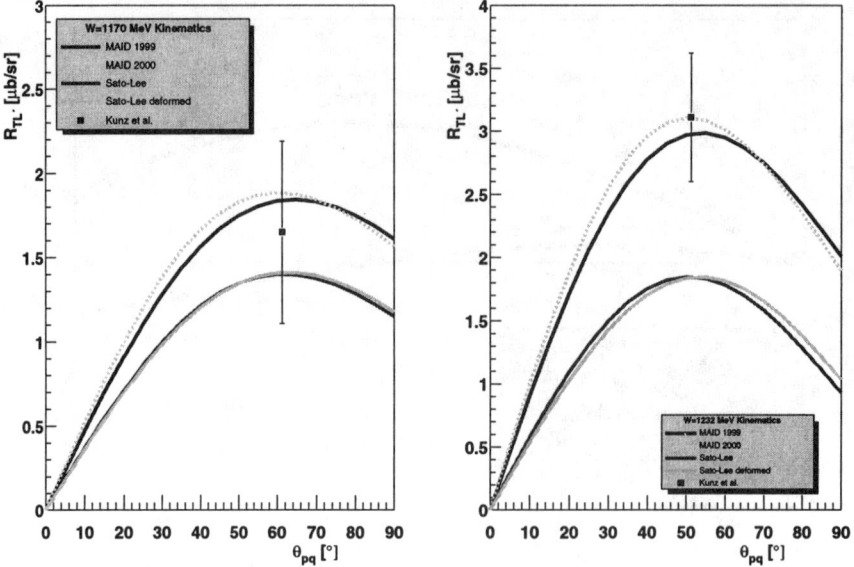

FIGURE 6. Preliminary result [29,28] of the imaginary part of the longitudinal-transverse (fifth) response at two values of invariant mass, as a function of the CM proton polar angle. The curves are theoretical calculations due to Refs. [8,15].

the systematic.

VIRTUAL COMPTON SCATTERING

As in all complex systems, extracting the polarizabilities (most commonly electric or magnetic) can provide rich information on the dynamical behavior of the system under investigation and though it better understanding of its structure. This has been the case in solid state, atomic and nuclear physics. In the study of hadronic systems, and in particular of the nucleon, the field has seen a tremendous growth in recent years. First, through the use of Real Compton Scattering (RCS) (using monochromatic or tagged photons) a thorough program at Illinois, Mainz and Saskatoon has provided us with precise data [32] on the electric and magnetic polarizabilities denoted as (α and β). In more recent years important developments both in theory and experiment have brought to maturity the use of Virtual Compton Scattering (VCS) in the study of the structure of the nucleon.

The use of virtual photons in the VCS process ($\gamma^* N \to \gamma N$) offers many advantages: the ability to study the process with varying wavelength and by taking advantage of the additional polarization state (longitudinal) offered by the virtual

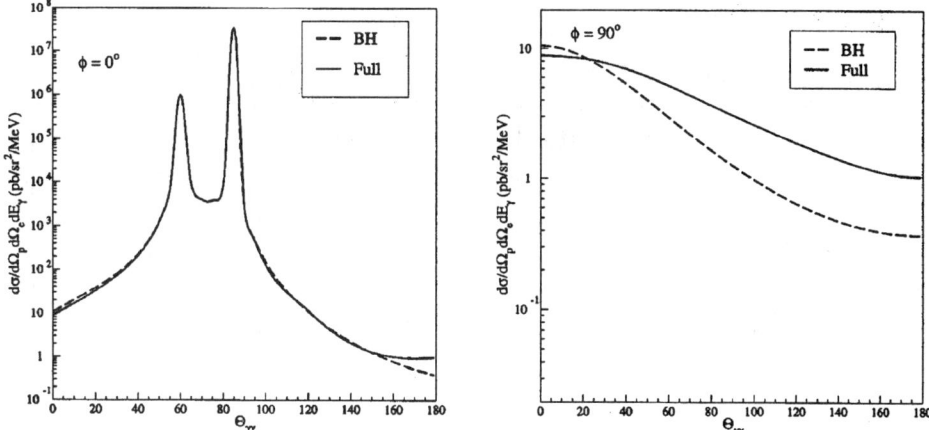

FIGURE 7. The VCS coincident cross section for Bates Kinematics ($q = 240$ MeV/c, $\theta_e = 90^0$). It can be seen that with in-plane kinematics, left frame, the BH amplitude overwhelms the cross -section. A far more favorable situation is achieved with out-of-plane kinematics, right frame.

photon. The VCS amplitude can then be analyzed in terms of structure coefficients that depend on the characteristic wavelength (q) of the process, the so called "Generalized Polarizabilities" (GPs). An impressive theoretical effort [33,34,37] has already demonstrated the very rich information that can be revealed through the study of GPs and their discriminating power among different nucleon models that attempt to describe the process.

However, the many benefits of VCS come at a heavy price. Experimentally the extraction of the physics of interest (i.e. of the GPs) is rendered almost impossible because of the Bethe - Heitler (B-H) process. The two processes, VCS and B-H, are coherent, and therefore they interfere; they are experimentally indistinguishable. In most kinematic sectors the B-H cross section dominates the VCS by several orders of magnitude. This can be seen in Fig. 7. It can be seen that in the usual experimental arrangement of in plane kinematics ($\phi = 0$), shown at the left frame of the figure, the B-H process overwhelms the cross section by several orders of magnitude, allowing a small window of opportunity at large scattering angles, away from the beam direction and angle of the scattered electron.

In a pioneering, tour de force experiment, the Saclay -Mainz collaboration (A1 -Collaboration) [35] has already demonstrated the feasibility of using VCS as a probe of nucleon structure. Actually using the very difficult in-plane kinematics (similar to those shown in Fig. 7) they were able to convincingly extract for the first time information on GPs. As it can be seen in Fig. 8 the information extracted convincingly shows that the data actually discriminate among the available theory pointing to a clear preference to Chiral perturbation Theory result.

The OOPS collaboration at Bates, has proposed [36] to pursue VCS measure-

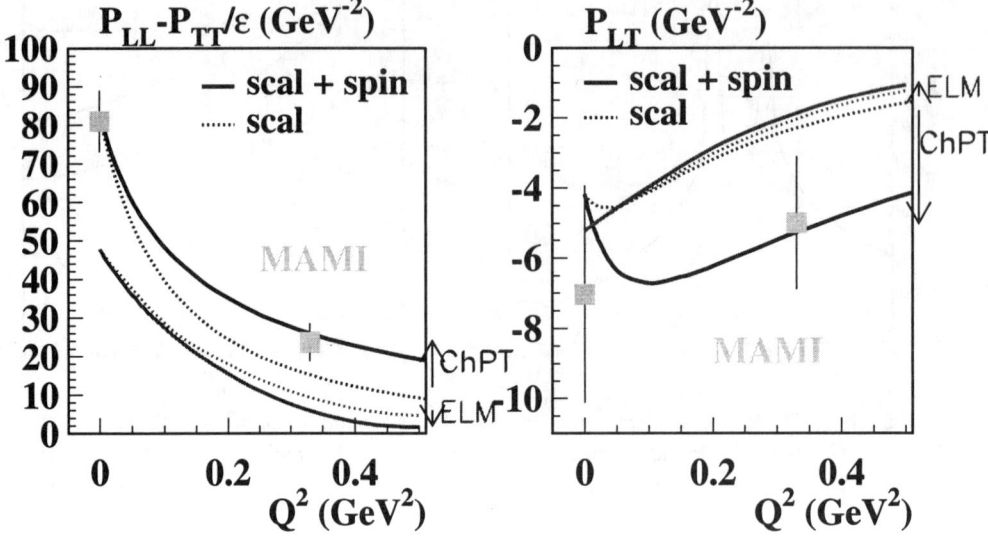

FIGURE 8. The VCS data from Mainz convincingly demonstrate the ability of this pioneering experiment to extract information on GPs and to discriminate among theoretical models. The data support the predictions of Chiral Perturbation Theory.

ments exploiting the unique out of plane capabilities of OOPS. In addition to the obvious advantage discussed above and shown in Fig. 7, of suppressing the B-H contribution by several orders of magnitude, out-of-plane detection offers the use of the out-of-plane angle (ϕ) as an added and very effective "knob' for the isolation of the GPs, in a similar fashion discussed in the case of the $N \to \Delta$ transition. In Fig. 9 the expected statistical accuracy of the planned measurements, which are expected to take place during next year, are shown together with the predictions of several theoretical approaches. As it can be seen the expected measurements should have no difficulty in distinguishing among the available theoretical predictions.

VCS IN THE RESONANCE REGION

As discussed earlier the motivation for theinvestigation of the $N \to \Delta$ in the $H(\vec{e},e'p)\gamma$ channel [24] is to complete the information on all decay channels of the $\Delta^+(1232)$. The use of Virtual Compton Scattering in the resonance region offers several distinct features, in comparison to the use of VCS for the determination of polarizabilities or to $(e,e\pi)$ for the study of nucleon resonances. The interference of the resonant amplitudes with those of Bethe Heitler can be used to enlarge the $\gamma^* N \to \Delta$ data base with different resonance-background interference observables,

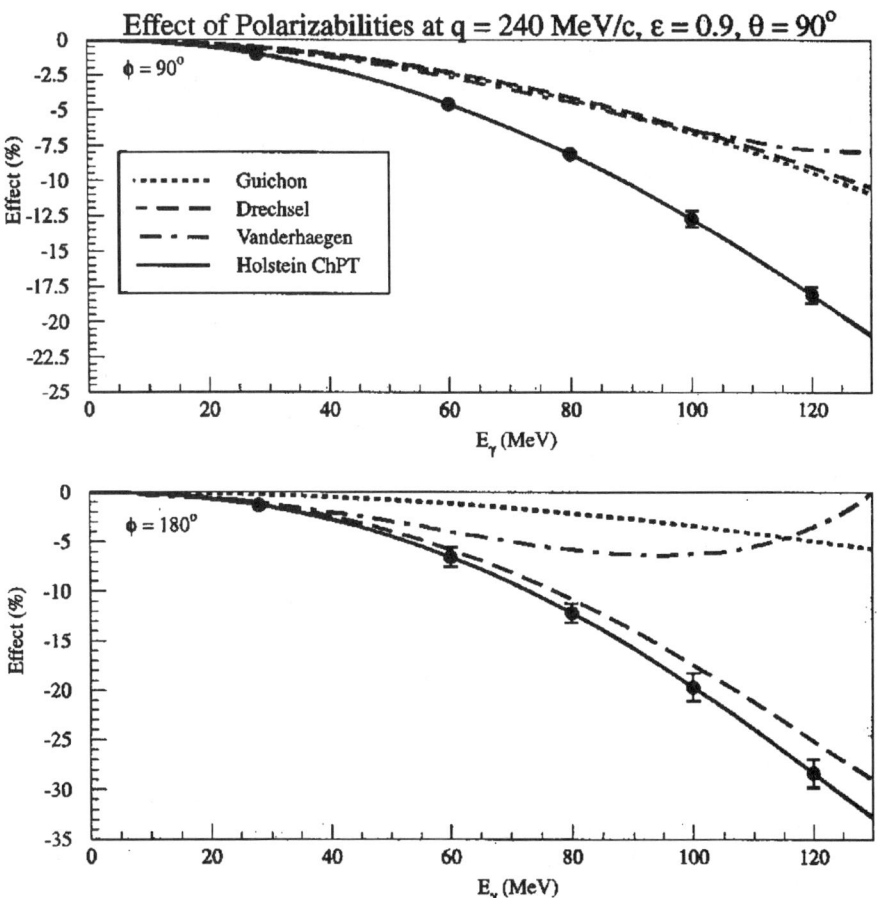

FIGURE 9. The kinematic parameters and expected statistical accuracy of the planned VCS measurements at Bates [36]. The anticipated measurements are arbitrarily placed on the chiral perturbation prediction [37]. The expected discriminatory power of the measurement is evident.

due to its purely electromagnetic nature. The characteristic shape of the resonant amplitude interfering with the featureless B-H amplitudes generates interesting oscillatory variations when explored as a function of invariant mass, W. This variations can in turn be used as signatures for exploring the resonance region. Two calculations for this reaction exist to date [38,39]. The one by Vanderhaeghen [39] predicts the helicity asymmetry to have an oscillation pattern, as a result of the resonance interfering with the Bethe-Heitler background. This distinct signal is of general interest in the study of other known resonances and potentially in the search for missing ones.

Although this measurement was first proposed a decade ago [22], it still remains a unexplored mainly due to the technical difficulties associated with the low cross section of this leptonic channel. At Bates this experiment will take place in the near future using high duty factor beam, which is absolutely necessary in order to suppress the background of accidental coincidences. It will be possible to measure the $p\gamma$ channel simultaneously with the $p\pi^0$ channel - due to the large momentum acceptance of the OOPS spectrometers - and to separate the two by reconstructing the missing mass spectrum. Similar investigations are under way at Jefferson, using the capabilities of Hall-A and CLAS.

CONCLUDING REMARKS

The refined experimental tools which are now well developed at the intermediate energy accelerator facilities, in terms of highly exclusive and precise coincident probes, are being used to provide important new insights in the structure of the nucleon.

The investigation of the long standing conjecture of nucleon deformation, through the electro-excitation of the first nucleon resonance, the $\Delta^+(1232)$ is being quantified through the isolation of quadrupole amplitudes in the $\gamma^* N \to \Delta$ transition. While the sensitivity of the new probes to such delicate effects has been demonstrated, the issue of model error in interpreting the data has emerged. The data that are now emerging have already have shown many of the shortcoming of the available phenomenology and have energized a theoretical effort to provide afar more sophisticated understanding of the issue.

The generalized polarizabilities of the nucleon studied through Virtual Compton Scattering (VCS) has been demonstrated in a most convincing way at the Mainz pioneering experiment. Intense experimental activity at Mainz, Bates and Jefferson, and parallel theoretical effort at many institutions indicates that this line of research promises to yield rich information in the coming years on the structure of the nucleon.

The Bates facility, renewed with fancy new instrumentation and an overhaul of its accelerator subsystems is continuing, as it has in the past twenty five years, to play a leading role this field. The scheduled availability of extracted polarized cw beams next years, will insure that this will continue to be the case for many years to come.

I am indebted to T. Botto, R. Gothe, N. dHose, J. Shaw, H. Schmieden, S. Stiliaris and C. Vellidis who have provided me with suggestions, critisism and in some cases their preliminary results shown in this presentation Partial support by the US Department of Energy, the Greek General Secretariat for Research and Technology and the Ministry of Education are gratefully acknowledged.

REFERENCES

1. S.L. Glashow, *Physica* **96A**, 27 (1979).
2. N. Isgur, G. Karl and R. Koniuk, *Phys. Rev.* **D25**, 2394 (1982).
3. S. Capstick and G. Karl, *Phys. Rev.* **D41**, 2767 (1990).
4. R.L. Crawford, *Nucl. Phys.* **B28**, 573 (1971).
5. R. Siddle et al., *Nucl. Phys.* **B35**, 93 (1971).
6. J.C. Alder et al., *Nucl. Phys.* **B46**, 573 (1972).
7. K. Batzner et al., *Nucl. Phys.* **B76**, 1 (1974).
8. T. Sato and T.-S.H. Lee, *Phys. Rev.* **C54**, 2660 (1996).
9. A. J. Buchmann, E. Hernández, and Amand Faessler, Phys. Rev. **C55**, 448 (1997).
10. G. Blanpied et al., *Phys. Rev.* **79**, 4337 (1997).
11. R. Beck et al., *Phys. Rev.* **78**, 606 (1997).
12. V. V. Frolov et al., Phys. Rev. Lett. **82**, 45 (1999).
13. C. N. Papanicolas, in *Topical Workshop on Excited Baryons*, edited by G. Adams, N. C. Mukhopadhyay and P. Stoler (World Scientific, Singapore, 1989).
14. J. M. Laget, *Nucl. Phys.* **A 481**, 765 (1988).
15. D. Drechsel et al., *Nucl. Phys.* **A645**, 145 (1999).
16. R. M. Davidson and N. C. Mukhopadhyay, Phys. Lett. **B353**, 131 (1995) and R. M. Davidson, private communication.
17. G.A. Warren et al., *Phys. Rev.* **C58**, 3722 (1998).
18. A. Bernstein et al., *Phys. Rev.* **C47**, 1274 (1993).
19. H. Schmieden, Proc. PANIC'99, Uppsala 1999 (nucl-ex/9909006) and private communication.
20. F. Kalleicher et al., *Z. Phys.* **A359**, 201 (1997).
21. A.S. Raskin and T.W. Donnelly, *Ann. Phys.* **191**, 78 (1989).
22. Bates Proposal 87-09, spokesman: C.N. Papanicolas (1987).
23. Bates Proposal 97-04, spokesmen: M.O. Distler, A.M. Bernstein (1997).
24. Bates Proposal 97-05, spokesmen: N.I. Kaloskamis, C.N. Papanicolas (1997).
25. S. Dolfini et al., *Nucl. Inst. Meth.* **A344**, 571 (1994).
26. J. Mandeville et al., *Nucl. Inst. Meth.* **A344**, 583 (1994).
27. C. Mertz et al., nucl-ex/9902012 (1999).
28. C. Kunz, PhD-Thesis, M.I.T. (2000).
29. N.I. Kaloskamis et al., to be published.
30. C. Vellidis, PhD-Thesis, under preparation, University of Athens (2000).
31. S. Mehrotra et al., *Nucl. Phys.* **A362**, 461 (1981).
32. B.E. MacGibbon et al., *Phys. Rev.* **C52**, 2097(1995)
33. P.A.M. Guichon, G.Q. Liu and A.W. Thomas, *Nucl. Phys.* **A591**, 606(1995)
34. D. Drechsel et al *Phys. Rev.* **C55**, 424 (1997); *Phys. Rev.* **C57**, 941 (1998).
35. N. D'Hose, private communication and J. Roche et al to be published.
36. Bates Proposal 97-03, J. Shaw, R. Miskimen spokesmen; private communication.
37. B. Holstein, contribution to this volume.
38. S. Kumano, *Nucl. Phys.* **A495**, 611 (1989).
39. M. Vanderhaeghen. Private communication and *Nucl. Phys.* **A595**, 219 (1995).

Electromagnetic Pion Production: From Yukawa to Goldstone

A. M. Bernstein

Physics Department and Laboratory for Nuclear Science
MIT, Cambridge, MA, USA

Abstract. The evolution of electromagnetic pion production is presented. The early experiments verified that the reaction mechanism of pion production in nuclei was understood; some open issues which can now be studied with CW electron beams are discussed. Emphasis is given to the present studies which measure meson production from the nucleon. The impact of spontaneously broken chiral symmetry and Goldstone's theorem on pion-nucleon physics is sketched. Present and future experiments to measure quark mass effects such as isospin breaking are described. The role of the new generation of beam and target polarization experiments are stressed, including such future possibilities at Bates as virtual photon tagging.

INTRODUCTION

In this paper I will sketch the evolution of electromagnetic pion production from approximately 25 years ago, when the main idea was its use as a probe of nuclear dynamics, to the present when we are testing effective field theory calculations in confinement scale QCD. In this time our understanding of the primary function of the pion has evolved considerably. The original idea of Yukawa was that the as yet undiscovered pion is the carrier of the long range part of the nucleon-nucleon interaction. Now we realize that the pion is a Goldstone Boson which is a manifestation of the spontaneously broken chiral symmetry of the QCD vacuum [1–3]. In this picture the pion would be massless in the chiral limit, where the masses of the up and down quarks are zero. The small mass of the pion is due to the non zero quark masses, $m_u \simeq 5$ MeV, $m_d \simeq 9$ MeV [4]. This view of the pion deepens the original Yukawa idea. Indeed a derivation (even approximate) of the pion mass, and of the πN coupling constant through the Goldberger-Treiman relation [1], constitutes a derivation of the long range part of the nucleon-nucleon potential.

Throughout the evolution from Yukawa to Goldstone, the processes involving pion production and interactions have remained central to probing the nature of the nucleon and complex nuclei. I will briefly trace this evolution to explain the shift in motivation of experiments and in their interpretation, with emphasis on the present status of the field. The development of this field involved the active participation of all of the major laboratories in the field: Bates, Saclay, Saskatoon, Mainz, and Tohuku, and has been

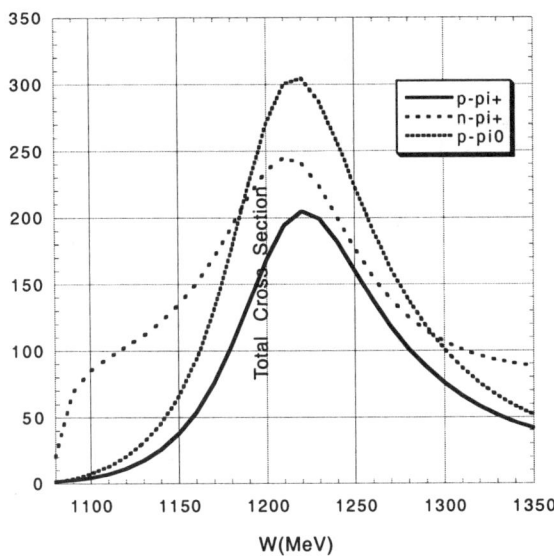

FIGURE 1. Total cross sections versus the CM energy W for $\pi^+ p$ elastic scattering in mb (solid line), the $\gamma p \to \pi^0 p$ reaction in μb (dotted line), and the $\gamma p \to \pi^+ n$ reaction in μb (dashed line).

extensively reviewed [5–7]. Here, to celebrate the occasion of the 25th year of Bates operation, its role will be presented primarily with emphasis on the future promise of polarized, internal target experiments.

Some of these ideas can be illustrated by considering the broad features of the $\pi^+ p$ and $\gamma p \to \pi^0 p, \pi^+ n$ reaction total cross sections in Fig. 1. The prominent Δ peak near W=total center of mass energy $\simeq$ 1232 MeV can be seen; this, and the differential cross section, data led to the identification of the Δ as a spin = isospin = 3/2 state [8]. We recognize the fact that p wave pions are coupled strongly to nucleons since the Δ resonance saturates the unitary bound, $\sigma(\pi^+ p) = 4\pi \lambdabar^2$. By contrast, in the s wave the interaction is weak as can be seen from the small $\pi^+ p$ cross section near threshold; this is a consequence of the fact that Goldstone Bosons do not interact at low energies [1–3]. The fact that the Δ peak seen in the $\pi^+ p$ and $\gamma p \to \pi^0 p$ total cross sections (Fig. 1) is so simple is due to the fact that for both cases the threshold cross section goes to zero in the chiral limit (and is therefore very small at low energies). This is in contrast to π^+ photoproduction where the low energy s wave, production amplitude is large, as can be seen in Fig. 1. Indeed, the first low energy theorem for charged pion photoproduction was proved by Kroll and Ruderman [9] and is in good agreement with experiment. This large electric dipole amplitude changes the shape of the $\gamma p \to \pi^+ n$ reaction total cross section considerably so that the Δ peak is not quite so clearly seen.

Twenty five years ago the broad facts of the πN interaction (ie the strong interaction in p wave and the weak interaction in the s wave) were well known. This is encoded in the phenomenological static model [10] which has gradient coupling of the pion field. We now have a much deeper understanding in terms of the general properties of Goldstone

Bosons. The gradient coupling is a signature of Goldstone's theorem [1–3]. The static model can now be derived [11] as the leading term in Chiral Perturbation Theory(ChPT) [1,12], which is an effective low energy representation of QCD. Indeed ChPT has been used to successfully calculate many observables in πN and $\gamma N \to \pi N$ reactions [13]. For a review of the status of this field see [14].

PHOTOPION NUCLEAR PHYSICS AT BATES

The first experiment performed at Bates in 1974 was on the threshold $^{12}C(\gamma, \pi^-)^{12}N$ reaction [15]. The immediate motivation was to study the effect of the Coulomb field on threshold photopion production. The general effect is to reduce(increase) the cross section for $\pi^+(\pi^-)$ production since the pion wave function is reduced (enhanced) in the nuclear volume [16]. The effect of the Coulomb field, an abrupt rise in the cross section at threshold, was clearly seen in the $^{12}C(\gamma, \pi^-)^{12}N$ reaction data [15]. At that time it was recognized that although the p wave part of the pion-nucleus optical potential could be calculated from first principles, the s wave part could not. This is probably related to the relative strengths of these interactions in the pion-nucleon interaction. It was our motivation that any reaction which enhanced the role of the s wave pion-nucleus interaction would be a good test of our understanding. We were able to conclude from the data that within the uncertainties of the reaction calculations, our understanding is at least qualitatively correct [15].

The experimental pion photoproduction program at Bates centered on understanding the reaction mechanism by testing the calculations. For this purpose we constructed two charged and one neutral pion spectrometers. The first charged pion spectrometer was fixed at $90°$ in the lab. in the $14°$ area and later we built MEPS (Medium Energy Pion Spectrometer) designed primarily by Ingvar Blomqvist. We succeeded in demonstrating that the dynamics of the photopion reaction were reasonably well described by the distorted wave impulse approximation as long as the nuclear wave functions were obtained empirically from magnetic electron scattering and the pion optical potential was obtained from fits to pion-nucleus scattering data. These reaction calculations are particularly accurate in the low and medium energy regions in situations where the pion mean free path is long and therefore the effects of absorption are not large. For charged pions, the dominant production mechanism is the theoretically well determined nucleon spin-flip, Kroll-Ruderman, term $\vec{\sigma} \cdot \vec{\epsilon}$ where $\vec{\sigma}$ is the nucleon spin and $\vec{\epsilon}$ is the photon polarization. This has been experimentally verified in the strong $0^+ \leftrightarrow 1^+$ spin flip (Gamow- Teller) transitions in the $^6Li(\gamma, \pi^+)^6He$ and $^{12}C(\gamma, \pi^+)^{12}B$ reactions [17]. For weaker spin-flip transitions, the other terms in the interaction Hamiltonian are more important. An interesting example of this can be seen in the cross sections for the severely hindered Gamow-Teller transition in the $^{14}N(\gamma, \pi^+)^{14}N$ reaction, measured using MEPS at Bates [18] and shown in Fig. 2. The sensitivity to the different terms in the interaction Hamiltonian [19] is demonstrated.

In the Δ region the distorted wave impulse approximation is not theoretically well based even though it can be used to fit the data. From a first principles point of view the

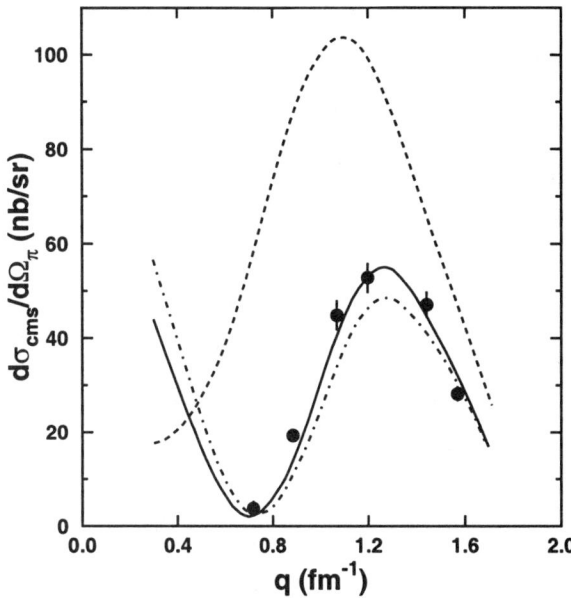

FIGURE 2. Differential cross section for the $^{14}N(\gamma, \pi^+)^{14}C$ reaction at E_γ=200 MeV compared to DWIA calculations using different forms of the production operator. The solid curve uses the full Blomqvist-Laget operator. The curve drops the s channel Δ contribution. The dashed curve only uses the Kroll-Ruderman term. For references see the text.

Δ-hole model [20] is preferable. This has been demonstrated in experiments at Bates for the coherent $^4He(\gamma, \pi^0)$ [21] and $^{16}O(\gamma, \pi^- p)$ [22] reactions. The results for the coherent reaction on ^4He are shown in Fig. 3 where the good agreement between the data and the Δ-hole calculations can be seen.

This initial Bates paper [15] stated the motivation to start the study of photo-pion physics as "Photomeson production in complex nuclei can be used as a probe of the nuclear mesonic field.... However due to experimental difficulties only a few experiments have been performed in which transitions to discrete nuclear states have been observed." Our program at Bates focussed on testing the calculations for this process in order to later use the charged photo-pion production reaction to study nuclear structure and the pion field. We never actually achieved these goals because we did not have the experimental tools to accomplish it. What we lacked was the high duty cycle electron beams which we now have. In fact a measurement of the predicted pion field in nuclei [23] is still a fundamental open issue in experimental nuclear physics. An early attempt with the low duty cycle electron accelerator at Saclay was performed [24] and a more current effort is being pursued at Jefferson Lab [25].

Sufficient work was done on photopion reactions in nuclei for its potential to be apparent. For example for the production of low energy charged pions the reaction proceeds primarily through the nucleon spin-flip Kroll-Ruderman term. This could be exploited to measure Gamow-Teller matrix elements in nuclei which are not only of intrinsic interest,

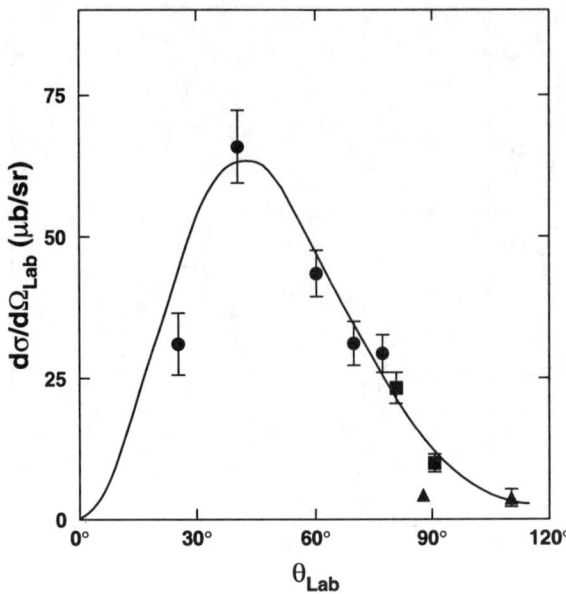

FIGURE 3. Differential cross section for the $^4He(\gamma,\pi^0)^4He$ reaction at E_γ=290 MeV compared to a Δ-hole model calculation. For references see the text.

but which bear on other fields such as neutrino detection and double beta decay. With high resolution tagged photon facilities these are subjects that can be studied in the future.

MEASUREMENTS OF THE THRESHOLD $\gamma P \rightarrow \pi^0 P$ REACTION AND LOW ENERGY THEOREMS

Electromagnetic pion production from the nucleon has seen major advances in the last decade. The excitement started when two experiments [26] claimed that the low energy theorems [27] based on current algebra and PCAC were incorrect (see [28,29] for reviews) and ended with improved experiments [30,31] and corrected theorems [28,13] based on ChPT.

In the 1930's Heisenberg started a long history of calculations of low energy phenomena by effective field theories with the scattering of light by atoms for energies which are low compared to the electron mass. Other well known examples are the low energy limit of Compton scattering [32] and charged pion photoproduction from the nucleon [9] (the Kroll-Ruderman theorem mentioned above). In analogy with gauge invariance in electrodynamics the four divergence of the axial field is proportional to m_π^2 which vanishes in the chiral limit (ie where the light quark masses and therefore m_π go to zero). Using this method, important results were derived such as the s wave $\pi - \pi$ and $\pi - N$ scattering lengths [33], and the low energy theorems for pion photoproduction [27].

However one should note that "the derivation of non-leading terms in the days of current algebra was more of an art than a science, often involving dangerous procedures like off shell extrapolations of amplitudes. The modern developments in this field have replaced the old notions by the effective field theory(EFT) of the standard model (SM), including the spontaneously broken chiral symmetry. This framework allows for a systematic expansion of amplitudes... in terms of momenta and meson masses. One recovers all of the old low energy theorems that are rightfully called theorems, but does not reproduce some of the old results that were based on unjustified assumptions not valid in the SM" [28]. The new calculations include the chiral corrections to the Kroll-Ruderman theorem and most important, low energy theorems for the p wave multipoles [13]. The upshot of all of this is that with the improved experiments and theory there is now good agreement as summarized in the Mainz workshop on chiral dynamics [14].

As an example the old low energy theorems predicted a value for the electric dipole amplitude $E_{0+}(\gamma p \rightarrow \pi^0 p) = -2.3$ [27], while the new ChPT theory calculations fit this value as a low energy constant (although reasonable estimates of this constant give a value which is consistent with experiment) [13,28], and experiment gives a value of -1.3 ± 0.2 [30,31](see Fig.5 and the discussion below). These are all in units of $10^{-3}/m_\pi$; this small amplitude ($\simeq$ milli-fermi's) indicates how technically demanding these experiments are. The extraction of the s wave multipole amplitude is not trivial since the p wave dominates relatively close to threshold. This is shown in Fig.4 for the $\gamma p \rightarrow \pi^0 p$ reaction at E_γ =158 MeV [30]. Although the s wave contribution to the differential cross section is negligible it is still possible to measure ReE_{0+} by the s-p interference (B) term in the differential cross section

$$\sigma(\theta) = (q/k)[A + Bcos\theta + Ccos^2\theta] \quad (1)$$

where q and k are the pion and photon center of mass momenta, and θ is the pion CM angle. The measured values of ReE_{0+} are shown in Fig.5. There is a relatively small disagreement between the Mainz [30] and Saskatoon [31] results; the errors shown in the figure are statistical only.

For the threshold $\gamma N \rightarrow \pi N$ reaction there are five amplitudes to measure: the real and imaginary part of the s wave multipole E_{0+} and the three p wave multipoles (these amplitudes are real close to threshold since the p wave πN phase shifts are very small). The unpolarized cross section measures three of these, Re E_{0+} and two linear combinations of the three p wave multipoles; these amplitudes are all in good agreement with ChPT [30,31]. However it must be pointed out that ChPT has three low energy constants which are fit to the data. Nevertheless the agreement is significant since the energy dependence of E_{0+} has been predicted without any parameters. For the p wave there is only one free parameter and three amplitudes. We have performed an experiment at Mainz with linearly polarized photons which, combined with the unpolarized cross section, will measure all three p wave amplitudes. This will a more stringent test of the p wave low energy theorems of ChPT [13]. This experiment should also be more accurate for the unpolarized cross section and resolve the small discrepancy between the older Mainz and Saskatoon data sets. To finish the complete determination of the five

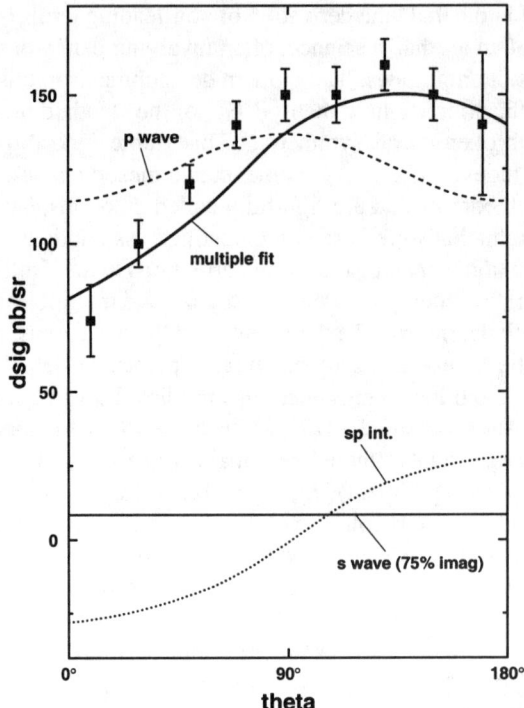

FIGURE 4. Differential cross section for the $\gamma p \to \pi^0 p$ reaction versus pion center of mass angle for E_γ=158 MeV. The errors are statistical only. The curves show the contributions of the s and p waves. Mainz data, see text for references.

multipole amplitudes will require an experiment with a polarized target. I shall return to this topic in the next section.

THE UNITARY CUSP IN THE $\gamma P \to \pi^0 P$ REACTION AND πN SCATTERING

The rapid energy dependence for Re E_{0+} exhibited in Fig.5 between the $\pi^0 p$ and $\pi^+ n$ thresholds at 144.68 and 151.44 MeV respectively is a unitary cusp. This is due to the interference of two flux paths, the direct $\gamma p \to \pi^0 p$ and two step $\gamma p \to \pi^+ n \to \pi^0 p$ reactions illustrated in Fig. 6. The formula for the energy dependence is [34]:

$$E_{0+}(\gamma p \to \pi^0 p) = e^{i\delta_0}[A_0 + i\beta q_+]$$
$$\beta = E_{0+}(\gamma p \to \pi^+ n) a_{cex}(\pi^+ n \to \pi^0 p) \qquad (2)$$

where δ_0 is the s wave $\pi^0 p$ phase shift, A_0 is a smooth function of the photon energy which equals $E_{0+}(\gamma p \to \pi^0 p)$ in the absence of the final state interaction, and q_+ is the

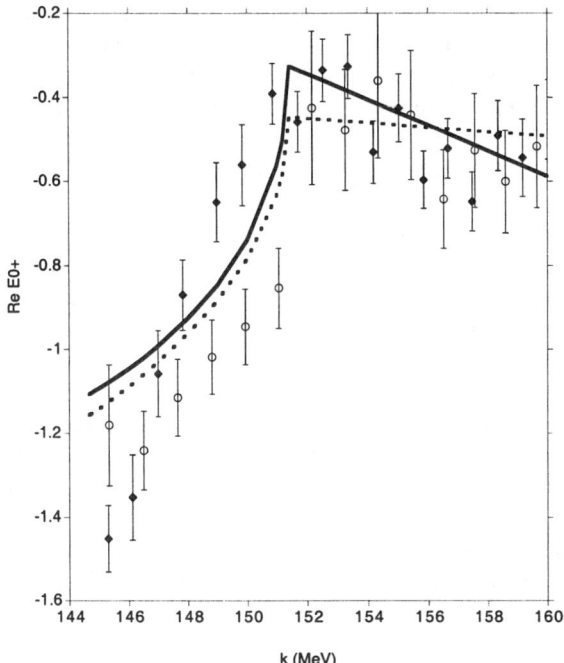

FIGURE 5. ReE_{0+} (in units of $10^{-3}/m_\pi$) for the $\gamma p \to \pi^0 p$ reaction versus photon energy k. The dashed dot curve is the ChPT fit [13] and the solid curve is the unitary fit. The solid diamonds (circles) are the Mainz (Saskatoon) points. The errors are statistical only. See text for references.

π^+ center of mass momentum in units of m_{π^+}. For photon energies $k_\gamma < k_T(\pi^+ n)$, the $\pi^+ n$ threshold energy, one must analytically continue $q_+ \to i|q_+|$. This switching of the amplitude from real to imaginary as the photon energy increases from below to above the secondary ($\pi^+ n$) threshold is a characteristic of a unitary cusp. For the specific form of Eq. 2 mild approximations have been made that the $\pi^0 p$ is much weaker than the $\pi^+ n$ channel, and that the s wave charge exchange phase is small [34]. If they are needed, more exact formulas for the cusp have been presented [34].

The rapid energy dependence of the unitary cusp can be considered to be analogous to a two slit interference pattern where one observes the interference between two flux paths,

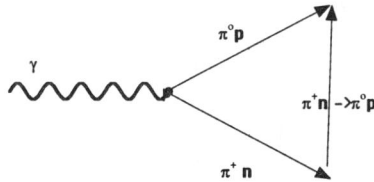

FIGURE 6. Schematic representation of the two flux paths in the $\gamma p \to \pi^0 p$ reaction.

in this case the one and two step reaction paths. From the data the value of $\beta = 3.3 \pm 0.5$ has been obtained [30,35]. The positive sign for β corresponds to the fact that the data lie below the smooth curves that would be obtained from the region above the $\pi^+ n$ threshold. If the data were to lie above that line then β would be negative. Thus, as in all interference phenomena, the sign of the interference amplitude is an observable.

It is interesting to note that both the power counting rules of ChPT and the constraints of unitarity lead to pion rescattering in the final state as a critical dynamical ingredient. Although the unitary and ChPT curves both agree with the data for $Re E_{0+}$ (Fig. 5) there is an important difference between them. The value of $\beta = 2.78$ [13,35] calculated for the $\gamma p \rightarrow \pi^0 p$ reaction is smaller then the one calculated using the separately predicted values of $E_{0+}(\gamma p \rightarrow \pi^+ n)$ and $a_{cex}(\pi^+ n \rightarrow \pi^0 p)$ of 3.51 ± 0.22 [34] This difference can be clearly seen in $Im E_{0+}$ (Fig. 7). The reason for this discrepancy, which was discussed by the ChPT authors [13], is due to the fact that the ChPT calculation is carried out to one loop which is not sufficient for $Im E_{0+}$. This is a general feature of ChPT in which the imaginary part of the amplitude is not calculated as accurately as the real part, and thus unitarity is only approximately satisfied at a given order. As will be discussed below, the difference between the unitary and one loop ChPT values of β can be observed in future experiments in which $Im E_{0+}$ will be measured directly.

The occurrence of the unitary cusp is isospin violating since it requires that the two thresholds not be degenerate (as required by isospin symmetry). Furthermore, as will discussed in the following section, a_{cex} is predicted to be isospin violating due to the mass difference of the up and down quarks [40,42,43]. This means that β is also isospin violating [34].

To accurately measure the magnitude of the unitary cusp and to exploit the connection between electromagnetic pion production and low energy πN interactions, one must measure $Im E_{0+}$. In photoproduction this requires experiments with polarized beams and/or targets [34]. To briefly demonstrate the power of polarized photo-pion experiments, two asymmetries are shown in Fig.9: Σ for linearly polarized photons with an unpolarized target; and T for unpolarized photons but with a target polarized normal to the reaction plane. The results presented in Fig. 9 use the p wave predictions of chiral perturbation theory [13] and the unitary fit to E_{0+} discussed above [30]. Σ is primarily sensitive to the p wave multipoles and since the unitary fit has essentially the same p wave multipoles as ChPT, the curves for ChPT and the unitary fit are almost identical. By contrast, T is sensitive to a linear combination of p wave multipoles times $Im E_{0+}$, and shows its rapid rise above the $\pi^+ n$ threshold. For T, the large difference in $Im E_{0+}$ between the unitary fit [30] and ChPT to one loop [13] should be straightforward to distinguish experimentally. One method to measure this, utilizing small angle electron scattering with polarized, internal, targets in a storage ring facility [36] will be discussed in the next section. Other possibilities for this measurement include the use of a laser backscattering source [37] or using an active, polarized target in a conventional tagged photon beam [38].

Finally I would like to point out that a measurement of β (Eq. 2) is equivalent to a measurement of $a_{cex}(\pi^+ n \rightarrow \pi^0 p)$. This demonstrates the fact that through the final state interaction one can perform πN scattering measurements with electromagnetic probes; the final states that are reached this way cannot be accessed with conventional

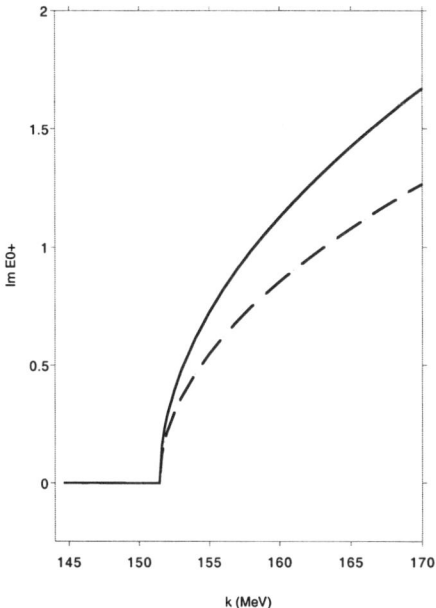

FIGURE 7. ImE_{0+} (in units of $10^{-3}/m_\pi$) for the $\gamma p \to \pi^0 p$ reaction versus photon energy k. The dashed dot curve is the ChPT calculation and the solid curve is the unitary calculation).

pion beams and targets. One can compare the value of $a_{cex}(\pi^+ n \to \pi^0 p)$ and compare it to the accurately measured value of $a_{cex}(\pi^- p \to \pi^0 n)$ [39], obtained from the width of the 1s state in pionic hydrogen, to see if there is any deviation from the pure isospin prediction that these values are equal. Predicted deviations from isospin conservation will be discussed below.

ISOSPIN VIOLATION AND QUARK MASS DIFFERENCES

Weinberg first showed that there is an isospin violating effect in the s wave πN scattering length $a(\pi N)$ [40] due to the up, down quark mass difference ($m_u \simeq 5 MeV, m_d \simeq 9 MeV$) [4]. This magnitude of this predicted effect, which occurs for $\pi^0 N$ scattering or charge exchange but not for $\pi^\pm N$ scattering, was estimated originally using Chiral Lagrangians [42] and more recently calculated by ChPT [43]. The estimated magnitude of this effect is the same (to within a factor of $\sqrt{2}$) in $\pi^0 N$ scattering and charge exchange reactions. However, since the magnitude of $a(\pi^0 N)$ is small, the relative magnitude of the isospin violating term is $\simeq 30\%$. By contrast, for charge exchange where the isospin conserving amplitude is larger, the relative isospin violation is estimated to be $\simeq 2$ to 3%. These fundamental predictions have never been experimentally tested. Indeed, as was discussed in the previous section, πN scattering lengths are hard to measure.

There have been two empirical analyses of πN scattering and charge exchange reactions at intermediate energies (pion lab kinetic energies between 30 and 70 MeV) that have

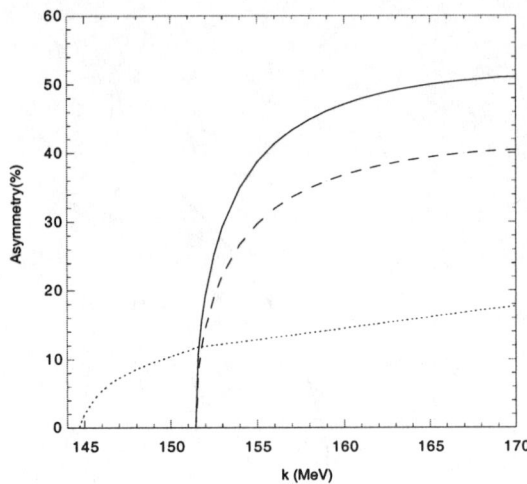

FIGURE 8. The polarized photon (Σ) and polarized target (T) asymmetries (in %) at a π^0 center of mass angle of 90° for the $\gamma p \to \pi^0 p$ reaction versus photon energy k. The lower curve is the ChPT calculation for Σ. The two curves for T which rise rapidly at the $\pi^+ n$ threshold at 151.4 MeV are the ChPT calculation (dashed curve) and the unitary calculation (solid curve).

concluded that isospin is not conserved. In particular they tested the triangle relation:

$$D \equiv f(\pi^- p \to \pi^0 n) - \frac{(f(\pi^+ p) - f(\pi^- p))}{\sqrt{2}} \tag{3}$$

where D is the difference between the $\pi^- p \to \pi^0 n$ scattering amplitude as observed experimentally and what would have been obtained from analyzing $\pi^+ p$ and $\pi^- p$ elastic scattering and isospin conservation. Both analyses [44,45] have obtained values of $D \simeq -.012 \pm 0.003 fm$ or $D/f_{exp}(\pi^- p \to \pi^0 n) \simeq 7\%$ This appears to be a significantly larger effect than was predicted at lower energies due to the up, down quark mass differences and electromagnetic effects [33,42,43] calculated in ChPT. Due to this relatively large isospin breaking effect, there was concern about the accuracy of the Coulomb effects used in these analyses. Subsequently, new, more accurate Coulomb calculations were performed [46] which do not significantly change the original result Therefore, to the extent that the πN scattering and charge exchange data base is accurate, a significant isospin violation has been observed. This is of sufficient importance that further experiments and analyses need to be performed to either verify or alter this conclusion.

Isospin violation in the πN system will also show up in electromagnetic pion production [47] as an observable violation of the Fermi-Watson theorem [48]; this is the relationship between the phase shifts in πN scattering and the phases of the electromagnetic pion production amplitudes. This theorem is based on time reversal invariance, unitarity, and isospin conservation. When this theorem was derived, quarks were not known. It was assumed that isospin violation was caused only by electromagnetic effects.

The Fermi-Watson theorem has been generalized to include isospin breaking but time reversal invariance and unitarity are still assumed [47]. Writing the multipole amplitudes for the photo- or electro-production of the πN channels in the isospin states I =1/2 and 3/2 as iM_{2I} one obtains [47]:

$$M_1 = e^{i\delta_1}[A_1 \cos \frac{\psi}{2} + iA_3 \sin \frac{\psi}{2}]$$
$$M_3 = e^{i\delta_3}[A_3 \cos \frac{\psi}{2} + iA_1 \sin \frac{\psi}{2}] \quad (4)$$

where $A_{2I=1,3}$ are real functions of the CM energy which can be identified as the multipole matrix elements M_{2I} in the absence of final state interactions and isospin breaking, δ_1, and δ_3, represent elastic πN scattering phase shifts in the I = 1/2 and 3/2 states, $\sin\psi$ represents the isospin violating term where ψ is a real number. Eq. 4 shows the violation of the Fermi-Watson theorem due to isospin breaking since the two isospin amplitudes are coupled. For $\psi \to 0$ the isospin violation vanishes and the Fermi-Watson theorem is recovered. Therefore the phases of the πN multipoles should be measured, and not calculated from the πN phase shifts (using the Fermi-Watson theorem) as they now are. To my knowledge only one such measurement of the phases of the photoproduction multipoles has been performed [49] and that did not have the required precision to determine the effects predicted here.

An isospin sensitive quantity is the polarized target (normal to the reaction plane) asymmetry: this is presented for the $ep \to e'\pi^+ n$ reaction at $Q^2 = 0.1 GeV^2$ for a center of mass energy W = 1120 MeV (pion lab kinetic energy $\simeq 40 MeV$) in Fig. 9. The calculation employed the isospin conserving amplitudes of the Mainz unitary model [50] and then coupled the pure isospin states using $\psi \simeq -0.01$ which was obtained [47] from the isospin violation found in πN scattering [44,45]. This asymmetry shows a significant effect due to isospin breaking, This effect is very similar at the photon point. Other isospin sensitive quantities include $A_{TL}(t)$ for the $ep \to e'\pi^+ n$ reaction and $A_{TL'}$ for the $ep \to e'\pi^0 p$ reaction.

Thus we have demonstrated the breakdown of the Fermi-Watson theorem due to isospin breaking effects in the πN System. The equations presented here are based on time reversal invariance and unitarity and are therefore very general. There are some significant, experimentally observable effects which can be used to make an independent determination of the magnitude of the isospin breaking utilizing a different reaction to observe this effect. This is of particular interest since the $\gamma^* p \to \pi^+ n, \pi^0 p$ reaction leads to charge states which are not accessible with conventional πN reactions. These isospin violating effects should occur in the intermediate energy region but also in the threshold region as was discussed in the previous section. In the next section a novel method is presented to make such measurements.

FUTURE POLARIZED INTERNAL TARGET EXPERIMENTS

In this section I want to point out new and fundamental experimental possibilities which will be made possible by detecting the small angle electron scattering from thin, pure, polarized, targets in a storage ring. This facility would utilize the polarized internal targets being built for BLAST in the Bates ring. The outgoing hadrons could either be detected

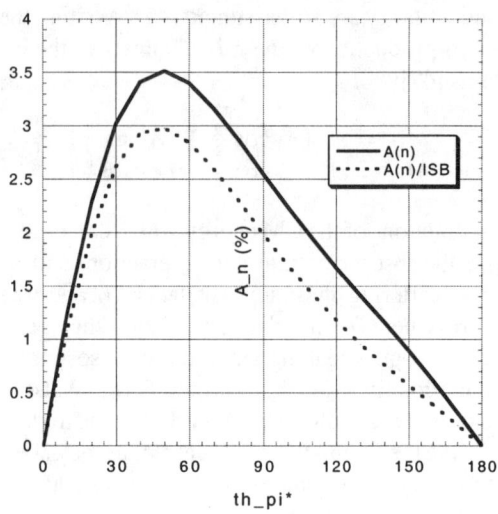

FIGURE 9. The polarized target (normal to the reaction plane) asymmetry for the $ep \to e'\pi^+ n$ reaction at $Q^2 = 0.1 GeV^2$ for a center of mass energy W = 1120 MeV (pion lab kinetic energy $\simeq 40 MeV$).

in BLAST or in special purpose detectors such as silicon strips. This would enable a complete set of experiments to be performed on the $\vec{\gamma}\vec{p} \to \pi^0 p$ reaction discussed above, in the threshold and intermediate energy regions. The possibilities also include the study of the quadrupole transitions in the $\vec{\gamma}\vec{p} \to \Delta$ transition. Space does not allow detailed discussion of many other possibilities. One, which is of great interest in studying the chiral structure of the nucleon, is the polarized Compton scattering $\vec{\gamma}\vec{p} \to \gamma p$ reaction. It should also be noted that such a facility would be capable of measuring all photo-hadron processes with polarized proton, deuteron, and ^3He targets and polarized photons. In particular the coherent $\vec{\gamma}\vec{D} \to \pi^0 D$ reaction can be accessed from threshold through the Δ region and could possibly produce important new results on the $\vec{\gamma}\vec{n} \to \pi^0 n$ amplitude.

The method of small angle electron scattering or virtual photon tagging has been proposed many years ago [51] and more recently has been proposed for use in the $\gamma p \to K^+ \Lambda$ reaction at CEBAF [52]. What is new about the present proposal is the utilization of full polarization observables for both the photon and target. The target polarization is made possible by the use of internal targets. For the photon one obtains circular polarization from longitudinally polarized electrons. The equivalent response function for linear polarized photons [53] is obtained by measuring the ϕ dependence of the cross section. Therefore this virtual tagging proposal goes much further in utilizing the polarized and windowless nature of internal targets.

A proposed method to perform small angle electron scattering in the Bates Storage Ring is presented in Fig.10 [54]. The change in the beam transport system in the vicinity of the target is shown. The crucial point is the magnetic chicane downstream from the target. For electrons which do not interact with the target the beam will be bent away and then returned to its original trajectory. For an electron which has lost energy and is

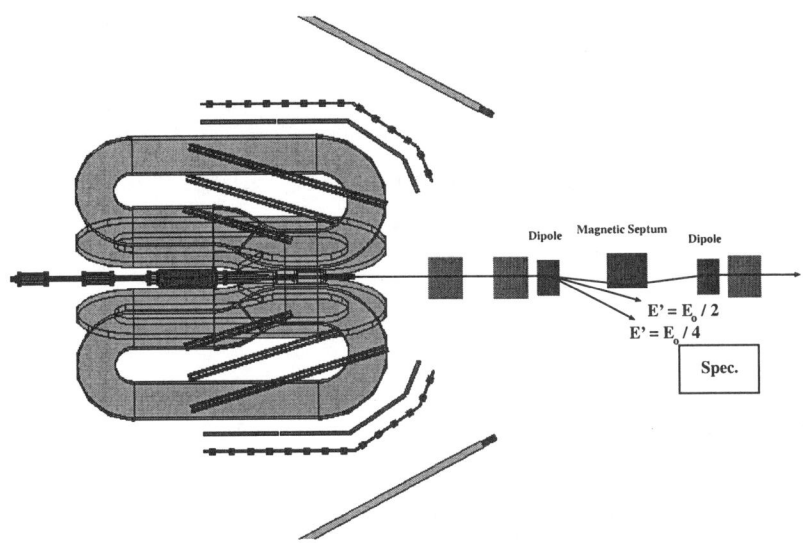

FIGURE 10. Schematic layout for a small angle electron facility in the Bates ring using the BLAST polarized internal target setup. The existing quadrupole magnets are shown as boxes and the proposed magnetic dipoles are labeled. Some representative electron trajectories are shown.

scattered at a small angle ($\leq 1^o$) the combination of the first two quadrupole and first dipole magnets will separate the electrons which have lost a significant amount of energy from the full energy electron beam. These pre-bending magnets must then be followed by an electron spectrometer. By placing a detector which has both good position and angular resolution, the energy and initial scattering and out of plane angles (θ_e, ϕ_e) can be determined. Thus important requirements for this system are good angular acceptance and resolution. The detailed design of this system must still be worked out.

The method of virtual photon has been originally proposed as a way to increase the count rates in photon induced reactions [51,52]. This proposed facility will also have this advantage. In a conventional tagged photon facility, the count rate is limited by the maximum rate capacity of the tagging ladder. In the virtual photon tagging facility proposed here, the count rate is limited by the ring luminosity. In Fig. 11 a comparison is made between the effective flux anticipated in the proposed internal tagging facility and that of a conventional tagger. It can be seen that one can expect a large count rate advantage due to virtual internal target tagging.

CONCLUSIONS

The evolution of photo- and electro-pion production from nuclear to nucleon targets has followed the paradigm shift from the Yukawa to the Goldstone view of the pion. In fact the long range part of the nucleon- nucleon interaction and the importance of the pion field in nuclei is as important as ever and remains an important frontier in physics. The

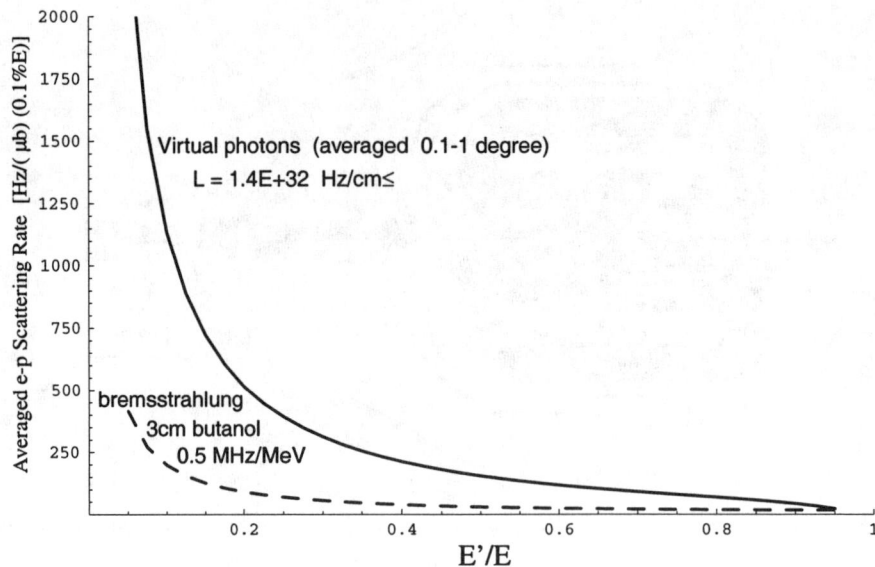

FIGURE 11. Effective luminosity for a virtual and real photon tagging as a function of the photon energy E' divided by the incoming electron energy E.

future role of Bates in studying chiral dynamics through electromagnetic pion production from the nucleon and nuclei is very promising with the use of internal, polarized, targets and the combination of BLAST and forward angle detectors.

ACKNOWLEDGMENTS

I would like to thank my many colleagues over the years for their sharing their efforts and insights. These include the organizers Bill Donnelly and Bill Turchinetz, and many others including Ingvar Blomqvist, Ed Booth, Reinhard Beck, Harald Merkel, and Marcello Pavan for their collaboration, Christophe Tzara for originally interesting me in photo-pion physics, Jean Marc Laget, Ernie Moniz, Justus Koch, Barry Holstein, and Ulf Meissner for many theoretical discussions.

REFERENCES

1. See e.g. *Dynamics of the Standard Model*, J. F. Donoghue, E. Golowich, and B. R. Holstein, Cambridge University Press (1992).
2. J. Goldstone, *Nuovo Cimento*, **19**, 154 (1961); Y. Nambu, *Phys. Rev. Lett.* **4**, 380 (1960).
3. H. Leutwyler, hep-ph 9409422 and hep-ph 9609466.

4. J. Gasser and H. Leutwyler, Phys. Reports 87, 77 (1982). H. Leutwyler, *Phys. Lett.* **B378**, 313 (1996) and hep-ph/9602255
5. *Photopion Nuclear Physics*, P. Stoler editor, Plenum Press (1979).
6. J. M. Laget, *Physics Reports*, **69**, 1 (1981).
7. A. M. Bernstein, Summary and Outlook in [5]; International School of Intermediate Energy Nuclear Physics, Verona, Italy, 1981, R. Bergere, C. Costa, and C. Schaerf editors, World Publishing (1982); AIP Conference Proceedings No. 183, American Institute of Physics (1985); Proceedings of the Seminar on Electromagnetic Interactions in Nuclei, Moscow, December 1988, G. M. Gurevitch editor, *Academy of Science*, USSR (1990).
8. B. H. Brandson and R. G. Moorhouse, *The Pion-Nucleon System*, Princeton University Press, N.Y. (1973).
9. N. M. Kroll and M. A. Ruderman, *Phys. Rev.* **93**, 233 (1954).
10. See e.g. E. M. Henley and W. Thirring, *Elementary Quantum Field Theory*, McGraw Hill, N.Y. (1962), Part III, Pion Physics.
11. H. Leutwyler, *Effective Field Theory of the Pion-Nucleon- Interaction*, $\pi - N$ Newsletter 15, 1 (1999).
12. S. Weinberg, *Physica*, A**96**, 327 (1979); J. Gasser and H. Leutwyler, *Ann. Phys.* **158**, 142 (1984); *Nucl. Phys.* B**250**, 465 and 517 (1985).
13. V. Bernard, N. Kaiser, and U. G. Meißner, *Int. J. Mod. Phys.* E4, 193 (1995); *Z. für Physik*, C**70**, 483 (1996); *Phys. Lett.* B**378**, 337 (1997); *Nucl. Phys.* A**607**, 379 (1996); *Phys. Rev. Lett.* **74**, 3752 (1995).
14. *Proceedings of the Workshops on Chiral Dynamics: Theory and Experiment*, Mainz, Germany, September 1997, A. M. Bernstein, D. Drechsel, and Th. Walcher editors, *Springer-Verlag, Lecture Notes in Physics*, Vol. **513** (1998); MIT, July 1994, A. M. Bernstein and B. Holstein editors, *Springer-Verlag, Lecture Notes in Physics*, Vol. **452** (1995).
15. A. M. Bernstein, N. Paras, W. Turchinetz, B. Chasan, and E. C. Booth, *Phys. Rev. Lett.* **37**, 819 (1976).
16. C. Tzara, *Nucl. Phys.* B**18**, 246 (1970).
17. K. Shoda *et al.*, *Phys. Lett.* **101**B, 124 (1981); Ch. Schmitt *et al.*, *Nucl. Phys.* A**395**, 435 (1983).
18. B. H. Cottman *et al.*, *Phys. Rev. Lett.* **55**, 684 (1985).
19. K. I. Blomqvist and J. M. Laget, *Nucl. Phys.* A**280**, 405 (1977).
20. J. H. Koch and E. J. Moniz, *Phys. Rev.* C**27**, 751 (1983).
21. D. R. Tieger *et al.*, *Phys. Rev. Lett.* **53**, 755 (1984).
22. L. D. Pham *et al.*, *Phys. Rev.* C**46**, 621 (1992).
23. R. J. Loucks, V. R. Pandharipande, and R. Schiavilla, *Phys. Rev.* C**9**, 342 (1994); R. J. Loucks and V. R. Pandharipande, *Phys. Rev.* C**54**, 32 (1996).
24. R. Gilman *et al.*, *Phys. Rev. Lett.* **64**, 622 (1990).
25. H. E. Jackson, *Bull. Am. Phys. Soc.* **43**, No. 6, 1546 (1998).
26. R. Beck *et al.*, *Phys. Rev. Lett.* **65**, 1841 (1990); E. Mazzacuato *et al.*, *Phys Rev. Lett.* **57**, 3144 (1986).
27. P. de Baenst, *Nucl. Phys.* B**24**, 633 (1970); A. I. Vainsthein and V. I. Zakaharov, *Sov. J. Nucl. Phys.* **12**, 333 (1971); and *Nucl. Phys.* B**36**, 589 (1972).
28. G. Ecker and U. G. Meissner, *Comm. Nucl. Part. Phys.* **21**, 347 (1995).
29. A. M. Bernstein and B. R. Holstein, *Comm. Nucl. Part. Phys.* **20**, 197 (1991).

30. A. M. Bernstein *et al.*, *Phys. Rev.* C**55**, 1509 (1997); M. Fuchs *et al.*, *Phys. Lett.* B**368**, 20 (1996).
31. J. C. Bergstrom *et al.*, *Phys. Rev.* C**53**, R1052 (1996).
32. F. Low, *Phys. Rev.* **96**, 1428 (1954); M. Gell-Mann and M. L. Goldberger, *ibid.* 1433.
33. S. Weinberg, *Phys. Rev. Lett.* **17**, 168 (1966).
34. A. M. Bernstein, *Phys. Lett.* B**442**, 20 (1998).
35. The values of the electric dipole multipole E_{0+} and β are quoted in units of $10^{-3}/m_\pi$.
36. A. M. Bernstein and M. M. Pavan in *Proceedings of the Second Workshop on Electronuclear Physics With Internal Targets and the BLAST Detector*, MIT, Cambridge, MA, May 1998, R. Alarcon and R. Milner editors, World Scientific Publishing Co., Pte. Ltd., (1999).
37. T. S. Carman *et al.*, *Nucl. Instr. and Meth.* A**378**, 1 (1996); V. N. Litvinenko *et al.*, *Phys. Rev. Lett.* **24**, 4569 (1997).
38. Mainz proposal A2/10-97, A. M. Bernstein, R. Beck, and M. Pavan, contact persons.
39. D. Sigg *et. al.*, *Phys. Rev. Lett.* **75**, 3245 (1995); *Nucl. Phys.* A**609**, 269 (1996); *ibid* A**617**, 526 (1997) A. Badertscher, in *Proceedings of the Workshop on Chiral Dynamics*, Mainz, Germany, September 1997 [14].
40. S. Weinberg, Transactions of the N.Y. Academy of Science Series II 38 (I. I. Rabi Festschrift), 185 (1977), and contribution to the MIT Workshop [14].
41. H. Leutwyler, contribution to the MIT Workshop [14].
42. U. van Kolck, Ph.D. thesis, University of Texas (1994), *unpublished and private communication*.
43. U. G. Meissner and S. Steininger, *Phys. Lett.* B**419**, 403 (1998); N. Fettes, U-G. Meissner, S. Steininger, *Phys. Lett.* B**451**, 233 (1999).
44. W. R. Gibbs, Li Ali, and W. B. Kaufmann, *Phys. Rev. Lett.* **74**, 3740 (1995).
45. E. Matsinos, *Phys. Rev.* C**58**, 3014 (1997).
46. A. Gashi, E. Matsinos, G. C. Oades, G. Rasche, and W. S. Woolcock, hep-ph/9903434, hep-ph/9902224, and hep-ph/9902207. E. Matsinos, *private communication*.
47. A. M. Bernstein, *Isospin Violation in πN Scattering and the Breakdown of the Fermi-Watson Theorem*, $\pi - N$ Newsletter **15**, 1 (1999).
48. E. Fermi, *Suppl. Nuovo Cimento*, **2**, 17 (1955); K. M. Watson, *Phys. Rev.* **95**, 228 (1954).
49. V. F. Grushin, in Photoproduction of Pions on Nucleons and Nuclei, *Proceedings of the Lebedev Physics Institute*, Academy of Sciences of the USSR, A. A. Komar, editor, Vol. 186 (Z1988), English Translation by Nova Press, N.Y. (1989).
50. D. Drechsel, O. Hanstein, S. S. Kamalov, and L. Tiator, *Nucl. Phys.* A**645**, 145 (1999); An on line version of the numerical results are available on the internet at http://www.kph.uni-mainz.de/T/maid/.
51. L. Hand and R. Wilson, SLAC Summer Study Report, SLAC-25 (1963).
52. C. E. Hyde-Wright, J. M. Finn, W. Bertozzi, CEBAF Summer Study, June 1995, p. 582.
53. D. Drechsel and L. Tiator, *J. Phys. G: Nucl. Part. Phys.* **18**, 449 (1992). A. S. Raskin and T. W. Donnelly, *Annals of Phys.* **191**, 78 (1989).
54. T. Zwart, *private communication*.

Effective Field Theory and χpt

Barry R. Holstein

Department of Physics and Astronomy
University of Massachusetts
Amherst, MA 01003

Abstract. A brief introduction to the subject of chiral perturbation theory (χpt) is given, including a discussion of effective field theory and application to the upcoming Bates virtual Compton scattering measurement.

I INTRODUCTION

We have gathered to celebrate the fact that Bates has been delivering beam successfully for twenty five years and to review some of the things which have been learned and which are still to be studied. One thing that *has* changed theoretically during this period is that we now have a new paradigm for analysis of low energy processes such as studied at Bates. I was a student in the 1960's and at that time our goal was to attempt to find a renormalizable field theory which describes all particle interactions with the same sort of success as quantum electrodynamics (QED). In 1967 we went part of the way with development of the Weinberg-Salam theory, which incorporated the weak interaction as a sibling to the electromagnetic. Because the interaction was weak it could be treated via the same perturbative techniques as could its electromagnetic kin and what has resulted is an extremely successful description of all weak and electromagnetic processes.

For the strong interactions a renormalizable picture has also been developed— quantum chromodynamics or QCD. The theory is, of course, deceptively simple on the surface. Indeed the form of the Lagrangian[1]

$$\mathcal{L}_{\text{QCD}} = \bar{q}(i\slashed{D} - m)q - \frac{1}{2}\text{tr}\, G_{\mu\nu}G^{\mu\nu}. \tag{3}$$

is elegant, and the theory is renormalizable. So why are we not satisfied? While at the very largest energies, asymptotic freedom allows the use of perturbative techniques, for those who are interested in making contact with low energy experimental findings there exist at least three fundamental difficulties:

i) QCD is written in terms of the "wrong" degrees of freedom—quarks and gluons—while low energy experiments are performed with hadronic bound states;

ii) the theory is non-linear due to gluon self interactions;

iii) the theory is one of strong coupling—$g^2/4\pi \sim 1$—so that perturbative methods are not practical.

Nevertheless, there has been a great deal of recent progress in making contact between theory and experiment using the technique of "effective field theory", which exploits the chiral symmetry of the QCD interaction. In order to understand how this is accomplished, we shall first review this idea of effective field theory in the simple context of quantum mechanics. Then we show how these ideas can be married via chiral perturbation theory and indicate applications at Bates.

II EFFECTIVE FIELD THEORY

The power of effective field theory is associated with the feature that there exist many situations in physics involving *two scales*, one heavy and one light. Then, provided one is working at energies small compared to the heavy scale, it is possible to fully describe the interactions in terms of an "effective" picture, which is written only in terms of the light degrees of freedom, but which fully includes the influence of the heavy mass scale through virtual effects. A number of very nice review articles on effective field theory can be found in ref. [1].

Before proceeding to QCD, however, it is useful to study this idea in the simpler context of ordinary quantum mechanics, in order to get familiar with the concept.

---

[1] Here the covariant derivative is

$$iD_\mu = i\partial_\mu - gA^a_\mu \frac{\lambda^a}{2}, \tag{1}$$

where λ^a (with $a = 1, \ldots, 8$) are the SU(3) Gell-Mann matrices, operating in color space, and the color-field tensor is defined by

$$G_{\mu\nu} = \partial_\mu A_\nu - \partial_\nu A_\mu - g[A_\mu, A_\nu], \tag{2}$$

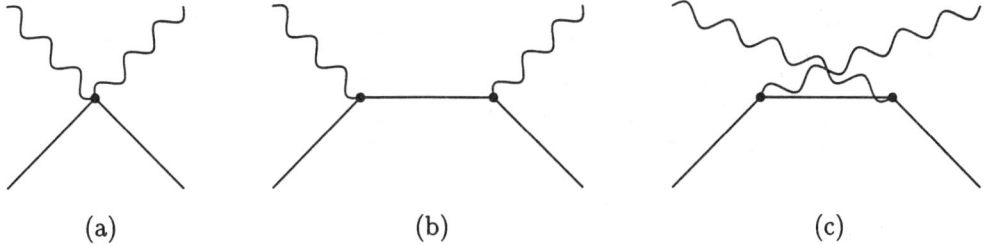

FIGURE 1. Feynman diagrams for nonrelativistic photonl-atom scattering.

Specifically, we examine the question of why the sky is blue, whose answer can be found in an analysis of the scattering of photons from the sun by atoms in the atmosphere—Compton scattering [2]. First we examine the problem using traditional quantum mechanics and consider elastic (Rayleigh) scattering from, for simplicity, single-electron (hydrogen) atoms. The appropriate Hamiltonian is then

$$H = \frac{(\vec{p} - e\vec{A})^2}{2m} + e\phi \qquad (4)$$

and the leading—$\mathcal{O}(e^2)$—amplitude for Compton scattering is found from calculating the diagrams shown in Figure 1, yielding the familiar Kramers-Heisenberg form

$$\text{Amp} = -\frac{e^2/m}{\sqrt{2\omega_i 2\omega_f}} \left[\hat{\epsilon}_i \cdot \hat{\epsilon}_f^* + \frac{1}{m} \sum_n \left(\frac{\hat{\epsilon}_f^* \cdot <0|\vec{p}e^{-i\vec{q}_f \cdot \vec{r}}|n> \hat{\epsilon}_i \cdot <n|\vec{p}e^{i\vec{q}_i \cdot \vec{r}}|0>}{\omega_i + E_0 - E_n} \right. \right.$$
$$\left. \left. + \frac{\hat{\epsilon}_i \cdot <0|\vec{p}e^{i\vec{q}_i \cdot \vec{r}}|n> \hat{\epsilon}_f^* \cdot <n|\vec{p}e^{-i\vec{q}_f \cdot \vec{r}}|0>}{E_0 - \omega_f - E_n} \right) \right] \qquad (5)$$

where $|0>$ represents the hydrogen ground state having binding energy E_0.

Here the leading component is the familiar ω-independent Thomson amplitude and would appear naively to lead to an energy-independent cross-section. However, this is *not* the case. Indeed, by expanding in ω and using a few quantum mechanical identities one can show that, provided that the energy of the photon is much smaller than a typical excitation energy—as is the case for optical photons, the cross section can be written as

$$\frac{d\sigma}{d\Omega} = \lambda^2 \omega^4 |\hat{\epsilon}_f^* \cdot \hat{\epsilon}_i|^2 \left(1 + \mathcal{O}\left(\frac{\omega^2}{(\Delta E)^2} \right) \right) \qquad (6)$$

where

$$\lambda = \alpha_{em} \sum \frac{2|z_{n0}|^2}{E_n - E_0} \qquad (7)$$

is the atomic electric polarizability, $\alpha_{em} = e^2/4\pi$ is the fine structure constant, and $\Delta E \sim m\alpha_{em}^2$ is a typical hydrogen excitation energy. We note that $\alpha_{em}\lambda \sim$

$a_0^2 \times \frac{\alpha_{em}}{\Delta E} \sim a_0^3$ is of order the atomic volume, as will be exploited below, and that the cross section itself has the characteristic ω^4 dependence which leads to the blueness of the sky—blue light scatters much more strongly than red [3].

Now while the above derivation is certainly correct, it requires somewhat detailed and lengthy quantum mechanical manipulations which obscure the relatively simple physics involved. One can avoid these problems by the use of effective field theory methods. The key point is that of scale. Since the incident photons have wavelengths $\lambda \sim 5000A$ much larger than the $\sim 1A$ atomic size, then at leading order the photon is insensitive to the presence of the atom, since the latter is electrically neutral. If χ represents the wavefunction of the atom then the effective leading order Hamiltonian is simply

$$H_{eff}^{(0)} = \chi^* \left(\frac{\vec{p}^2}{2m} + e\phi \right) \chi \tag{8}$$

and there is *no* interaction with the field. In higher orders, there *can* exist such atom-field interactions and this is where the effective Hamiltonian comes in to play. In order to construct the effective interaction, we demand certain general principles—this Hamiltonian must satisfy fundamental symmetry requirements. In particular H_{eff} must be gauge invariant, must be a scalar under rotations, and must be even under both parity and time reversal transformations. Also, since we are dealing with Compton scattering, H_{eff} should be quadratic in the vector potential. Actually, from the requirement of gauge invariance, it is clear that the effective interaction can utilize $\vec{A}$ only via the electric and magnetic fields, rather than the vector potential itself—

$$\vec{E} = -\vec{\nabla}\phi - \frac{\partial}{\partial t}\vec{A}, \qquad \vec{B} = \vec{\nabla} \times \vec{A} \tag{9}$$

since these are invariant under a gauge transformation

$$\phi \to \phi + \frac{\partial}{\partial t}\Lambda, \qquad \vec{A} \to \vec{A} - \vec{\nabla}\Lambda \tag{10}$$

while the vector and/or scalar potentials are not. The lowest order interaction then can involve only the rotational invariants $\vec{E}^2, \vec{B}^2$ and $\vec{E} \cdot \vec{B}$. However, under spatial inversion—$\vec{r} \to -\vec{r}$—electric and magnetic fields behave oppositely—$\vec{E} \to -\vec{E}$ while $\vec{B} \to \vec{B}$—so that parity invariance rules out any dependence on $\vec{E} \cdot \vec{B}$. Likewise under time reversal invariance $\vec{E} \to \vec{E}, \vec{B} \to -\vec{B}$ so such a term is also T-odd. The simplest such effective Hamiltonian must then have the form

$$H_{eff}^{(1)} = \chi^*\chi[-\frac{1}{2}c_E\vec{E}^2 - \frac{1}{2}c_B\vec{B}^2] \tag{11}$$

(Terms involving time or spatial derivatives are much smaller.) We know from electrodynamics that $\frac{1}{2}(\vec{E}^2 + \vec{B}^2)$ represents the field energy per unit volume, so by dimensional arguments, in order to represent an energy in Eq. 11, c_E, c_B must have dimensions of volume. Also, since the photon has such a long wavelength, there is no penetration of the atom, so only classical scattering is allowed. The relevant scale must then be atomic size so that we can write

$$c_E = k_E a_0^3, \qquad c_B = k_B a_0^3 \tag{12}$$

where we anticipate $k_E, k_B \sim \mathcal{O}(1)$. Finally, since for photons with polarization $\hat{\epsilon}$ and four-momentum q_μ we identify $\vec{A}(x) = \hat{\epsilon}\exp(-iq\cdot x)$, then from Eq. 9, $|\vec{E}| \sim \omega$, $|\vec{B}| \sim |\vec{k}| = \omega$ and

$$\frac{d\sigma}{d\Omega} \propto |<f|H_{eff}|i>|^2 \sim \omega^4 a_0^6 \tag{13}$$

as found in the previous section via detailed calculation. This is a nice example of the power of simple effective field theory arguments.

III APPLICATION TO QCD: CHIRAL PERTURBATION THEORY

Now let's apply these ideas to the case of QCD. In this case the invariance we wish to exploit is "chiral symmetry." The idea of "chirality" is defined by the operators

$$\Gamma_{L,R} = \frac{1}{2}(1 \pm \gamma_5) = \frac{1}{2}\begin{pmatrix} 1 & \mp 1 \\ \mp 1 & 1 \end{pmatrix} \tag{14}$$

which project "left-" and "right-handed" components of the Dirac wavefunction via

$$\psi_L = \Gamma_L \psi \qquad \psi_R = \Gamma_R \psi \quad \text{with} \quad \psi = \psi_L + \psi_R \tag{15}$$

In terms of these chirality states the quark component of the QCD Lagrangian can be written as

$$\bar{q}(i\not{D} - m)q = \bar{q}_L i\not{D} q_L + \bar{q}_R i\not{D} q_R - \bar{q}_L m q_R - \bar{q}_R m q_L \tag{16}$$

The reason that these chirality states are called left- and right-handed is that in the limit $m \to 0$ they coincide with quark *helicity* projection operators. With this background, we note that QCD, in the mathematical limit as $m \to 0$ has the structure

$$\mathcal{L}_{\text{QCD}} \xrightarrow{m=0} \bar{q}_L i\not{D} q_L + \bar{q}_R i\not{D} q_R \tag{17}$$

and is invariant under *independent* global left- and right-handed rotations

$$q_L \to \exp(i \sum_j \lambda_j \alpha_j) q_L, \qquad q_R \to \exp(i \sum_j \lambda_j \beta_j) q_R \qquad (18)$$

This invariance is called $SU(3)_L \otimes SU(3)_R$ or chiral $SU(3) \times SU(3)$. Continuing to neglect the light quark masses, we see that in a chiral symmetric world one would expect to have sixteen—eight left-handed and eight right-handed—conserved Noether currents

$$\bar{q}_L \gamma_\mu \frac{1}{2} \lambda_i q_L, \qquad \bar{q}_R \gamma_\mu \frac{1}{2} \lambda_i q_R \qquad (19)$$

Equivalently, by taking the sum and difference we would have eight conserved vector and eight conserved axial vector currents

$$V_\mu^i = \bar{q} \gamma_\mu \frac{1}{2} \lambda_i q, \qquad A_\mu^i = \bar{q} \gamma_\mu \gamma_5 \frac{1}{2} \lambda_i q \qquad (20)$$

In the vector case, this is just a simple generalization of isospin ($SU(2)$) invariance to the case of $SU(3)$. There exist *eight* ($3^2 - 1$) time-independent charges

$$F_i = \int d^3 x V_0^i(\vec{x}, t) \qquad (21)$$

and there exist various supermultiplets of particles having identical spin-parity and (approximately) the same mass in the configurations—singlet, octet, decuplet, *etc.* demanded by $SU(3)$-invariance.

If chiral symmetry were realized in the conventional fashion one would expect there also to exist corresponding nearly degenerate same spin but *opposite* parity states generated by the action of the time-independent axial charges $F_i^5 = \int d^3 x A_0^i(\vec{x}, t)$ on these states. However, it is known that the axial symmetry is broken spontaneously, whereby Goldstone's theorem requires the existence of eight massless pseudoscalar bosons, which couple derivatively to the rest of the universe [4]. Of course, in the real world such massless 0^- states do not exist, because in the real world exact chiral invariance is broken by the small quark mass terms which we have neglected up to this point. Thus what we have are eight very light (but not massless) pseudo-Goldstone bosons which make up the pseudoscalar octet. Since such states are lighter than their other hadronic counterparts, we have a situation wherein effective field theory can be applied—provided one is working at energy-momenta small compared to the ~ 1 GeV scale which is typical of hadrons, one can describe the interactions of the pseudoscalar mesons using an effective Lagrangian. Actually this has been known since the 1960's, where a good deal of work was done with a *lowest order* effective chiral Lagrangian [5]

$$\mathcal{L}_2 = \frac{F_\pi^2}{4} \text{Tr}(\partial_\mu U \partial^\mu U^\dagger) + \frac{m_\pi^2}{4} F_\pi^2 \text{Tr}(U + U^\dagger). \qquad (22)$$

where the subscript 2 indicates that we are working at two-derivative order or one power of chiral symmetry breaking—*i.e.* m_π^2. Here $U \equiv \exp(\sum \lambda_i \phi_i / F_\pi)$, where

$F_\pi = 92.4$ is the pion decay constant. This Lagrangian is *unique*—if we expand to lowest order in $\vec{\phi}$

$$\mathrm{Tr}\partial_\mu U \partial^\mu U^\dagger = \mathrm{Tr}\frac{i}{F_\pi}\vec{\tau}\cdot\partial_\mu\vec{\phi} \times \frac{-i}{F_\pi}\vec{\tau}\cdot\partial^\mu\vec{\phi} = \frac{2}{F_\pi^2}\partial_\mu\vec{\phi}\cdot\partial^\mu\vec{\phi}$$

$$\mathrm{Tr}(U+U^\dagger) = \mathrm{Tr}(2 - \frac{1}{F_\pi^2}\vec{\tau}\cdot\vec{\phi}\vec{\tau}\cdot\vec{\phi}) = \mathrm{const.} - \frac{2}{F_\pi^2}\vec{\phi}\cdot\vec{\phi} \qquad (23)$$

we reproduce the free pion Lagrangian, as required,

At the SU(3) level, including an appropriately generalized chiral symmetry breaking term, there is even predictive power—one has

$$\frac{F_\pi^2}{4}\mathrm{Tr}\partial_\mu U \partial^\mu U^\dagger = \frac{1}{2}\sum_{j=1}^{8}\partial_\mu\phi_j\partial^\mu\phi_j + \cdots \qquad (24)$$

$$\frac{F_\pi^2}{4}\mathrm{Tr}2B_0 m(U+U^\dagger) = \mathrm{const.} - \frac{1}{2}(m_u+m_d)B_0\sum_{j=1}^{3}\phi_j^2$$
$$-\frac{1}{4}(m_u+m_d+2m_s)B_0\sum_{j=4}^{7}\phi_j^2 - \frac{1}{6}(m_u+m_d+4m_s)B_0\phi_8^2 + \cdots \qquad (25)$$

where B_0 is a constant and m is the quark mass matrix. We can then identify the meson masses as

$$m_\pi^2 = 2\hat{m}B_0$$
$$m_K^2 = (\hat{m}+m_s)B_0$$
$$m_\eta^2 = \frac{2}{3}(\hat{m}+2m_s)B_0, \qquad (26)$$

where $\hat{m} = \frac{1}{2}(m_u+m_d)$ is the mean light quark mass. This system of three equations is *overdetermined*, and we find by simple algebra

$$3m_\eta^2 + m_\pi^2 - 4m_K^2 = 0 . \qquad (27)$$

which is the Gell-Mann-Okubo mass relation and is well-satisfied experimentally [6]. Expanding to fourth order in the fields we also reproduce the well-known and experimentally successful Weinberg $\pi\pi$ scattering lengths [7]

$$a_0^0 = \frac{7m_\pi^2}{32\pi F_\pi^2}, \quad a_0^2 = -\frac{m_\pi^2}{16\pi F_\pi^2}, \quad a_1^1 = \frac{m_\pi^2}{24\pi F_\pi^2} \qquad (28)$$

However, when one attempts to go beyond tree level in order to unitarize the results, divergences arise and that is where the field stopped at the end of the

1960's. The solution, as pointed out ten years later by Weinberg [8] and carried out by Gasser and Leutwyler [9], is to absorb these divergences in phenomenological constants, just as done in QED. A new wrinkle in this case is that the theory is nonrenormalizabile in that the forms of the divergences are *different* from the terms that one started with. That means that the form of the counterterms that are used to absorb these divergences must also be different, and Gasser and Leutwyler wrote down the most general counterterm Lagrangian that one can have at one loop, which involves *four-derivative* interactions

$$\mathcal{L}_4 = \sum_{i=1}^{10} L_i \mathcal{O}_i = L_1 \Big[\text{tr}(D_\mu U D^\mu U^\dagger)\Big]^2 + L_2 \text{tr}(D_\mu U D_\nu U^\dagger) \cdot \text{tr}(D^\mu U D^\nu U^\dagger)$$
$$+ L_3 \text{tr}(D_\mu U D^\mu U^\dagger D_\nu U D^\nu U^\dagger) + L_4 \text{tr}(D_\mu U D^\mu U^\dagger)\text{tr}(\chi U^\dagger + U\chi^\dagger)$$
$$+ L_5 \text{tr}\left(D_\mu U D^\mu U^\dagger \left(\chi U^\dagger + U\chi^\dagger\right)\right) + L_6 \Big[\text{tr}\left(\chi U^\dagger + U\chi^\dagger\right)\Big]^2$$
$$+ L_7 \Big[\text{tr}\left(\chi^\dagger U - U\chi^\dagger\right)\Big]^2 + L_8 \text{tr}\left(\chi U^\dagger \chi U^\dagger + U\chi^\dagger U\chi^\dagger\right)$$
$$+ iL_9 \text{tr}\left(F^L_{\mu\nu} D^\mu U D^\nu U^\dagger + F^R_{\mu\nu} D^\mu U^\dagger D^\nu U\right) + L_{10} \text{tr}\left(F^L_{\mu\nu} U F^{R\mu\nu} U^\dagger\right)$$
(29)

where the covariant derivative is defined via

$$D_\mu U = \partial_\mu U + \{A_\mu, U\} + [V_\mu, U] \quad (30)$$

the constants $L_i, i = 1, 2, \ldots 10$ are arbitrary (not determined from chiral symmetry alone) and $F^L_{\mu\nu}, F^R_{\mu\nu}$ are external field strength tensors defined via

$$F^{L,R}_{\mu\nu} = \partial_\mu F^{L,R}_\nu - \partial_\nu F^{L,R}_\mu - i[F^{L,R}_\mu, F^{L,R}_\nu], \qquad F^{L,R}_\mu = V_\mu \pm A_\mu. \quad (31)$$

Now just as in the case of QED the bare parameters L_i which appear in this Lagrangian are not physical quantities. Instead the experimentally relevant (renormalized) values of these parameters are obtained by appending to these bare values the divergent one-loop contributions—

$$L_i^r = L_i - \frac{\gamma_i}{32\pi^2}\left[\frac{-2}{\epsilon} - \ln(4\pi) + \gamma - 1\right] \quad (32)$$

By comparing predictions with experiment, Gasser and Leutwyler were able to determine empirical values for each of these ten parameters. Typical results are shown in Table 1, together with the way in which they were determined. The important question to ask at this point is why stop at order four derivatives? Clearly if two-loop amplitudes from $\mathcal{L}_2$ or one-loop corrections from $\mathcal{L}_4$ are calculated, divergences will arise which are of six-derivative character. Why not include these? The answer is that the chiral procedure represents an expansion in energy-momentum. Corrections to the lowest order (tree level) predictions from one-loop corrections from

Coefficient	Value	Origin
L_1^r	0.65 ± 0.28	$\pi\pi$ scattering
L_2^r	1.89 ± 0.26	and
L_3^r	-3.06 ± 0.92	$K_{\ell 4}$ decay
L_5^r	2.3 ± 0.2	F_K/F_π
L_9^r	7.1 ± 0.3	π charge radius
L_{10}^r	-5.6 ± 0.3	$\pi \to e\nu\gamma$

TABLE 1. Gasser-Leutwyler counterterms and the means by which they are determined.

Reaction	Quantity	Theory	Experiment
$\pi^+ \to e^+\nu_e\gamma$	$h_V(m_\pi^{-1})$	0.027	0.029 ± 0.017 [11]
$\pi^+ \to e^+\nu_e e^+ e^-$	r_V/h_V	2.6	2.3 ± 0.6 [11]
$\gamma\pi^+ \to \gamma\pi^+$	$(\alpha_E + \beta_M)\,(10^{-4}\,\text{fm}^3)$	0	1.4 ± 3.1 [12]
	$\alpha_E\,(10^{-4}\,\text{fm}^3)$	2.8	6.8 ± 1.4 [13]
			12 ± 20 [14]
			2.1 ± 1.1 [15]

TABLE 2. Chiral Predictions and data in radiative pion processes.

$\mathcal{L}_2$ or tree level contributions from $\mathcal{L}_4$ are $\mathcal{O}(E^2/\Lambda_\chi^2)$ where $\Lambda_\chi \sim 4\pi F_\pi \sim 1$ GeV is the chiral scale [10]. Thus chiral perturbation theory is a *low energy* procedure. It is only to the extent that the energy is small compared to the chiral scale that it makes sense to truncate the expansion at the one-loop (four-derivative) level. Realistically this means that we deal with processes involving $E < 500$ MeV, and for such reactions the procedure is found to work very well.

In fact Gasser and Leutwyler, besides giving the form of the $\mathcal{O}(p^4)$ chiral Lagrangian, have also performed the one loop integration and have written the result in a simple algebraic form. Users merely need to look up the result in their paper and, despite having ten phenomenological constants the theory is quite predictive. An example is shown in Table 2, where predictions are given involving quantities which arise using just two of the constants—L_9, L_{10}. The table also reveals an interesting dilemma—one solid chiral prediction, that for the charged pion polarizability, is possibly violated, although this is far from clear since there are three experimental results here, only one of which is in disagreement. This represents a serious challenge to the chiral predictions (and therefore to QCD!) and should be the focus of future experimental work. However, there are no Bates implications and, because of space limitations, we shall have to be content to stop here. Interested readers, however, can find applications to this and other systems in a number of review articles [16].

IV χPT AND BATES

For application at Bates it is important to note that the same ideas can be applied within the sector of meson-nucleon interactions, although with a bit more difficulty. Again much work has been done in this regard [17], but there remain important challenges [18]. Writing the lowest order chiral Lagrangian at the SU(2) level is straightforward—

$$\mathcal{L}_{\pi N} = \bar{N}(i\slashed{D} - m_N + \frac{g_A}{2}\slashed{u}\gamma_5)N \tag{33}$$

where g_A is the usual nucleon axial coupling in the chiral limit, the covariant derivative $D_\mu = \partial_\mu + \Gamma_\mu$ is given by

$$\Gamma_\mu = \frac{1}{2}[u^\dagger, \partial_\mu u] - \frac{i}{2}u^\dagger(V_\mu + A_\mu)u - \frac{i}{2}u(V_\mu - A_\mu)u^\dagger, \tag{34}$$

and u_μ represents the axial structure

$$u_\mu = iu^\dagger \nabla_\mu U u^\dagger \tag{35}$$

Expanding to lowest order we find

$$\mathcal{L}_{\pi N} = \bar{N}(i\slashed{\partial} - m_N)N + g_A \bar{N}\gamma^\mu \gamma_5 \frac{1}{2}\vec{\tau} N \cdot (\frac{i}{F_\pi}\partial_\mu \vec{\pi} + 2\vec{A}_\mu)$$
$$- \frac{1}{4F_\pi^2}\bar{N}\gamma^\mu \vec{\tau} N \cdot \vec{\pi} \times \partial_\mu \vec{\pi} + \ldots \tag{36}$$

which yields the Goldberger-Treiman relation, connecting strong and weak couplings of the nucleon system [19]

$$F_\pi g_{\pi NN} = m_N g_A \tag{37}$$

Using the present best values for these quantities, we find

$$92.4\text{MeV} \times 13.05 = 1206\text{MeV} \quad \text{vs.} \quad 1189\text{MeV} = 939\text{MeV} \times 1.266 \tag{38}$$

and the agreement to better than two percent strongly confirms the validity of chiral symmetry in the nucleon sector. Actually the Goldberger–Treiman relation is only strictly true at the unphysical point $g_{\pi NN}(q^2 = 0)$ and one *expects* about a 1% discrepancy to exist. An interesting "wrinkle" in this regard is the use of the so-called Dashen-Weinstein relation, which takes into account lowest order SU(3) symmetry breaking, to predict this discrepancy in terms of corresponding numbers in the strangeness changing sector [20].

Another successful application at tree level involves threshold charged pion photoproduction and the Kroll-Ruderman term [21], which arises from the feature that, since the pion must be derivatively coupled, there exists a $\bar{N}N\pi^\pm \gamma$ contact

Quantity	Expt.
$E_{0+}(\gamma p \to \pi^+ n)$	$(+27.9 \pm 0.5) \times 10^{-3}/m_\pi$ [23]
	$(+28.8 \pm 0.7) \times 10^{-3}/m_\pi$ [24]
	$(+27.6 \pm 0.3) \times 10^{-3}/m_\pi$ [25]
$E_{0+}(\gamma n \to \pi^- p)$	$(-31.4 \pm 1.3) \times 10^{-3}/m_\pi$ [23]
	$(-32.2 \pm 1.2) \times 10^{-3}/m_\pi$ [26]
	$(-31.5 \pm 0.8) \times 10^{-3}/m_\pi$ [27]

TABLE 3. Experimental values for E_{0+} multipoles in charged pion photoproduction.

interaction which dominates threshold charged pion photoproduction. Here what is measured is the s-wave or E_{0+} multipole, defined via

$$\text{Amp} = 4\pi(1+\mu)E_{0+}\vec{\sigma}\cdot\hat{\epsilon} + \ldots \qquad (39)$$

where $\mu = m_\pi/M$. The chiral symmetry prediction is [22]

$$E_{0+} = \pm\frac{1}{4\pi(1+\mu)}\frac{eg_A}{\sqrt{2}F_\pi}(1\mp\frac{\mu}{2}) = \frac{eg_A}{4\sqrt{2}F_\pi}\begin{pmatrix} 1-\frac{3}{2}\mu & \pi^+ \\ -1+\frac{1}{2}\mu & \pi^- \end{pmatrix}$$
$$= \begin{cases} +26.3 \times 10^{-3}/m_\pi & \pi^+ n \\ -31.3 \times 10^{-3}/m_\pi & \pi^- p \end{cases}, \qquad (40)$$

which is in excellent agreement with the present experimental results, as shown in Table 3.

However, any realistic approach must also involve loop calculations as well as the use of a Foldy-Wouthuysen transformation in order to assure proper power counting. This approach goes under the name of heavy baryon chiral perturbation theory (HBχpt) and interested readers can find a compendium of such results in the review article [28]. For our purposes we shall have to be content to examine just two applications. One is neutral pion photoproduction. In this case the Kroll-Ruderman term is absent and the chiral expansion of the E_{0+} threshold amplitude begins at order μ and a heavy baryon HBχpt calculation by Bernard, Kaiser, and Meissner found an important loop contribution which had been omitted in the previous PCAC/based approach [29]. The correct chiral prediction at $\mathcal{O}(\mu^2)$ was found to be [30]

$$E_{0+} = \frac{eg_A}{8\pi M}\mu\{1 - [\frac{1}{2}(3+\kappa_p) + (\frac{M}{4F_\pi})^2]\mu + \mathcal{O}(\mu^2)\} \qquad (41)$$

where the term in M^2 signifies the "new" chiral loop contribution. However, comparison with experiment is tricky because of the existence of isotopic spin breaking in the pion and nucleon masses, so that there are *two* thresholds—one for $\pi^0 p$ and the second for $\pi^+ n$—only 7 MeV apart. When the physical masses of the pions

	theory	expt.
$E_{0+}(\pi^0 p)(\times 10^{-3}/m_\pi)$	-1.2	-1.31 ± 0.08 [31]
		-1.32 ± 0.11 [32]
$E_{0+}(\pi^0 n)(\times 10^{-3}/m_\pi)$	2.1	1.9 ± 0.3 [33]
$P_1/|\vec{q}|(\pi p)(\times \text{GeV}^{-2})$	0.48	0.47 ± 0.01 [31]
		0.41 ± 0.03 [32]

TABLE 4. Threshold parameters for neutral pion photoproduction.

are used recent data from both Mainz and from Saskatoon agree with the chiral prediction. However, there are concerns about the convergence of the chiral expansion, which reads $E_{0+} = C(1 - 1.26 + 0.59 + ...)$. There also exist chiral predictions for threshold p-wave amplitudes which are in good agreement with experiment, as shown in Table 4, and for which the convergence is exprcted to be rapid.

Finally exists a chiral symmetry prediction for the reaction $\gamma n \to \pi^0 n$

$$E_{0+} = -\frac{eg_A}{8\pi M}\mu^2\{\frac{1}{2}\kappa_n + (\frac{M}{4F_\pi})^2\} + ... = 2.13 \times 10^{-3}/m_\pi \tag{42}$$

However, the experimental measurement of such an amplitude involves considerable challenge, and must be accomplished either by use of a deuterium target with the difficult subtraction of the proton contribution and of meson exchange contributions or by use of a ^{3}He target. Neither of these are straightforward although some limited data already exist [33].

Our final example involves an experiment at Bates—measurement of the *generalized* proton *polarizability* via virtual Compton scattering. First recall from section 2 the concept of polarizability as the constant of proportionality between an applied electric or magnetizing field and the resultant induced electric or magnetic dipole moment—

$$\vec{p} = 4\pi\alpha_E \vec{E}, \qquad \vec{\mu} = 4\pi\beta_M \vec{H} \tag{43}$$

The corresponding interaction energy is

$$E = -\frac{1}{2}4\pi\alpha_E E^2 - \frac{1}{2}4\pi\beta_M H^2 \tag{44}$$

which, upon quantization, leads to a proton Compton scattering cross section

$$\frac{d\sigma}{d\Omega} = \left(\frac{\alpha_{em}}{m}\right)^2 \left(\frac{\omega'}{\omega}\right)^2 [\frac{1}{2}(1 + \cos^2\theta)$$
$$- \frac{m\omega\omega'}{\alpha_{em}}[\frac{1}{2}(\alpha_E + \beta_M)(1 + \cos\theta)^2 + \frac{1}{2}(\alpha_E - \beta_M)(1 - \cos\theta)^2 + ...]. \tag{45}$$

It is clear from Eq.(45) that, from careful measurement of the differential scattering cross section, extraction of these structure dependent polarizability terms is possible provided that

i) the energy is large enough that these terms are significant compared to the leading Thomson piece and

ii) that the energy is not so large that higher order corrections become important

and this has been accomplished recently at SAL and MAMI, yielding [34]

$$\alpha_E^{exp} = (12.1 \pm 0.8 \pm 0.5) \times 10^{-4} \text{fm}^3, \qquad \beta_M^{exp} = (2.1 \mp 0.8 \mp 0.5) \times 10^{-4} \text{fm}^3 \qquad (46)$$

A chiral one loop calculation has also been performed by Bernard, Kaiser, and Meissner and yields a result in good agreement with these measurements [35]

$$\alpha_E^{theo} = 10\beta_M^{theo} = \frac{5e^2 g_A^2}{384\pi^2 F_\pi^2 m_\pi} = 12.2 \times 10^{-4} \text{fm}^3 \qquad (47)$$

The idea of *generalized* polarizability can be understood from the analogous venue of electron scattering wherein measurement of the charge form factor as a function of $\bar{q}^2$ leads, when Fourier transformed, to a picture of the *local* charge density within the system. In the same way the virtual Compton scattering process— $\gamma^* + p \to \gamma + p$ can provide a measurement of the $\bar{q}^2$-dependent electric and magnetic polarizabilities, whose Fourier transform provides a picture of the *local polarization density* within the proton. On the theoretical side our group has performed a one loop HBχpt calculation and has produced a closed from expression for the predicted polarizabilities [36]

$$\bar{\alpha}_E^{(3)}(\bar{q}) = \frac{e^2 g_A^2 m_\pi}{64\pi^2 F_\pi^2} \frac{4 + 2\frac{\bar{q}^2}{m_\pi^2} - \left(8 - 2\frac{\bar{q}^2}{m_\pi^2} - \frac{\bar{q}^4}{m_\pi^4}\right) \frac{m_\pi}{\bar{q}} \arctan \frac{\bar{q}}{2m_\pi}}{\bar{q}^2 \left(4 + \frac{\bar{q}^2}{m_\pi^2}\right)},$$

$$\bar{\beta}_M^{(3)}(\bar{q}) = \frac{e^2 g_A^2 m_\pi}{128\pi^2 F_\pi^2} \frac{-\left(4 + 2\frac{\bar{q}^2}{m_\pi^2}\right) + \left(8 + 6\frac{\bar{q}^2}{m_\pi^2} + \frac{\bar{q}^4}{m_\pi^4}\right) \frac{m_\pi}{\bar{q}} \arctan \frac{\bar{q}}{2m_\pi}}{\bar{q}^2 \left(4 + \frac{\bar{q}^2}{m_\pi^2}\right)}. \qquad (48)$$

In the electric case the structure is about what would be expected—a gradual falloff of $\alpha_E(\bar{q})$ from the real photon point with scale $r_p \sim m_\pi$. However, the magnetic generalized polarizability is predicted to *rise* before this general falloff occurs— chiral symmetry requires the presence of both a paramagnetic and a diamagnetic component to the proton. Both predictions have received some support in a soon to be announced (and tour de force) MAMI measurement at $\bar{q} = 600$ MeV [37]. However, since parallel kinematics were employed in the experiment the desired generalized polarizabilities had to be identified on top of an enormous Bethe-Heitler background. The Bates measurement, to be performed by the OOPS collaboration next spring, will take place at $\bar{q} = 240$ MeV and will use the cababilities of the OOPS detector system to provide a 90 degree out of plane measurement, which should be *much* less sensitive to the Bethe-Heitler blowtorch. We anxiously await the results.

V CONCLUSION

In a short paper it is not possible to give any sense of the range of phenomena to which the concept of effective field theory as manifested via chiral perturbation theory has been applied, and interested readers can find many further applications in [16] and [28]. Nevertheless, we have tried to convey the relatively direct connection of such predictions to the underlying QCD interaction and the feature that in this way QCD itself can be tested at Bates.

Acknowlegement

It is a pleasure to acknowledge the hospitality of MIT/Bates and the organizers of this meeting. This work was supported in part by the National Science Foundation.

REFERENCES

1. See, *e.g.* A. Manohar, "Effective Field Theories," in **Schladming 1966: Perturbative and Nonperturbative Aspects of Quantum Field Theory**, hep-ph/9606222; D. Kaplan, "Effective Field Theories," in Proc. 7th Summer School in Nuclear Physics, nucl-th/9506035,; H. Georgi, "Effective Field Theory," in Ann. Rev. Nucl Sci. **43**, 209 (1995).
2. B.R. Holstein, Am. J. Phys. **67**, 422 (1999).
3. A corresponding classical physics discussion is given in R.P Feynman, R.B. Leighton, and M. Sands, **The Feynman Lecures on Physics**, Addison-Wesley, Reading, MA, (1963) Vol. I, Ch. 32.
4. J. Goldstone, Nuovo Cim. **19**, 154 (1961); J. Goldstone, A. Salam, and S. Weinberg, Phys. Rev. **127**, 965 (1962).
5. S. Gasiorowicz and D.A. Geffen, Rev. Mod. Phys. **41**, 531 (1969).
6. M. Gell-Mann, CalTech Rept. **CTSL-20** (1961); S. Okubo, Prog. Theo. Phys. **27**, 949 (1962).
7. S. Weinberg, Phys. Rev. Lett. **17** 616 (1966).
8. S. Weinberg, Physica **A96**, 327 (1979).
9. J. Gasser and H. Leutwyler, Ann. Phys. (NY) **158**, 142 (1984); Nucl. Phys. **B250**, 465 (1985).
10. A. Manohar and H. Georgi, Nucl. Phys. **B234**, 189 (1984); J.F. Donoghue, E. Golowich and B.R. Holstein, Phys. Rev. **D30**, 587 (1984).
11. Particle Data Group, Phys. Rev. **D54**, 1 (1996).
12. Yu. M. Antipov et al., Z. Phys. **C26**, 495 (1985).
13. Yu. M. Antipov et al., Phys. Lett. **B121**, 445 (1983).
14. T.A. Aibergenov et al., Czech. J. Phys. **36**, 948 (1986).
15. D. Babusci et al., Phys. Lett. **B277**, 158 (1992).
16. See, *e.g.* B.R. Holstein, Int.J. Mod. Phys. **A7**, 7873 (1993); H. Leutwyler, in **Perspectives in the Standard Model**, eds. R.K. Ellis, C.T. Hill, and J.D. Lykken, World Scientific, Singapore (1992); J. Gasser, in Advanced School on Effective Theories, eds. F. Cornet and M.J. Herrero, World Scientific, Singapore (1997); H.

Leutwyler, in **Selected Topics in Nonperturbative QCD**, eds. A. DiGiacomo and D. Diakonov, IOS Press, Amsterdam (1996).
17. J. Gasser, M. Sainio, and A. Svarc, Nucl. Phys. **B307**, 779 (1988).
18. V. Bernard, N. Kaiser, and U.G. Meissner, Int. J. Mod. Phys. **E4**, 193 (1995).
19. M. Goldberger and S.B. Treiman, Phys. Rev. **110**, 1478 (1958).
20. R. Dashen and M. Weinstein, Phys. Rev. **188**, 2330 (1969); B.R. Holstein, "Nucleon Axial Matrix Elements," Few-Body Systems Suppl. **11**, 116 (1999); J.L. Goity, R. Lewis, and M. Schvelinger, "The Goldberger-Treiman Discrepancy in SU(3)," Phys. Lett. **B454**, 115 (1999).
21. N. Kroll and M.A. Ruderman, Phys. Rev. **93**, 233 (1954).
22. P. deBaenst, Nucl. Phys. **B24**, 613 (1970).
23. J.P. Burg, Ann. De Phys. (Paris) **10**, 363 (1965).
24. M.J. Adamovitch et al., Sov. J. Nucl. Phys. **2**, 95 (1966).
25. J. Bergstrom, private communication.
26. E.L. Goldwasser et al., *Proc. XII Int. Conf. on High Energy Physics, Dubna, 1964*, ed. Ya.-A Smorodinsky, Atomizdat, Moscow (1966).
27. M. Kovash, πN Newsletter **12**, 51 (1997).
28. V. Bernard, U.-G. Meissner, and N. Kaiser, Int. J. Mod. Phys. **E4**, 193 (1995).
29. P. deBaenst, Nucl. Phys. **B24**, 633 (1970); A.M. Bernstein and B.R. Holstein, Comm. Nucl. Part. Phys. **20**, 197 (1991).
30. V. Bernard, J. Gasser, N. Kaiser and Ulf-G. Meissner, Phys. Lett. **B268**, 291 (1991).
31. M. Fuchs et al., Phys. Lett. **B368**, 20 (1996).
32. J.C. Bergstrom et al., Phys. Rev. **C53**, R1052 (1996).
33. P. Argan et al., Phys. Lett. **B206**, 4 (1988).
34. F.J. Federspiel et al., Phys. Rev. Lett. **67**, 1511 (1991); A. L. Hallin et al., Phys. Rev. **C48**, 1497 (1993); A. Zieger et al., Phys. Lett. **B278**, 34 (1992); B.R. MacGibbon et al., Phys. Rev. **C52**, 2097 (1995).
35. V. Bernard, N. Kaiser, and U.-G. Meissner, Phys. Rev. Lett. **67**, 1515 (1991).
36. T.R. Hemmert, B.R. Holstein, G. Knoechlein, and D. Drechsel, hep-ph/9910036.
37. S. Kerhoas et al., Few Body Syst. Supp. **10**, 523 (1999).

V. PARITY-VIOLATING ELECTRON SCATTERING AND A LOOK FORWARD

Chair: R.D. McKeown

Session V

Parity Violation I
Then and Now

Paul A. Souder[1]

Syracuse University, Syracuse, NY 13244

Abstract. This contribution, together with the following one by E. J. Beise, will cover the topic of experiments using polarized electrons to study the parity-violating electroweak interference effects. The role of Bates in the evolution of the field is emphasized. The initial motivation of these experimental efforts was tests of the Standard Model. Examples that I will describe include deep inelastic scattering from deuterium at SLAC and elastic scattering from ^{12}C at Bates. Recently, the focus has shifted to the search for strange form factors in the elastic scattering from hydrogen. I will present recent results on the subject from the HAPPEX collaboration at JLab. Finally, future experiments that will measure very small asymmetries at SLAC and JLab are described.

INTRODUCTION

The idea of using polarized electrons to measure the parity-violating asymmetry

$$A^{PV} = \frac{\sigma_R - \sigma_L}{\sigma_R + \sigma_L} \quad (1)$$

in the scattering of polarized electrons was first proposed by Zeldovich in 1957 [1]. By assuming that the effect was due to the weak interactions, he predicted that the asymmetries would be on the order of $10^{-4}Q^2$ where the momentum transfer Q^2 is given in $(GeV/c)^2$. It took about 20 years for the first experimental paper to be published in the field. However, the field is now developing rapidly. As can be seen in Figure 1, a number of experiments have been published or are in various stages of progress [2] - [12].

This flourishing of the field is due to progress in both theory and experiment. On the theoretical side, the Standard Model with radiative corrections has been developed. In addition, the possible role of strange quarks in the nucleon has been clarified. Finally, the theory necessary to interpret scattering from heavy nuclei in terms of the neutron distributions has been developed.

---

[1] Work supported by the DOE under contract number DE-FG02-84ER40146

Progress in experimental techniques also has been impressive. Intense, highly polarized electron sources are available on most electron accelerators. In addition, methods have been developed to measure the tiny ($10^{-4} - 10^{-7}$) asymmetries that arise in important experiments.

STANDARD MODEL

The theory of parity violation in electron scattering is given in a recent review [13]. Here, I will review some of the basic ideas. The general cross section is given by the square of the scattering amplitude:

$$\frac{d\sigma}{d\Omega} = |f^{L(R)}(Q^2)|^2. \qquad (2)$$

The superscripts $L(R)$ refer to whether the beam is polarized with left(right) helicity.

The scattering amplitude is the sum of contributions of the following form that arise from each relevant interaction:

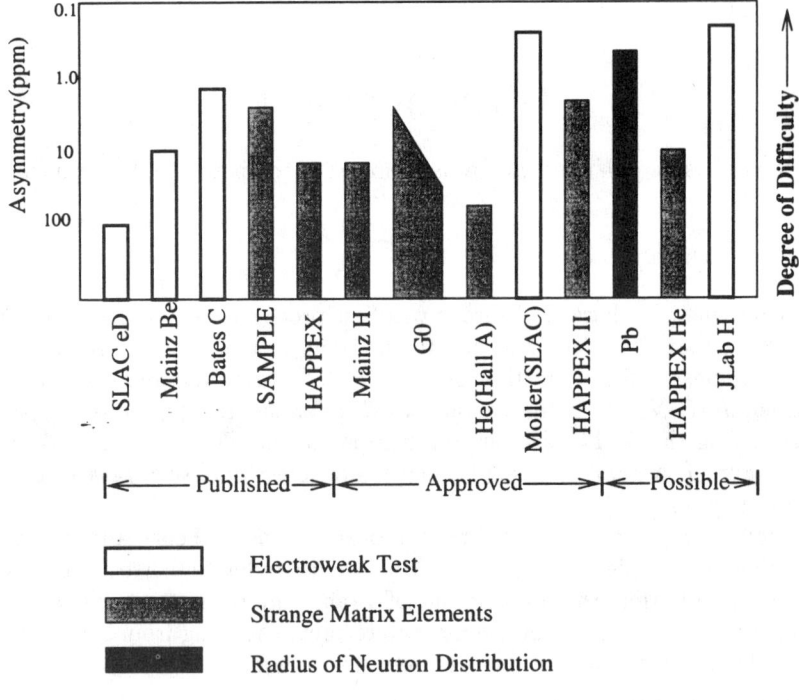

FIGURE 1. History of Polarized Electron Parity Experiments

$$f^{L(R)}(Q^2) \sim \frac{g_b^{L(R)} g_t}{Q^2 + M^2} \quad (3)$$

Here g_b is the charge of the beam and g_t is the charge of the target. M is the mass of the exchanged particle. For electron scattering, the electromagnetic and weak amplitudes are are relevant. For electromagnetic scattering (f_γ), the charge is independent of helicity and $M = 0$. For weak scattering (f_Z), the charge depends on helicity. Then $g^L - g^R \equiv 2g^A \neq 0$, and the cross section depends on helicity. Also $M = M_Z$. For $M_Z^2 \gg Q^2$,

$$A^{PV} = \frac{d\sigma_R - d\sigma_L}{d\sigma_R + d\sigma_L} \sim \frac{(f_Z^R - f_Z^L)}{2f_\gamma} \quad (4)$$

In the Standard Model, g_A is relatively large for the electron. Thus polarized electron scattering is a practical probe of the weak interactions.

Most targets are complex states of bound quarks, and the theory must account for this complication. For elastic scattering, the result is that the scattering amplitude is of the form

$$f = a_u F_u + a_d F_d + a_s F_s \quad \text{(Elastic Scattering)} \quad (5)$$

where the $F_1(Q^2)$ are form factors, one corresponding to each relevant quark flavor, and the a_i are constants determined by the Standard Model. Similar results apply for deep inelastic scattering.

In order to test the Standard Model with polarized electrons, there are two approaches. One is to use a target without quarks, namely the electron. This is the strategy of ref. [10,31]. The other strategy is to choose kinematics where the strange quarks are negligible and choose an isoscalar target so that $F_u = F_d$ and the form factors cancel in the asymmetry. That is the method used for the ^{12}C experiment at Bates [4] and effectively the method used for the SLAC eD experiment [2].

Today, the Standard Model has been well tested. Many experiments at LEP and Fermilab have been in impressive agreement with the predictions of the theory [14]. However, there are a few slight discrepancies, such as the latest result from Cs [15], and it is possible that there are new interactions that go beyond the scope of the Standard Model [16]. Examples include

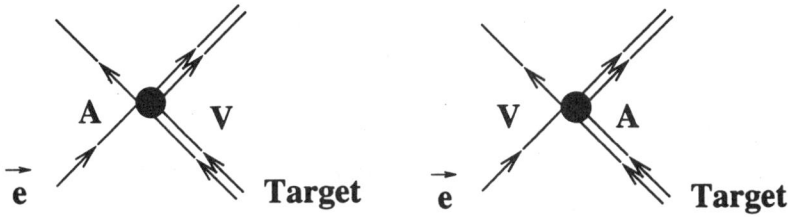

FIGURE 2. Feynman diagrams for contact interactions

TABLE 1. Parameters for contact interactions

	$A - V$	$V - A$
$u+d$	$\tilde{\gamma}$	$\tilde{\delta}$
$u-d$	$\tilde{\alpha}$	$\tilde{\beta}$
e	$\tilde{\epsilon}$	

1. Extra Z bosons

2. Lepto-quarks

3. Substructure of quarks or leptons

In order to parameterize the new interactions at low energies, four-Fermi current-current interactions shown in Figure 2 are appropriate. There are different effects for different processes. A convenient set of parameters is given in Table 1. Using these parameters, it is easy to compare the sensitivities of different experiments and describe what new limits may be set by a particular result.

A second motivation for experiments is to use electroweak interference to study hadronic structure. Targets are chosen where the form factors in Equation 5 contribute to the asymmetry instead of cancelling. Examples discussed below include measuring strange form factors or measuring the neutron radius of heavy nuclei.

EARLY EXPERIMENTS

The first published experiment with a non-zero result was performed at SLAC [2]. The reaction measured was deep inelastic scattering from deuterium. The experiment convincingly demonstrated that the Z_0 violates parity. In addition, it determined that $\sin^2 \theta_W = 0.224 \pm 0.014$, a result that was competitive with the best measurements of $\sin^2 \theta_W$ then available.

The SLAC eD experiment established many of the techniques that are used today for parity measurements:

1. Use of GaAs as a polarized injector.

2. Rapid random reversal of the helicity of the beam.

3. Integration of signals.

4. Precision beam monitors.

5. Sign reversals, including optical means.

These experimental techniques were further refined at Bates [4]. A different reaction, elastic scattering from ^{12}C, required the measurement of a raw asymmetry of less that 1 ppm, a factor of 50 smaller than the SLAC asymmetry. The new developments include:

1. Development of an intense polarized beam.

2. Intensity feedback.

3. On-line calibration of systematic corrections.

Figure 3 shows the problem of the correlation of the beam intensity with helicity. The top right panel shows the intensity asymmetry versus time that was typical without feedback at Bates. This asymmetry, being unstable and much larger than the experimental asymmetry, was unacceptable. However, as seen on the left panel, the intensity asymmetry is linear in one of the voltages on the Pockels cell, and there is a value that nulls the effect. By measuring the slope of the curve, the change in the voltage required to null the asymmetry could be determined. The intensity asymmetry after the feedback was applied, shown in the bottom right panel, was negligible.

The ^{12}C experiment at Bates measured the model-independent parameter $\tilde{\gamma} = 0.14 \pm 0.03$. This quantity is different from what the SLAC experiment determined: $(\tilde{\alpha} + 0.3\tilde{\gamma} + 0.2\tilde{\beta} + 0.1\tilde{\delta}) = -0.52 \pm 0.06$. Presently, the most precise data on $\tilde{\gamma}$ are obtained from atomic physics experiments [15] using the Cs atom: $(\tilde{\gamma}+ = 0.6\tilde{\alpha})$ is known to $\pm 0.6\%$.

The general experimental methods developed for the SLAC and Bates experiments and used at other laboratories are summarized in Figure 4. Polarized electrons are produced by laser light shining on the GaAs source. The helicity of the light and hence the helicity of the electrons is determined by the voltage on the Pockels cell in the laser beam. A half-wave plate inserted in the laser beam serves as a simple and effective method to reverse the experimental sign of the asymmetry. This procedure checks for systematic errors.

The polarized electrons are accelerated and focused onto a target by a beam

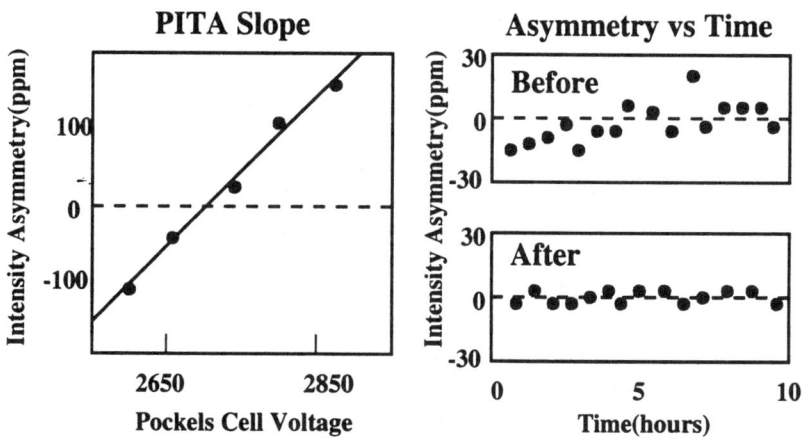

FIGURE 3. Intensity Feedback

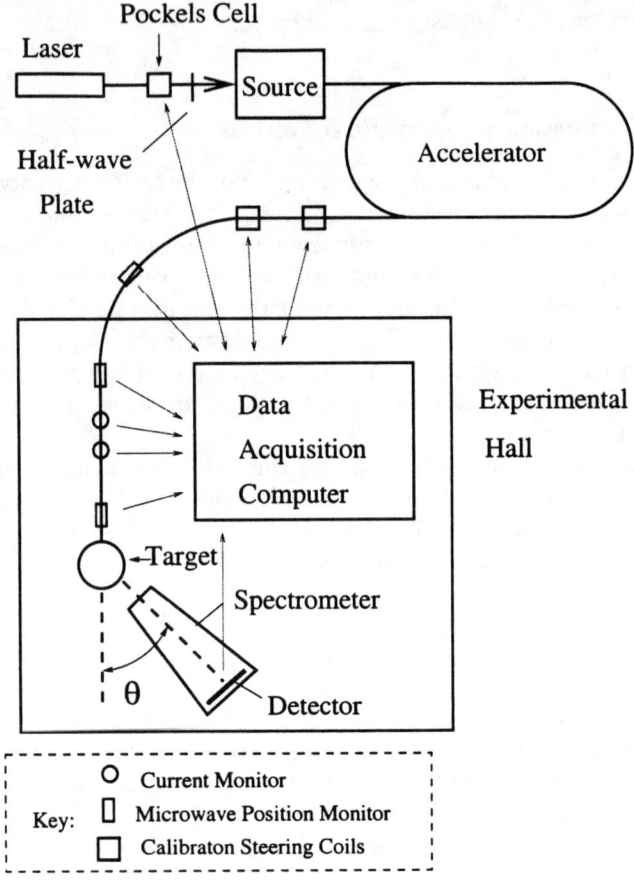

FIGURE 4. Generic Polarized Electron Parity Experiment

transport line. The transport line contains precise microwave position monitors that detect any helicity correlations in beam parameters such as position and energy. The position monitor in the curved part of the beam is sensitive to the beam energy. Current monitors are used in the intensity feedback and also normalize the cross section data. The spectrometers select scattered electrons with the desired kinematics and transport them to the detector. A computer monitors all the signals and records the data. In addition, the computer controls the Pockels cell to null the intensity asymmetry and controls the coils used to calibrate the sensitivity both of the monitors and of the spectrometers to beam parameters.

The helicity of the beam may be reversed rapidly. At Bates, with the high pulse rate, the helicity is reversed 300 times per second. For the continuous wave beam at JLab, the rate of reversal is chosen to be 15 per second. For each pulse, the flux in the detector D and the intensity I in the beam monitor are determined. For each pair of pulses with opposite helicity, an asymmetry

$$A_{pair} = \frac{D^R/I^R - D^L/I^L}{D^R/I^R + D^L/I^L} \qquad (6)$$

The raw experimental asymmetry A_{raw} is typically the average of $10^7 - 10^8$ values of A_{pair}.

THE HAPPEX EXPERIMENT

With the advent of the "spin crisis," [17] the primary motivation for performing polarized electron parity experiments became the search for strange form factors of the proton [18,19]. Two such experiments have been published recently, SAMPLE [5] at Bates and HAPPEX [6] in Hall A at JLab. The SAMPLE experiment, together with more details on the theory of strange form factors, will be discussed in the following contribution. Here I will discuss the HAPPEX experiment, both in the context of how the benefits gained from experience at Bates and also in the context of new techniques that have been developed at JLab. In addition, preliminary results on strange form factors will be given.

The HAPPEX experiment took place in two runs, the first in 1998 and the second in 1999. The 1998 run used the standard bulk GaAs crystal for the polarized source. The polarization was ~40% and the current on target was ~ $100\mu A$. The result was published in 1999 [6]. For the 1999 data, a strained GaAs crystal was used which produced a beam with ~70% polarization and a current of typically $40\mu A$.

The kinematics for the HAPPEX experiment corresponds to elastic scattering with a beam energy of 3.4 GeV and a nominal scattering angle of 12.5°. Two high resolution spectrometers served to focus the elastic events onto detectors made of layers of lead and lucite. The spectrometers performed well; unwanted inelastic events only contributed 0.2% to the signal.

The statistical properties of the data were ideal. For example, Figure 5 shows the distribution of the 23 million values of A_{pair} from the 1998 run. The distribution is Gaussian over seven decades. Data from the 1999 run look similar.

For the 1998 run, helicity correlations in the beam parameters were negligible. The intensity difference, which was controlled by the feedback method developed for the ^{12}C experiment at Bates, was less than 1 ppm. Position differences at the target were on the order of only a few nm. The helicity-correlated energy difference was also negligible; the position difference was <30 nm at a point on the beam line where the dispersion was ~ 5m.

The 1999 run presented some additional challenges. First, the strained GaAs crystal that produces the higher polarization also has a large analyzing power for linearly polarized light and thus tends to cause large helicity correlations in the beam parameters. Second, the experiment was run simultaneously with an intense beam delivered to a second experimental hall, Hall C.

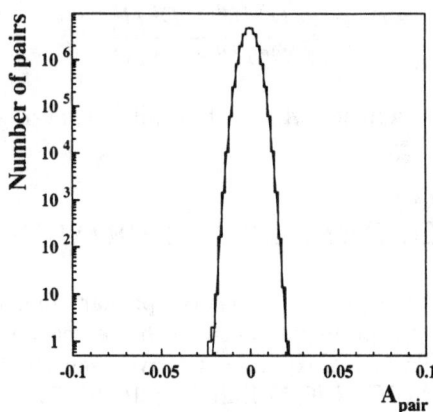

FIGURE 5. Distribution of asymmetries from individual pairs of pulses.

Additional hardware was required to cope with these new problems. A half wave plate that can be rotated remotely was inserted into the laser beam. An angle could always be found so that the intensity asymmetry versus Pockels cell voltage curve was as shown in Figure 3. Then the intensity feedback method developed at Bates worked well.

After the intensity asymmetry in Hall A was nulled, the beam to Hall C developed a large asymmetry which induced position differences in the Hall A beam. A second feedback system was installed to separately null the intensity asymmetry in Hall C. With these changes, the position differences in the Hall A beam were reduced to reasonable values, typically 20nm. However, these values are much larger than those with the bulk GaAs source in 1998.

The contribution that these larger position differences made to the asymmetry was measured by the calibration technique developed at Bates for the ^{12}C experiment. The position and angle of the beam was modulated with coils as shown in Figure 4. In addition, the energy of the beam was modulated. The response of the beam monitors and the detectors to the modulation was measured, and the result was used to determine the correction to the asymmetry. These calibration measurements were made simultaneously with data taking, assuring that the corrections apply to the apparatus under running conditions and saving beam time. As shown in Figure 6, the net contribution to the asymmetry was still negligible.

An important systematic check, first used at SLAC, is to insert a half-wave plate in the laser beam. This reverses the circular polarization of the laser light and hence the sign of the asymmetry, while leaving many possible systematic effects unchanged. At JLab, we grouped the data into sets of 24-48 hours duration. Alternating sets had the half-wave plate inserted. The raw asymmetry as a function

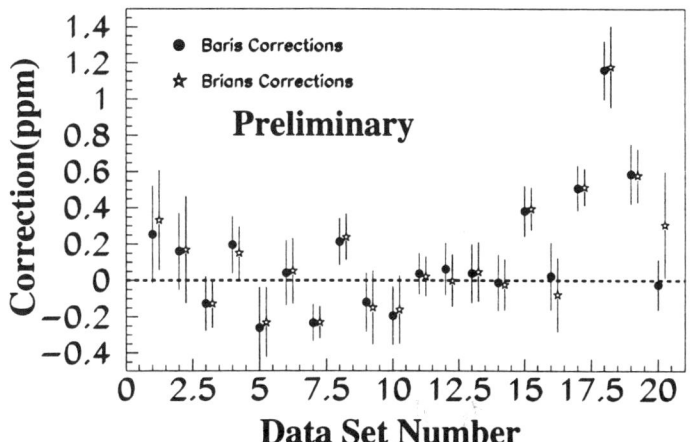

FIGURE 6. Corrections to the asymmetry due to helicity correlations in the beam parameters. The average correction is 0.03 ppm. The analysis was performed by two groups independently yielding similar results.

of data set is given in Figure 7. The asymmetry is constant but reverses sign as expected. The behavior of the data is consistent with the expected statistical fluctuations.

During each data set, the beam polarization P_e was measured by Moller scattering just upstream of the target. The experimental asymmetry was then computed according to $A_{exp} = A_{raw}/P_e$. The result is given in Table 2. A small correction was made for 1.5% of the events coming from the Al end windows on the target and the 0.2% contribution from inelastic events. The dominant systematic errors include a 3.2% uncertainty in the beam polarization and a 2% uncertainty in the average value for Q^2.

The preliminary experimental asymmetry for the combined 1998 and 1999 is A_{exp}=-14.5 ppm with a 6.7% statistical and a 3.8% systematic error. To search for strange matrix elements, we compare to the theoretical expression:

$$A^{PV}(\vec{e}P) \approx -\frac{G_F Q^2}{4\pi\alpha\sqrt{2}} \times \left\{ g_V - \frac{[\varepsilon G_E^p(G_E^n + G_E^s) + \tau G_M^p(G_M^n + G_M^s)]}{[\varepsilon(G_E^p)^2 + \tau(G_M^p)^2]} \right\} \quad (7)$$

where $g_V = 1 - 4\sin^2\theta_W$. The electric(magnetic) Sachs form factors are given by $G_{E(M)}$ and the superscript $n(p)$ refers to the neutron(proton). The strange form factors have the superscript s.

We define A_{ns} to be the theoretical asymmetry for $G_E^s = G_M^s = 0$. Any difference between A_{exp} and A_{ns} is evidence for the presence of strange form factors.

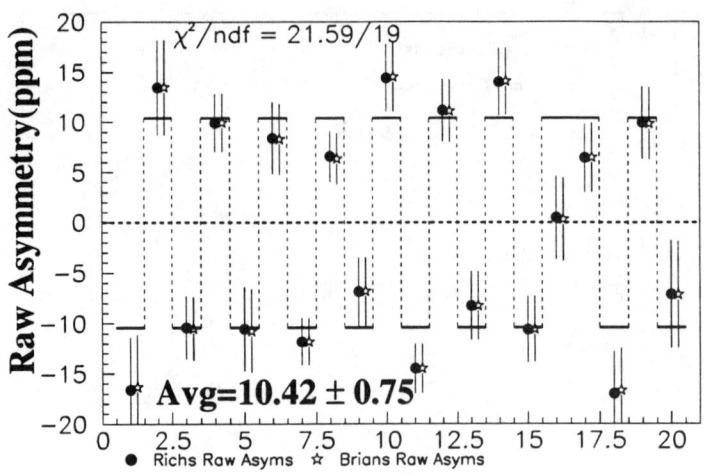

FIGURE 7. Averages of half wave plate data sets

The value for A_{ns} in Equation 7 also depends upon the data used for the electromagnetic form factors. As is apparent from the review by C. W. DeJager in these proceedings, the field of measurements of these form factors is developing rapidly. I will not review the field, but simply show how choosing from among various possible data sets influences the implications of the HAPPEX data. The difference between our result and A_{ns} is shown in Figure 8 for three choices of the form factors. The simplest approach (Dipole-Galster) is to assume the dipole approximation for G_E^p, G_M^p, and G_M^n and the Galster parameterization [20] for G_E^n.

TABLE 2. Preliminary results from HAPPEX

Quantity	Value	Error(%) (Stat.)	Error(%) (Syst.)
$A_{raw}(99)$	-14.7 ppm	7.2	
P_{beam}	68-74%		3.2
Q^2	0.465(GeV/c)^2		2.0
Background	1.7%		
ΔA_{Bg}	0.13 ppm		0.6
$A_{exp}(99)$	-14.6 ppm	7.2	3.8
$A_{exp}(98)$	-14.5 ppm	13.4	
$A_{exp}^{eff}(98)$	-14.2 ppm	15.1	
$A_{exp}(\text{All})$	-14.5 ppm	6.7	3.8

The value for A_{ns} changes by about half of our error if data on G_M^n from Mainz [23] replace the dipole approximation (Mainz-Galster). Alternatively, one might choose data on G_M^n from Bonn [21] and data on G_E^n from Saclay [22] (Bonn-Saclay). If the latter data are accurate, there is some hint that strange matrix elements may be nonzero. Fortunately, new data on these form factors will be available soon.

Also shown are the sizes of contributions from the strange form factors predicted by various theoretical calculations [24–27]. For the Jaffe and Musolf predictions, no Q^2 dependence is given, and we have taken the liberty to assume a Q^2 dependence as suggested in ref. [13] with $\lambda_s = 0$. The largest of the estimations are ruled out, but improvements in the form factor data are needed to provide a more precise search.

Selected Future Experiments

I will conclude by discussing four experiments planned for the future. These experiments have diverse physics motivations, but are similar in that they require the accurate measurement of very small asymmetries.

Two of the experiments are extensions of the HAPPEX experiment and focus on searching for strange form factors. The first, called HAPPEX II, will measure elastic scattering from hydrogen, but at $Q^2 \sim 0.1 (\text{GeV}/c)^2$. The motivation is

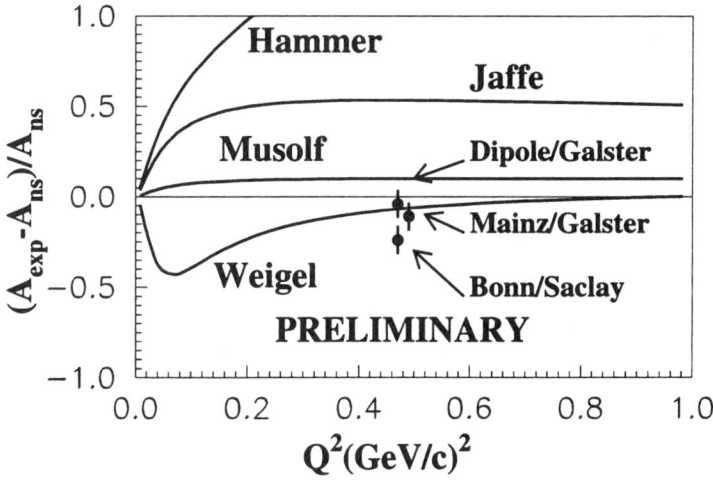

FIGURE 8. Preliminary result of the entire HAPPEX experiment for different assumptions about the nucleon electromagnetic form factors. Also shown are typical theoretical predictions.

the possibility that the strange form factors are large at small Q^2, but fall off significantly at the HAPPEX kinematics. The new kinematics will be reached by using new septum magnets that allow the spectrometers in Hall A to reach angles as small as 6°. This experiment was recently approved at JLab.

The nucleus ^{4}He also makes an attractive target. The asymmetry is sensitive only to G_E^s, whereas for hydrogen the asymmetry depends upon a combination of G_E^s and G_M^s. Combining measurements of He and H at the same Q^2 provides a method of measuring G_E^s and G_M^s separately as shown in Figure 9. At these low values of Q^2, G_E^s provides a measure of the strange radius of the nucleon $\langle r_s^2 \rangle$ and $G_M^s \approx \mu_s$, the strange contribution to the nucleon magnetic moment. Also shown in Figure 9 are various theoretical predictions [25,26,28,29].

Another important issue in nuclear physics is the precise determination of the radius of the neutron distribution in heavy nuclei. An experiment has been proposed to measure A^{PV} for elastic scattering from ^{208}Pb [12]. This will provide data on the neutron radius as easy to interpret as electromagnetic scattering data is for the

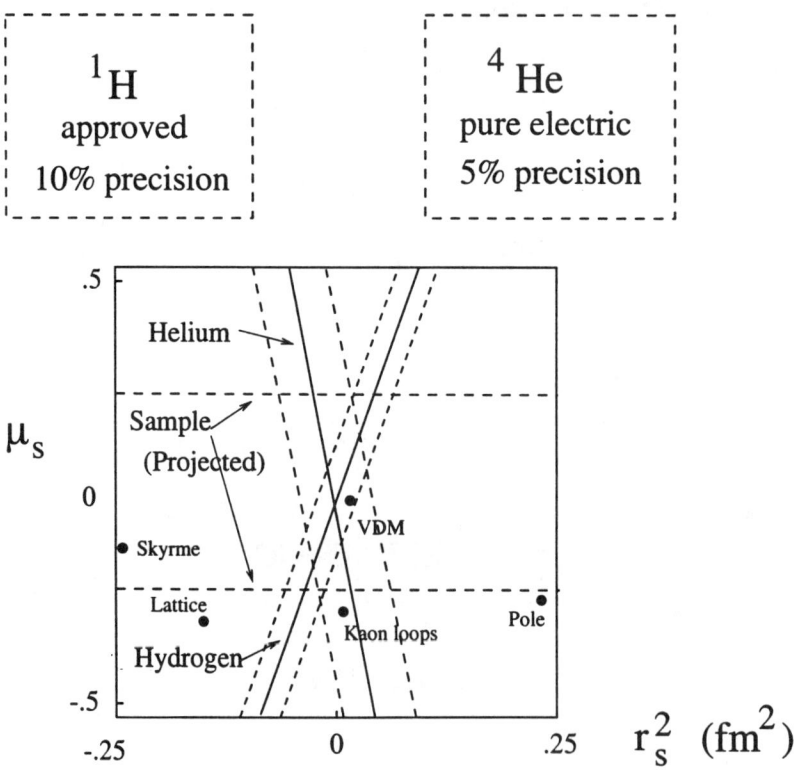

FIGURE 9. Projected results from future HAPPEX experiments from H and He at $\theta = 6°$. Also shown are representative theoretical predictions for strangeness in the nucleon.

charge radius. The basic idea is that the Z couples mainly to neutrons in the same way that the photon couples to protons. Recently, the theory of distorted waves for spin-dependent electron scattering has been worked out [30] and the experiment appears feasible. Given our knowledge of nuclear shapes, a single kinematic point will be sufficient. The point planned is $\theta = 6°$ and E=850 MeV. The asymmetry will be measured to $\sim \pm 3\%$ and will provide a 1% measure of the radius of the neutron distribution.

A new experiment at SLAC, E158 [10], is presently being installed. The goal of the experiment is to perform the most precise measurement of $\sin^2 \theta_W$ away from the Z pole. The result will determine the running of $\sin^2 \theta_W$, a central effect for gauge theories, but one that has not been measured conclusively for the weak interactions. Alternatively, the experiment is sensitive to physics beyond the Standard Model. A new force, such as a neutral current mediated by a new Z boson with a mass as high as 1 TeV, would influence the result. Another possibility is compositeness of the electron characterized by a new strong interaction with a scale $\Lambda < 15$ TeV.

The experiment is a challenging one, with a predicted asymmetry $A^{PV} = 0.32$ ppm. A 10 μA 48 GeV beam will scatter from a 1.5 m long liquid hydrogen target. A spectrometer based on quadrupoles will accept most of the possible solid angle for the desired kinematics. The plan is to obtain first data in the year 2001 and ultimately achieve a precision of 7% in the asymmetry. The success of the ^{12}C experiment at Bates in minimizing the systematic errors in a polarized electron parity experiment is one of the important reasons why this difficult experiment is considered practical.

REFERENCES

1. Ia. B. Zeldovich, *Zh. Eksp. Teor. Fiz.*, **33**, 1531 (1957) [*Sov. Phys. -JETP*, **6** 1184 (1958).
2. C. Y. Prescott, et al., *Phys. Lett. B* **77**, 347 (1978)); *Phys. Lett. B* **84**, 524 (1979)).
3. W. Heil, et al., *Nucl. Phys. B* **327**, 1 (1989).
4. P. A. Souder, et al., *Phys. Rev. Lett.* **65**, 694 (1990).
5. B. Mueller, et al., *Phys. Rev. Lett.* **78**, 3824 (1997).
6. K. Aniol, et al., *Phys. Rev. Lett.* **82**, 1096 (1999).
7. Mainz propoosal A4/1-93 (D. von Harrach, spokesperson).
8. JLab experiment 91-017 (D. Beck, spokesperson).
9. JLab experiment 91-004 (E. J. Beise, spokesperson).
10. SLAC experiment E158 (K. S. Kumar, spokesperson, E. W. Hughes and P. A. Souder, deputy spokespersons).
11. JLab experiment (K. Kumar and D. Lhuillier, spokespersons).
12. JLab experiment 99-012 (R. Michaels and P. A. Souder, spokespersons).
13. M. J. Musolf, et al., Phys. Rep. **239**, 1 (1994) and references therein.
14. J. L. Rosner, preprint hep-ph/9907524; submitted to Phys. Rev. D.
15. S. C. Bennett and C. E. Weiman, *Phys. Rev. Lett.* **82**, 2484 (1999).
16. P. Langacker, M. Luo, and A, K, Mann, *Rev. Mod. Phys.* **64** 87 (1992).

17. J. Ashman, et al., *Phys. Lett. B* **206**, 364 (1988), *Nucl. Phys. B* **328**, 1 (1989).
18. D. B. Kaplan and A. Manohar, *Nucl. Phys. B* **310**, 527 (1988).
19. R. D. McKeown, *Phys. Lett. B* **219**, 140 (1989).
20. S. Galster, et al., *Nucl. Phys. B* **32**, 221 (1971).
21. E. E. W. Bruins, et al., *Phys. Rev. Lett.* **75**, 21 (1995).
22. S. Platchkov, et al., *Nucl. Phys. A* **510**, 740 (1990).
23. H. Anklin, et al., *Phys. Lett. B* **428**, 248 (1998).
24. H. -W. Hammer, Ulf-G. Meissner, and D. Drechsel, *Phys. Lett. B* **367**, 323 (1996).
25. M. J. Musolf and M. Burkhardt, *Z. Phys. C* **61**, 433 (1994).
26. R. L. Jaffe, *Phys. Lett. B* **229**, 275 (1989).
27. H. Weigel, et al., *Phys. Lett. B* **353**, 20 (1995).
28. N. W. Park, J; Schecter, and H. Weigel, *Phys. Rev. D* **43**, 869 (1991).
29. S. J. Dong, K. F. Liu, and A. G. Williams, *Phys. Rev. D* **58**, 074504 (1998).
30. C. J. Horowitz, *Phys. Rev. C* **57**, 3430 (1998).
31. K. S. Kumar, E. W. Hughes, R. Holmes, and P. A. Souder, *Mod. Phys. Lett.* **A10**, 2979 (1995).

Parity Violation and Hadron Structure

E. J. Beise, for the SAMPLE Collaboration

Dept. of Physics, University of Maryland, College Park, MD

Abstract. Parity violation has played an important role in electron scattering since the 1970's when the first experiments were performed at SLAC. Since the late 1980's the focus of experiments has shifted from probing the neutral weak interaction itself to using the exchange of a Z-boson to access new information about the structure of hadrons. A brief survey of experiments that investigate this physics is presented, with a focus on the recently completed SAMPLE experiment at the MIT-Bates Laboratory.

INTRODUCTION

The history of parity violation at the MIT-Bates Laboratory is almost as long as the history of beam delivery at Bates. In the mid-1970's, soon after the completion of the SLAC experiment [1] that studied the neutral weak interaction in deep-inelastic scattering, a parity violation experiment was proposed for Bates in which elastic electron scattering from ^{12}C [2] would be used to probe the neutral weak interaction in a regime not yet studied by the SLAC or atomic parity violation programs. The goal of the experiment was to put limits on the value of $\sin^2\theta_W$. To do so required measurement of a parity violating asymmetry well below one part per million. In order to carry out the experiment, Souder and collaborators implemented several of the SLAC techniques, including polarized beam technology, helicity control and fast precise beam parameter determination. These techniques, described in more detail in ref. [3] of these proceedings, have now become a standard part of the present day program of parity violation measurements at Bates and JLAB.

In the 1980's a parity violation experiment was also carried out at Mainz; quasielastic electron scattering from ^{9}Be [4] The aim of this measurement was to put constraints on the hadronic axial-vector coupling of the e-N interaction. The Mainz collaborators used an open-geometry air Cerenkov detector to achieve a large solid angle for the backward scattered electrons. Because all electrons above 20 MeV were accepted in the detector, the measured asymmetry included not only quasielastic scattering events but also electrons resulting from pion electron production in the dip and Δ regions, resulting in corrections to the quasielastic asymmetry of about 20-25%. Nonetheless, the success of the Mainz experiment demonstrated that a large open-geometry detector could be used in the presence of a high current electron beam with acceptable levels of background.

In 1988, Kaplan and Manohar [5] showed that measurement of the neutral weak interaction between leptons and hadrons would lead to new information on the contribution of strange quarks to the electromagnetic structure of the proton, thus providing a unique window into the proton's quark-antiquark sea. This led to a proposal by McKeown and Beck [6] to use parity violating electron scattering from hydrogen to measure $\bar{s}s$ contributions to the proton's magnetic moment. The experimental design relied on the foundations laid by the earlier Mainz and Bates experiments.

PARITY VIOLATING ELECTRON SCATTERING

The electromagnetic and weak form factors of the proton can be constructed as a sum of individual quark pieces multiplied by coupling constants given by the Standard Model. The neutral weak vector form factors can then be written in terms of the EM form factors and a contribution from strange quarks. For the proton,

$$G_{E,M}^Z = \left(1 - 4\sin^2\theta_W\right)\left(1 + R_V^p\right)G_{E,M}^p - \left(1 + R_V^n\right)G_{E,M}^n - G_{E,M}^s$$

The factors R_V^i are weak radiative corrections that must be applied to account for higher order processes. They have been calculated in ref. [7], and are on the order of a few percent. With the exception of G_E^n, the electromagnetic form factors are determined with very good precision, and the only undetermined quantities in $G_{E,M}^Z$ are the strange quark contributions $G_{E,M}^s$. Thus, a measurement of the weak vector form factors allows a complete decomposition of the proton's electromagnetic structure into contributions from different quark flavors (up, down and strange).

In addition to the weak vector form factors, the axial vector coupling leads to a third form factor G_A^Z.

$$G_A^Z = -\left(1 + R_A^1\right)G_A + R_A^0 + G_A^s$$

In this expression, G_A is the charged current nucleon form factor: its $Q^2=0$ value is determined from neutron β decay. The strange axial form factor G_A^s is the same observable as the Δs deduced from deep-inelastic electron scattering, the contribution of strange quarks to the proton's spin. As in the case of the weak vector form factors, there are higher order terms to consider. The axial radiative corrections $R_A^{0,1}$ are potentially large and very uncertain. They contain contributions from "box diagrams" with both a Z and photon exchange, and from "anapole" type contributions where the electron-nucleon interaction is electromagnetic in nature but there is also a weak interaction between two quarks in the nucleon. An estimate of the radiative corrections to G_A^Z were made in 1990 by Musolf and Holstein [7], based on the weak NN coupling parameters. Since these couplings are also not fully determined experimentally, the estimate used the reasonable ranges indicated in [8].

For elastic scattering from a free proton, the parity-violating asymmetry is

$$A_p = \left[\frac{G_F Q^2}{4\pi\alpha\sqrt{2}}\right] \frac{\left[\varepsilon G_E^p G_E^Z + \tau G_M^p G_M^Z - \left(1 - 4\sin^2\theta_W\right)\varepsilon' G_M^p G_A^Z\right]}{\varepsilon(G_E^p)^2 + \tau(G_M^p)^2},$$

where ε, τ and $\varepsilon' = \sqrt{(1-\varepsilon^2)\tau(1+\tau)}$ are kinematic factors. The first two terms of A_p dominate at forward angles and the latter two terms contribute at backward angles. Although the term containing G_A^Z is suppressed by the factor $(1-4\sin^2\theta_W)$, it still contributes about 20% to the asymmetry under kinematical conditions such as those in the SAMPLE experiment. As a result, the uncertainty in G_A^Z puts a theoretical limit on the extraction of quantitative information about G_M^s from a measurement on the proton alone. Fortunately, independent experimental information on G_A^Z can be extracted by performing the same experiment with deuterium [9]. In the "static" approximation, the deuteron asymmetry is an incoherent sum of contributions from the proton and neutron weighted by the unpolarized cross sections:

$$A_d = \frac{\sigma_p A_p + \sigma_n A_n}{\sigma_d}.$$

The resulting deuteron asymmetry is very insensitive to s-quark contributions but has approximately the same sensitivity to the isovector components of G_A^Z. Hadjimichael, Poulis and Donnelly investigated the dependence of A_d on the structure of the deuteron [10] and found that is insensitive to corrections to the static model at the level of 1-2%. There are also contributions from elastic e-d scattering and from electrodisintegration, which are also estimated to be less than a few percent.

THE SAMPLE EXPERIMENT

The SAMPLE experiment at Bates measured A_p and A_d at ($130° < \theta < 170°$) and $E_{lab} = 200$ MeV, resulting in $Q^2 = 0.1$ (GeV/c)^2, by scattering polarized electrons from unpolarized hydrogen and deuterium. In the case of hydrogen, the expected asymmetry with no contribution from strange quarks is about -7×10^{-6} (or -7 ppm), the deuterium asymmetry is expected to be about 40% larger.

The majority of the data taking for the two experiments was carried out at Bates in the summers of 1998 and 1999 using a 200 MeV polarized electron beam incident on hydrogen (in 1998) and deuterium (in 1999) targets. Scattered electrons were detected by a large solid angle air Čerenkov detector consisting of ten mirrors that image the target onto ten 8 inch photomultiplier tubes. Typically about 40 μA of beam was incident on the target, and in order to dissipate the heat deposited by the beam (approximately 700 Watts in the case of deuterium), the liquid was rapidly circulated through a heat exchanger. The incident electron beam (2.7 mA peak) was pulsed at 600 Hz, and because of the resulting high counting rate the phototube signals were integrated over the 25 µsec beam pulse and normalized to the charge in each burst. Various other beam monitors were also integrated and digitized for every 25 µsec long beam pulse in order to monitor helicity correlations in the beam. The parity-violating asymmetry A was determined from the asymmetries in ratios of integrated detector signal to beam intensity for left- and right-handed beam pulses. Background in the detector was measured by closing shutters in front of the phototubes, with empty target runs, and by reducing the peak current by a large factor so that individually scattered electrons could be detected in coincidence with Čerenkov photons incident on the phototube (the latter technique is described in detail in ref. [14]).

The Bates polarized electron source uses photoemission from unstrained GaAs by circularly polarized laser light, resulting in an electron beam which is 35% polarized. The laser beam helicity for each pulse is determined by a $\lambda/4$ Pockels cell and is randomly chosen for each of 10 consecutive beam pulses; the complement helicities are then delivered for the next 10 pulses. For SAMPLE, the asymmetry in the normalized detector yields was computed for "pulse pairs" separated by 1/60 of a second. In addition, a half-wave plate was periodically inserted upstream of the Pockels cell to reverse the polarization of the beam independent of all electronic signals. (This configuration is denoted as $\lambda/2 = $ "IN" as opposed to $\lambda/2 = $ "OUT".) During each running period, the IN/OUT configuration was reversed every few days to minimize false asymmetries and test for systematic errors.

Helicity correlations of various parameters of the electron beam were monitored continuously. These parameters include the beam intensity, position and angle at the target in both transverse dimensions (x and y), the beam energy, and the beam "halo". Two forward angle lucite Čerenkov counters were also implemented at $\sim 12°$ to monitor luminosity and test for helicity dependence. These monitors detected low Q^2 elastic scattering at forward angles and other soft electromagnetic radiation and should show negligible parity violating asymmetry.

As mentioned earlier, much of the electronics and beam control for the SAMPLE experiment was based on that developed for the earlier ^{12}C elastic scattering measurement, including an active feedback system to reduce any asymmetry introduced by helicity correlations in the beam intensity. In 1998, this feedback was implemented with an independent Pockels cell located between linear polarizers to separate this function from the Pockels cell that controlled the helicity (HPC). The HPC was also repositioned to be downstream of all laser transport elements. These changes resulted in improved stability of the laser beam position under helicity reversal as compared with data taken in 1996. In addition, a feedback system was implemented to reduce the remaining helicity-correlated beam position asymmetry [11]. This was accomplished using a tilted glass plate in the laser beam path and a piezoelectric transducer. By adjusting the tilt of this glass plate with helicity reversal, the first order beam position asymmetry is reduced, resulting in improved quality of the data. For example, the helicity correlated vertical beam shift at the target was reduced from ~ 200 nm to typically < 20 nm.

In the summer 1998, 110 C of beam were delivered to the hydrogen target. In figure 1a, the helicity correlated vertical position differences at the target, for the IN (triangles) and OUT (circles) states of the $\lambda/2$ plate, are shown. Figure 1b contains the raw measured asymmetry for the unweighted sum of all 10 mirror signals. The first few weeks of data were taken before the position feedback system was fully implemented, and significant position asymmetries were present. A linear regression technique is used to remove such effects from the data [12]. The procedure involves six beam parameters: x, y, θ_x, θ_y, energy, and intensity and the correlations between them. It was determined to be very effective in removing correlations in the luminosity monitors, which had large corrections relative to the measured asymmetry. The total correction to the detector asymmetry, however, was only about 5% of the raw asymmetry or 0.06 ppm.

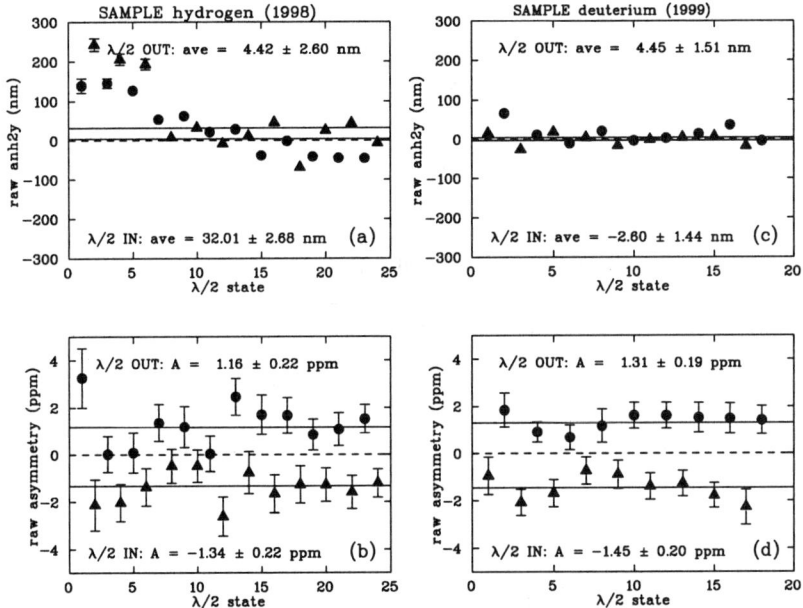

FIGURE 1. (a) Helicity correlated vertical position differences (a) and the raw asymmetry (b) measured during the 1998 SAMPLE hydrogen run. Panels (c) and (d) contain the same information for the 1999 deuterium run. In (b) and (d) the raw asymmetry is computed from the sum of the 10 mirror yields.

The elastic scattering asymmetry was determined from a weighted sum of the 10 individual mirror asymmetries after each was separately corrected for all effects, including background dilution and beam helicity correlations. The resulting asymmetry is

$$A = -4.92 \pm 0.61 \pm 0.73 \,\text{ppm} \tag{1}$$

where the first uncertainty is statistical and the second is the estimated systematic error, dominated by uncertainties in the background dilution. This value [13] is in good agreement with our previous reported measurement [14].

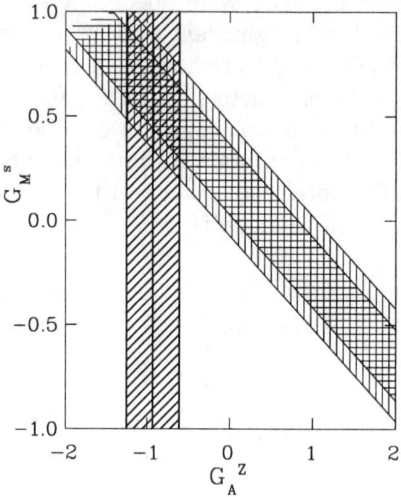

FIGURE 2. Error band of G_M^s for the allowed region (shaded) corresponding to the SAMPLE hydrogen measurement. The inner hatched region includes the statistical error, the outer represents the systematic uncertainty added in quadrature, and the vertical band corresponds to the calculated value of G_A^Z (see text).

At the mean kinematics of the experiment ($Q^2 = 0.1$ (GeV/c)^2 and θ=146.1°), the theoretical asymmetry is

$$A_p = -5.61 + 3.49 G_M^s + 1.55 G_A^Z \qquad (2)$$

Combining the theoretical value with our measured result gives the graphical display in Fig. 2, along with G_A^Z (vertical band) computed from taking the Musolf and Holstein values of R_A^0 and R_A^1, with their errors added quadratically. Combining this band for G_A^Z with our measurement implies a substantially positive value of G_M^s. As noted in recent papers [18,20] most model calculations tend to produce negative values of $\mu_s \equiv G_M^s(Q^2 = 0)$, typically about -0.3. The SAMPLE measurement implies that the computed negative value of G_A^Z is inconsistent with $G_M^s < 0$.

In the summer of 1999, 170 C of polarized beam were delivered to a deuterium target in order to gain experimental information about G_A^Z. Figures 1c and 1d show the deuterium equivalent plots to the hydrogen data. Analysis of these data is still underway, but when completed the deuterium asymmetry should result in a nearly vertical band on figure 2.

PARITY VIOLATION AT JLAB AND MAINZ

At higher beam energies and forward angles it is possible to measure the Q^2 dependence of the strange form factors, and PV measurements of this type are underway at Jefferson Lab [15] and Mainz [16]. The HAPPEX collaboration has recently completed a measurement of A_p at forward angles and $Q^2=0.5$ (GeV/c)^2. The first published results [17] gave a result consistent with no contribution from strange quarks:

$$G_E^s + 0.39 G_M^s = 0.023 \pm 0.034 \pm 0.022 \pm 0.026$$

where the last error is due to the uncertainty in G_E^n. In the most recent HAPPEX run, a strained photocathode was used for the first time to deliver polarized beam to a parity violation experiment, resulting in a factor of two smaller statistical error. The experiment is described in more detail in [3].

At Mainz, a large solid angle detector using a 1022 PBF$_2$-crystal calorimeter is presently under construction [16]. The detectors are designed to have high count rate capabilities and sufficient energy resolution to separate elastic events from the large inelastic background from pion electroproduction. The kinematics of the Mainz experiment are comparable to those of HAPPEX; they will determine the combination $G_E^s + 0.22 G_M^s$ and $Q^2=0.23$(GeV/c)^2.

Although the HAPPEX data is consistent with no contribution from strange quarks, HAPPEX alone does not constrain the individual form factors. At least two models, one based on heavy baryon chiral perturbation theory [19] and one lattice QCD calculation [20], predict that at $Q^2 > 0$, G_E^s and G_M^s have opposite sign. In order to distinguish between these models and the implied result from HAPPEX, G_E^s and G_M^s must be determined independently over a wide range of momentum transfer. The G0 experiment at Jefferson Lab is designed specifically for this purpose, with a large solid angle spectrometer to detect protons from forward angle e-p scattering or electrons from backward angle scattering, up to a momentum transfer of 1 (GeV/c)^2. The experiment uses a superconducting toroidal magnet surrounding a 20 cm liquid hydrogen target, and an array of scintillators with specialized electronics to count scattered particles at a rate of up to a few MHz. Construction is underway, and it is anticipated that the experimental program will begin within the next few years.

Parity violation experiments at Bates have helped lay the foundation for an exciting experimental program that uses the neutral weak interaction to probe hadron structure. After a decade of work we are seeing the first quantitative results of this work from SAMPLE and HAPPEX, and we can look forward to continued progress in this area from experiments now underway at Mainz and Jefferson Lab.

Much of the work describe in this paper was supported by the National Science Foundation and the Department of Energy. The author's work is supported by NSF grant PHY-9971819 and the NSF Young Investigator award PHY-9457906.

REFERENCES

1. C.Y. Prescott *et al.*, Phys. Lett. **B77**, 347 (1978) and C.Y. Prescott *et al.*, Phys. Lett. **B84**, 524 (1979).
2. Bates proposal 77-14, P. Souder and and S. Kowalski, spokespersons. P.A. Souder *et al.*, Phys. Rev. Lett. **65**, 694 (1990).
3. P.A. Souder, these proceedings.
4. W. Heil *et al.*, Nucl. Phys. **B327**, 1 (1989).
5. D. Kaplan and A. Manohar, Nucl. Phys. B **310**, 527 (1988).
6. R. D. McKeown, Phys. Lett. **B219**, 140 (1989), D. H Beck, Phys. Rev. **D39**, 3248 (1989). Bates experiment 89-06, R.D. McKeown and D. Beck, spokespersons.
7. M. J. Musolf and B. R. Holstein, Phys. Lett. **B242**, 461 (1990); M. J. Musolf, *et al.*, Phys. Rep. **239**, 1 (1994).
8. B. Desplanques, J.F. Donoghue and B.R. Holstein, Ann. Phys. (NY) 124, 449 (1980).
9. Bates experiment 94-11 (M. Pitt and E. J. Beise, spokespersons).
10. E. Hadjimichael, G.I. Poulis and T.W. Donnelly, Phys. Rev. **C45**, 2666 (1992).
11. T. Averett, *et al.*, Nucl. Inst. Meth. **A438**, 246 (1999).
12. B. A. Mueller, Ph.D. Thesis, California Institute of Technology, (1997), unpublished.
13. D.T. Spayde, *et al.*, submitted to Phys. Rev. Lett., Sept. 1999, preprint nucl-ex/9909010.
14. B. A. Mueller, *et al.*, Phys. Rev. Lett. 78, 3824 (1997).
15. Experiments E91-010 (P. Souder, *et al.*), E91-017 (D. Beck, *et al.*), and E91-004 (E. Beise, *et al.*).
16. Mainz experiment A4/1-93 (D. von Harrach, *et al.*). F. Maas, private communication.
17. K.A. Aniol, *et al.*, Phys. Rev. Lett. **82**, 1096 (1998).
18. R.D. McKeown, in *Parity Violation in Atoms and Polarized Electron Scattering*, ed. by B. Frois and M.A.-Bouchiat, World Scientific, 423 (1999).
19. T.R. Hemmert, B. Kubis, and U.-G. Meissner, Phys. Rev. **C60**, 045501 (1999).
20. S. J. Dong, K. F. Liu, and A. G. Williams, Phys. Rev. **D58**, 074504 (1998).

Electrons, New Physics, and the Future of Parity-Violation

M.J. Ramsey-Musolf [a,b]

[a] *Department of Physics, University of Connecticut, Storrs, CT 06269 USA*
[b] *Theory Group, Thomas Jefferson National Laboratory, Newport News, VA 23606 USA*

Abstract. The study of parity-violation in semi-leptonic processes has yielded important insights into the structure of the Standard Model and the substructure of the nucleon. I discuss the future of semi-leptonic parity-violation and the role it might play in uncovering physics beyond the Standard Model.

I INTRODUCTION

In addition to celebrating the silver anniversary this year of electron scattering at the MIT-Bates Laboratory, we may also mark the passing of 25 years for another sub-field of physics: parity-violation (PV) in semi-leptonic neutral current interactions. Since the MIT-Bates Laboratory has made important contributions to the field of neutral current PV, it seems highly appropriate to consider the future of the field at this Symposium. Paul Souder has discussed in detail the history of neutral current PV in electron scattering, and Betsy Beise has summarized the present program of strange-quark searches here at MIT-Bates, the Jefferson Lab, and Mainz. Consequently, I will focus on the future: where the field might go once the current round of parity-violating electron scattering (PVES) experiments are completed. I will also broaden the topic to include PV in atoms. Historically, atomic parity-violation (APV) has been at the forefront of the field, and it will undoubtedly continue to hold such a position in the future. In discussing the future, I hope to convey the following three points: (a) the forefront of neutral current PV will consist of searches for physics beyond the Standard Model; (b) APV and PVES can play complementary roles in this search for "new physics"; and (c) parity-violating, low-energy semi-leptonic processes and high-energy collider searches can, in principle, provide complementary insights as to what may lie beyond the Standard Model (SM). My guess is that this situation will persist for the better part of the next decade, until the LHC begins to produce significant physics results.

Before considering the next decade, it is useful to look back briefly a the last quarter century. One may trace the birth of this field to the Bouchiats, who proposed in 1974 that studying PV atomic processes might produce evidence for the weak neutral currents of the SM in the semi-leptonic sector [1]. The Bouchiats suggested a clever technique for enhancing the signal for these tiny neutral currents so that they might be observed in table top experiments. This technique, called "Stark mixing", relies on the interference of a Stark-induced mixing of opposite parity states in an atom and the mixing caused by weak neutral currents. In effect, the Stark-induced amplitude functions as a lever arm to magnify the importance of the neutral current amplitude. The importance of this idea cannot be overstated. Following the Bouchiats' proposal, a number of groups endeavored to search for weak neutral currents in APV, using either the Stark-mixing idea or by studying the rotation of plane-polarized light as it passes through a gas of atoms (see, e.g., Ref. [2] and references therein). In fact, the recent, very precise result for cesium APV reported by the Boulder group was obtained using a variation on the Bouchiats' original Stark-mixing idea [3]. The result of these APV experiments has been to confirm the SM prediction for the structure of the weak neutral current in the low-energy domain at the few percent level. Given the scope of effort involved in testing the SM in high-energy collider experiments, the results of the APV measurements represent a significant triumph for table top physics.

Among noteworthy collider experiments are those involving semi-leptonic PVES. Results from the SLAC deep-inelastic PVES experiment on deuterium were reported in the late 1970's [4]. These results also confirmed the structure of the semi-leptonic weak neutral currents of the SM and yielded a value for the weak mixing angle with nine uncertainty. About a decade later, the collaboration at Mainz reported results on a quasi-elastic PVES experiment involving a ^{8}Be target [5]. This experiment tested a different combination of the neutral current parameters than tested by the SLAC experiment. Shortly after the appearance of the Mainz result, the results of the elastic PVES experiment on ^{12}C performed at Bates were reported [6]. Again, the results of the carbon experiment complemented those from quasi-elastic and deep inelastic measurements and confirmed the predictions of the SM. As discussed in more detail by Paul Souder, an important benefit of these PVES experiments was the development of experimental expertise and technology that is crucial to the success of the present program and the future prospects of PVES.

Turning back to APV, the Boulder group's result for cesium dominates the present landscape. The group reports an experimental error of less than 0.4 %. As with the earlier APV and PVES experiments, the goal of the cesium measurement was to test the SM. The cesium results deviate from the SM prediction by about 1.5%, representing a 2.5σ difference. The potentially serious consequences of this deviation call for a repeat measurement. To that end, the Bouchiat group in Paris is currently involved in another Stark-mixing experiment with cesium, although the experimental uncertainty is not projected to be as small as in the Boulder measurement.

Over the last decade, the emphasis of PVES has shifted away from SM tests to the study of hadron structure. As Paul Souder and Betsy Beise discussed, a well-defined program of measurements to determine the nucleon's strange quark vector current form factors is underway [7]. Instead of studying the structure of the lepton-quark weak neutral current interaction, these experiments rely on the present knowledge of that interaction in order to learn something new about the sea-quark structure of the nucleon. Results from the MIT-Bates backward angle experiments on the proton and deuterium have been reported by the SAMPLE collaboration [8], and the results of a forward angle measurement have been published by the HAPPEX collaboration at the Jefferson Lab [9]. The list of approved strange-quark experiments also includes the G0 experiment at the Jefferson, a Hall-C experiment on ^{4}He, and an experiment on the proton at the MAMI facility in Mainz [7]. The HAPPEX collaboration has also been approved to run another forward angle proton measurement at Q^2 similar to that of the SAMPLE experiment. In addition, the G0 detector will be used to measure the axial vector $N \to \Delta$ transition form factor.

For both PVES and APV, the next generation of experiments are on the horizon. The groups in Seattle and Berkeley have undertaken measurements of APV observables for several atoms along the chain of isotopes. As I discuss below, *ratios* of such observables are less sensitive to atomic theory uncertainties than is the APV observable for a single isotope. One hopes that such measurements may provide an even more precise tool for uncovering new physics than the Boulder cesium experiment. In order to realize this goal, however, one requires a new level of insight into *nuclear* structure than required for the interpretation of a single isotope APV measurement. In the case of PVES, the interest of future experiments seems to be returning to studying the weak neutral current interaction at the elementary fermion level. To that end, a purely leptonic experiment involving PV Møller scattering has been approved for SLAC [10]. Similarly, a letter of intent to perform a precise, forward angle PV $\vec{e}p$ experiment at the Jefferson Lab has appeared [11]. Finally, the Jefferson Lab PAC is considering a proposal to carry out elastic PVES with a ^{208}Pb target [12]. This experiment would provide the most precise information we have to date on the distribution of neutrons in a nucleus, something of considerable interest to nuclear structure physicists. At the same time, the ^{208}Pb experiment may provide enough nuclear structure information to help with the interpretation of the APV isotope ratio studies in terms of new electroweak physics. In this respect, the lead experiment would solidify a unique marriage of table top and collider efforts having important consequences for atomic, nuclear, and particle physics.

In the remainder of this discussion, I consider these future APV and PVES experiments in detail. First, I review the motivation for searching for new physics at low-energies. I subsequently review the basics of the relevant PV observables and show how precise measurements of these observables can provide a window on physics at the TeV scale. I give a few examples of new physics scenarios that can be tested by low-energy PV and consider a possible connection with nuclear β-decay. Finally, I discuss the relationship between the APV isotope ratio studies, the nuclear neutron distribution $\rho_n(r)$, and the PVES experiment on ^{208}Pb. For an in-depth discussion of these issues, I refer the reader to Refs. [13,14]

II SEARCHING FOR NEW PHYSICS

Although there exist a plethora of data confirming the electroweak sector of the Standard Model at the few $\times 0.1\%$ level, there also exist strong conceptual reasons to believe that the SM is only a piece of some larger framework. A nice perspective from which to view the reasons for this belief is the so-called high-energy desert. The high-energy desert is the region in mass scale ranging from the weak scale $M_{\text{WEAK}} \sim 250$ GeV up to the Planck scale $M_P \sim 1/\sqrt{8\pi G_{\text{NEWTON}}} = 2.4 \times 10^{18}$ GeV. The conceptual shortcomings of the SM appear at both edges of this desert. First, at the high-energy end, the SM does not appear to produce unification of the electroweak and strong interactions at any scale. If one perturbatively runs the $SU(3)_C$, $SU(2)_L$, and $U(1)_Y$ couplings up from the weak scale, they never meet at a common point. This lack of unification is undesirable, particularly if one believes a common framework ought to describe the electroweak, strong, and gravitational interactions.

At the low-energy ($\mu \ll M_{\text{WEAK}}$) edge of the desert, the SM is similarly less than satisfying. The most obvious shortcoming is the presence of 19 independent parameters (in the limit of zero neutrino mass) which must be determined from experiment. In addition, the violation of discrete symmetries, such as parity and CP, is put in by hand. The SM does not explain why nature violates these symmetries; it simply incorporates them into a unified framework. Similarly, the quantization of electric charge must be put in by hand; it does not follow naturally (at tree-level in the theory) as does, say, the quantization of isospin charge [15]. A particularly serious challenge for the SM is to account for the wide range of mass scales in the SM spectrum. A related aspect of this "hierarchy problem" has to do with quadratic divergences appearing in the renormalization of the Higgs mass. The presence of these divergences lead one to wonder why the Higgs mass should turn out to be at or below the weak scale without the aid of some fine tuning of electroweak parameters. In short, the SM leaves open many questions regarding the various mass scales governing low-energy physics.

Despite the phenomenological successes of the SM, then, one has good reason to believe there must exist some larger framework which contains the $SU(3)_C \times SU(2)_L \times U(1)_Y$ theory and which, presumably, provides answers to the conceptual puzzles of the SM. Hence, there exists intense interest these days in the search for new physics. In considering what such new physics might be, one faces two broad questions: (a) Which new physics scenarios are most viable, both conceptually and phenomenologically? (b) What are the mass scales associated with a given scenario? In the remainder of this discussion, I will illustrate the insight parity-violating processes involving electrons might play.

III PV OBSERVABLES

The basic quantity of interest in considering neutral current PV is the so-called weak charge, Q_W. This quantity is the weak neutral current analog of the EM charge. It gives the strength of the vector current coupling of Z^0-boson to an elementary fermion or system of fermions. For our purposes, it is useful to write the weak charge as

$$Q_W = Q_W(\text{SM}) + \Delta Q_W(\text{NEW}) + \Delta Q_W(\text{MB}) \tag{1}$$

where the first term, $Q_W(\text{SM})$ is the contribution to the weak charge from the SM. This contribution can be computed precisely and compared with an experimental value for Q_W. Any significant deviation would signal non-zero values for the remaining terms. Of these, $\Delta Q_W(\text{NEW})$ represents contributions from possible physics beyond the SM, while $\Delta Q_W(\text{MB})$ denotes contributions from conventional many-body effects, such as strong interactions among quarks which interfere with the Z^0-quark interaction. The extent to which we can reliably compute the latter determines the confidence with which we can learn about ΔQ_w(NEW) from a given measurement.

Presently, the most precise determination of Q_w has been obtained with APV in cesium. In an APV process, the weak neutral current interaction between the electron and nucleus generates a PV atomic Hamiltonian which mixes states of opposite parity in the atom:

$$\mathcal{H}_w^{PV} = \mathcal{H}_w^{PV}(\text{NSID}) + \mathcal{H}_w^{PV}(\text{NSD}) . \tag{2}$$

Here, "NSID" and "NSD" denote, respectively, nuclear spin-independent and nuclear spin-dependent components of the interaction. The for mer arises from the product of axial vector electron and vector nuclear currents, whereas the latter arises from a $V(e) \times A(\text{nucleus})$ structure. These terms can be separated by measuring PV transitions between different hyperfine levels. The NSID term contains Q_w. The physics of the NSD term, which includes the effects of the nuclear anapole moment, is also interesting, though I will not consider it further here (for a general discussion, see Ref. [16]).

As pointed out by the Bouchiats, the small parity-mixing effects caused by $\mathcal{H}_w^{PV}$ can be enhanced by applying an electric field, which also causes states of opposite parity to mix. Reversing the direction of the applied field can isolate terms in the transition rate which depend on the interference of the Stark and weak interaction amplitudes. In the end, one extracts a ratio such as

$$|A_{PV}| / |A_{STARK}| = \xi Q_w \tag{3}$$

where ξ is an atomic structure-dependent constant that must be computed by atomic theorists. Thus, any errors associated with atomic theory will propagate into uncertainties in Q_w (we might associate these conventional, atomic structure uncertainties with ΔQ_w(MB)). In fact, the dominant uncertainty in the present value for Q_w of cesium is from atomic theory.

An alternate method for determining Q_w is with PVES. In PVES, one scatters longitudinally polarized electrons from a target and compares the cross sections when the electron helicity is flipped. Any non-zero difference results from an interference of the PV neutral current electron-nucleus scattering amplitude and the more familiar, parity-conserving electromagnetic amplitude. The observable of interest in this case is the "left-right" asymmetry

$$A_{LR} = \frac{N_+ - N_-}{N_+ + N_-} = a_0 Q^2 \left\{ \frac{Q_w}{Q_{EM}} + F(q) \right\} . \tag{4}$$

Here, $N_\pm$ denote the number of electrons detected for a given helicity of the incident beam; a_0 is a constant whose scale is set by the Fermi and EM fine structure constants; Q^2 is the square of the momentum transfer; Q_{EM} is the electromagnetic charge of the target; and $F(q)$ is a term which depends on hadronic or nuclear form factors. In principle, one can separate the effects of $F(q)$ from those of Q_w by exploiting the kinematic dependence of the former. The goal of the present strange-quark program is to determined the contribution made by strange quarks to $F(q)$.

TABLE 1. Present and prospective limits on low-energy electroweak observables. First three lines give present weak charge limits for cesium and prospective precision for the SLAC Möller experiment and possible Jefferson Lab experiment. Fourth line gives prospective isotope ratio limits for APV studies in Seattle and Berkeley. Fifth line gives present results for $|V_{ud}|^2$ for nine superallowed β-decays. Following line gives present and prospective limits on the muon anomalous magnetic moment. Final three lines give upper bounds on the permanent EDM's of the electron, neutron, and neutral mercury atom. Experimental errors are denoted by (E) and theoretical uncertainties by (T).

Observable	Quantity	Present Value	Source
Weak Charge	$(Q_W^{EX} - Q_W^{SM})/Q_W^{WM}$	$-0.016 \pm 0.0038(E) \pm 0.005(T)$	Cesium APV
		$? \pm 0.07(E) \pm 0.03(T)$	PV $\bar{e}e$
		$? \pm 0.03(E) \pm 0.03(T)$	PV $\bar{e}p$
Isotope Ratios	$(\mathcal{R}_{EX} - \mathcal{R}_{SM})/\mathcal{R}_{SM}$	$? \pm 0.001(E) \pm 0.004(T)$	APV on Ba,Yb
CKM Matrix	$|V_{ud}|^2_{EX} - |V_{ud}|^2_{SM}$	-0.0028 ± 0.0013	$0^+ \to 0^+$ β-decay
Muon M.M.	$\kappa_\mu^{EX} - \kappa_\mu^{SM}$	$(750 \pm 733) \times 10^{-11}$	Present
		$(? \pm 250) \times 10^{-11}$	BNL E821
EDM	$|d|$	$\leq 4 \times 10^{-28}$ e – cm	electron
		$\leq 0.97 \times 10^{-25}$ e – cm	neutron
		$\leq 9 \times 10^{-28}$ e – cm	^{199}Hg

It is interesting to compare present and prospective determinations of Q_w with those of other low-energy electroweak observables. In Table I, I list several of interest.

The top line in Table I gives the present limits on the agreement of the cesium weak charge with the SM prediction. The Boulder group finds 2.5σ deviation (about 1.5%) from the SM value. The following rows give the expected precision on the weak charge of the electron expected in the SLAC Möller experiment and the weak charge of the proton in a prospective Jefferson Lab experiment. It is worth noting that the electron and proton weak charges are suppressed at tree level by $(1-4\sin^2\theta_W) \approx 0.1$; the electron weak charge is further suppressed by SM radiative corrections [17]. Crudely speaking, then, a 10% determination of the proton or electron weak charge is equivalent to a 1% determination of the weak charge of the cesium atom. The fourth line gives the expected precision for the isotope ratio measurements at Berkeley and Seattle. Note that the prospective experimental error is much smaller than the present theoretical uncertainty – a point I address at the end of this discussion.

The other entries in Table I include the anomalous magnetic moment of the muon, the permanent electric dipole moments (EDM's) of the electron, neutron, and mercury atom; and the square of the $u - d$ matrix element of the CKM matrix. Thus far, one has no evidence of a non-vanishing permanent EDM or of a muon anomalous moment which differs from the SM prediction. In the case of $|V_{ud}|^2$, however, an average over the results of nine superallowed, Fermi nuclear β-decays yields a deviation from the requirements of CKM unitarity at the 2.2σ level (about 0.3%) [18,19]. It is intriguing that both semi-leptonic observables – Q_w from cesium APV and $|V_{ud}|^2$ from superallowed β-decay – have the same relative sign for the experimental deviation from the SM prediction. If this discrepancy is due to new physics, this common sign may point to a common new physics scenario, as I discuss below.

IV PV AND NEW PHYSICS

Before considering specific scenarios for physics beyond the SM, it is useful to consider the generic sensitivity of PV observable to such scenarios. In doing so, I follow the discussion of Ref. [13] restrict my attention to those scenarios which generate new effective, four-fermion interactions. Specifically, I write the PV fermion-fermion interaction as

$$\mathcal{L} = \mathcal{L}^{PV}_{S.M.} + \mathcal{L}^{PV}_{NEW} \tag{5}$$

where

$$\mathcal{L}^{PV}_{S.M.} = \frac{G_F}{2\sqrt{2}} g^e_A \bar{e}\gamma_\mu\gamma_5 e \sum_f g^f_V \bar{f}\gamma^\mu f \tag{6}$$

gives the SM contribution and

$$\mathcal{L}^{PV}_{NEW} = \frac{4\pi\kappa^2}{\Lambda^2} \bar{e}\gamma_\mu\gamma_5 e \sum_f h^f_V \bar{f}\gamma^\mu f \tag{7}$$

is the contribution from some new physics. Here, g^e_A axial vector electron-Z^0 coupling and g^f_V is the vector current coupling of the Z^0 to fermion f. In Eq. (7), Λ denotes the mass scale associated with the new physics and κ^2 parameterizes the overall strength of the interaction. The h^f_V give the scenario-specific couplings of the electron axial vector cu current to the vector current of fermion f. If the SM interaction in Eq. (6) determines the SM value of Q_W, the the fractional shift induced by the new interaction in Eq. (7) is

$$\frac{\Delta Q_W}{Q_W(SM)} = \frac{8\sqrt{2}\pi}{\Lambda^2 G_F} \tag{8}$$

assuming $g^e_A g^f_V$ and h^f_V have commensurate magnitudes. If an experiment is sensitive to shifts on the order of $\Delta Q_W/Q_W(SM) \sim 0.01$, then Eq. (8) implies one is probing new physics at the $\Lambda \sim 20\kappa$ TeV scale. For new physics of a strong-interaction character, one expects $\kappa^2 \sim 1$, while for new gauge interactions one expects $\kappa^2 \sim \alpha$. In either case, high-precision PV measurements are incredibly powerful probes of physics at the TeV scale.

It is instructive to consider how these general features apply in the case of specific new physics scenarios. One of the most interesting such scenarios is that of extended gauge symmetry. The basic of extended gauge symmetry is that the SM group structure is embedded in some larger group G. The full symmetry of G may break down spontaneously at one or more scales M_X above the weak scale, leaving the $SU(3)_C \times SU(2)_L \times U(1)_Y$

symmetry of the SM intact at $M_{\rm WEAK}$. In principle, the gauge bosons associated with the additional symmetries of G will acquire masses commensurate with the symmetry breaking scales M_X. If one of these scales is not too much larger than $M_{\rm WEAK}$, then the additional massive gauge bosons could generate small effects in low-energy processes.

In addition to its phenomenological implications, extended gauge symmetry can provide resolution to some of the rough edges of the SM. For example, if G contains an $SU(2)_R$ subgroup, then one has a natural explanation for PV at low-energies. At some high scale, one has exact parity symmetry. However, if the scale of symmetry breaking associated with the right-handed sector is much larger than $M_{\rm WEAK}$, the right-handed gauge bosons will be too heavy to compete effectively with the SM gauge bosons, so that low-energy processes favor the left-handed sector. Similarly, the electromagnetic charge can appear as a generator of G, in which case its quantization is natural. Even the apparent lack of SM coupling unification can be resolved by extended gauge symmetry. The presence of additional symmetry breaking scales implies that the running of the couplings will change as one crosses each scale. Thus, there exists sufficient room within different extended gauge group scenarios to bring about coupling unification near the expected grand unified scale.

Here, I concentrate on the neutral current phenomenology of extended gauge symmetry. Specifically, I consider a scenario in which spontaneous symmetry breaking of G yields a second neutral gauge boson Z' with mass not too different from the weak scale. To make life simple, I also consider the case in which this Z' does not mix with the SM Z^0. If it did mix, its effects would show up strongly in the Z-pole observables. In fact, the latter severely constrain the mass of a Z' that does mix with the Z^0 [20]. In the language of Eq. (7), we have for this scenario $\kappa^2 = \alpha'$, the fine-structure constant associated with the Z' interaction; $\Lambda = M_{Z'}$; and the h_V^f to be specified by a particular scenario.

Given the experimental precisions listed in Table I, how sensitive would the different measurements be to extended gauge symmetry-induced new interactions? A detailed summary is given in Ref. [13]. Here, I quote a few illustrative examples. Extended gauge symmetry scenarios which fit naturally into the framework of heterotic strings live in a group called E_6. The factors of E_6 include two U(1) groups called $U(1)_\chi$ and $U(1)_\psi$. The neutral gauge boson associated with the $U(1)_\chi$ would show up particularly strongly in low-energy PV if it had a sufficiently low mass; the Z_ψ, on the other hand, does not contribute to PV amplitudes at tree-level. Let G_χ denotes the Fermi constant associated with the interactions of the Z_χ. We may characterize the sensitivity of various PV observable in terms of the ratio $r_\chi = G_\chi/G_F$. The present cesium APV is able to discern effects of the scale $r_\chi \sim 0.003$ or larger. The sensitivities of the SLAC Möller experiment, the proposed Jefferson Lab PV ep experiment, and the isotope ratio measurements are comparable. We can turn this statement about Fermi constants into mass limits by assuming the break down of E_6 to the SM $\times$ $U(1)_\chi$ occurs in one step, so that the coupling associated with the new U(1) group is maximal. In this case, the cesium APV, isotope ratio, and PVES measurements would probe M_{Z_χ} at about the one TeV level or better. In contrast, the sensitivity of the cesium measurement to a neutral right-handed gauge boson vastly exceeds the corresponding sensitivities of the isotope ratio and PVES measurements. Thus, the use of different low-energy PV measurements could prove useful in sorting out among competing scenarios.

It is also interesting to compare the sensitivities of low-energy PV and high-energy collider experiments. In terms of mass limits, the "reach" of the present and prospective PV experiments exceeds that of the Tevatron by almost a factor of two. Even an up-graded Tevatron (Tev33) would only achieve comparable sensitivities. One must wait until the LHC has taken sufficient data before the PV sensitivities will be surpassed. In fact, the information provided by colliders and the PV measurements is complementary. The colliders are primarily sensitive to the mass scale associated with the new gauge boson relatively insensitive to the coupling strength g' or detailed structure of the fermion-Z' coupling. The PV observables, in contrast, depend on $(g'/M_{Z'})^2$ (κ/Λ in the language of Eq. (7)) and on the effective couplings fermion-Z' couplings (h_V^f in Eq. (7)).

To illustrate, I again consider E_6 theories [21]. The phenomenology of neutral E_6 gauge bosons is essentially governed by three parameters: $M_{Z'}$; a parameter λ_g which governs the overall coupling strength g' and whose value depends on the number of symmetry breaking steps leading to a massive Z'; and an "extended" weak mixing angle ϕ which describes the structure of the additional "low-energy" U(1) group. Specifically, if Z_χ and Z_ψ are the gauge bosons associated with the U(1)$_\chi$ and U(1)$_\psi$ groups, respectively, then a general neutral E_6 gauge boson can be written as

$$Z' = \cos\phi Z_\psi + \sin\phi Z_\chi . \tag{9}$$

The couplings h_V^f of this Z' to electrons and light quarks are given by

$$h_V^u = 0 \tag{10}$$

$$h_V^d = -h_V^e = \left[\sin^2\phi - \sqrt{15}\sin\phi\cos\phi/3\right]/20 . \tag{11}$$

Note that for $\phi = 0$ or π, $Z' = Z_\psi$ and all of the PV couplings vanish. The d-quark and electron couplings also vanish for $\phi = \phi_c = \tan^{-1}(\sqrt{5/3})$ and have opposite signs for ϕ on either side of ϕ_c. Thus, the net effect of the Z' on Q_W can be either positive or negative, depending on the value of ϕ. The present present cesium APV results favor $\phi > \phi_c$, if an E_6 gauge boson is responsible for the observed deviation from the SM value for Q_W. This kind of information about the structure of the extended gauge sector is difficult to obtain from high-energy collider limits.

It is also amusing to combine information obtained from colliders and low-energy experiments. To do so, let's assume the E_6 gauge boson is responsible for the deviation of the cesium Q_W from the SM value (about a two σ effect). Under this assumption, one has a relationship between $M_{Z'}$, λ_g, and ϕ. A second condition derives from the CDF lower bounds, which are roughly 600 GeV with little dependence on the value of ϕ. Combining the two pieces of information, one obtains

$$600 \text{ GeV} \lesssim M_{Z'} \lesssim 1.15\lambda_g \text{ TeV} \tag{12}$$

where $\lambda_g \leq 1$. This range is already rather narrow. If a future up-graded Tevatron found no evidence for extra neutral gauge bosons with a mass less than about one TeV, then a low-mass Z' would be ruled out as the culprit behind the cesium APV result.

Another popular extension of the Standard Model is supersymmetry. The literature on SUSY extensions of the SM is legion, so I will not discuss SUSY models in detail. The appeal of SUSY includes its solution to the hierarchy problem associated with mass renormalization. In addition, the gauge couplings in the minimal supersymmetric standard model (MSSM) unify at the GUT scale when run perturbatively up from the weak scale. Whether this coupling unification is fortuitous or reflects deeper physics can be debated. It is, nevertheless, intriguing. One important characteristic of the MSSM as far as low-energy phenomenology is concerned involves a quantity called R-parity. The R-parity quantum number is defined as

$$P_R = (-1)^{3(B-L)+2S} \tag{13}$$

where B, L, and S denote the baryon number, lepton number, and spin, respectively, of a given particle. Every SM particle has $P_R = 1$ while each superpartner has $P_R = -1$. The MSSM conserves total P_R, which implies that every interaction involves an even number of superpartners. As a result, superpartners cannot appear in low-energy processes involving SM particles at tree-level. They only contribute through loops. Their effects are correspondingly suppressed by loop factors, making them hard to see at low-energies.

It is possible, however, to write down simple extensions of the MSSM in which P_R is not conserved. In such B and/or L-violating theories, superpartner effects can appear at tree-level. To illustrate, consider a purely leptonic R parity-violating SUSY model. The relevant Lagrangian is [22]

$$\mathcal{L}_{\text{RPV}} = \lambda_{ijk}(\tilde{e}_R^k)^* (\bar{\nu}_L^i)^c e_L^j + \text{h.c.} \tag{14}$$

where $\tilde{e}_R^k$ denotes the bosonic superpartner of a right-handed charged lepton of generation k (the other superscripts denote generation). Since the interaction contains three leptons, L (and P_R) are not conserved. Tree-level exchange of the $\tilde{e}_R^k$ between lepton currents can generate new four-fermion effective interactions, such as the following interaction relevant to μ-decay:

$$\mathcal{L}_{\text{EFF}} = -(\lambda_{12k}/\sqrt{2}M_{\phi_{kR}^c})^2 \bar{e}_L \gamma_\alpha \nu_L^e \bar{\nu}_L^\mu \gamma^\alpha \mu_L . \tag{15}$$

The interaction of Eq. (14) may provide a partial explanation for both the cesium APV result and the apparent CKM unitarity violation inferred from the superallowed β-decays. The reason has to do with the Fermi constant. Both the β-decay amplitude and the PV amplitude of Eq. (6) are written in terms of the Fermi constant. The reason is that these amplitudes depend on g^2/M_w^2, which can be related to the Fermi constant as measured in μ-decay. At tree-level, this relationship is given by

$$\frac{g^2}{8M_w^2} = \frac{G_F}{\sqrt{2}} . \tag{16}$$

Because of the precision with which μ-decay is measured, Eq. (16) must be modified to account for electroweak radiative corrections:

$$\frac{g^2}{8M_W^2}(1+\Delta r) = \frac{G_\mu}{\sqrt{2}} \qquad (17)$$

where Δr contains the radiative corrections. Suppose now some new physics, such as the interaction of Eq. (15), contributes to μ-decay. Then one must further modify Eq. (17) as

$$\frac{g^2}{8M_W^2}(1+\Delta r + \Delta_\mu^{NEW}) = \frac{G_\mu}{\sqrt{2}} \qquad (18)$$

where Δ_μ^{NEW} gives the corrections from the new interaction. When writing down the amplitude for β-decay or PV, one needs g^2/M_W^2 in terms of G_μ:

$$\frac{g^2}{8M_W^2} = \frac{G_\mu}{\sqrt{2}}(1 - \Delta r - \Delta_\mu^{NEW}) \qquad (19)$$

to first order in the small corrections.

To make contact with the semi-leptonic observables, it is useful to consider the effective Fermi constants G_F^β and G_F^{PV} which govern them. In terms of other quantities, these effective Fermi constants are

$$G_F^\beta = G_\mu(1 - \Delta r + \Delta r_\beta - \Delta_\mu^{NEW} + \Delta_\beta^{NEW})|V_{ud}|^2 \qquad (20)$$
$$G_F^{PV} = G_\mu(1 - \Delta r + \Delta r_{PV} - \Delta_\mu^{NEW} + \Delta_{PV}^{NEW})Q_W \qquad (21)$$

where Δr_β and Δr_{PV} denote SM radiative corrections to the β-decay and PV amplitudes, respectively, and Δ_β^{NEW} and Δ_{PV}^{NEW} are the corresponding contributions from new interactions.

The results of from the superallowed decays and cesium APV imply

$$G_F^{\beta,EX}/G_F^{\beta,SM} < 1 \qquad (22)$$
$$G_F^{PV,EX}/G_F^{PV,SM} < 1 \qquad (23)$$

where the EX and SM superscripts denote the experimental and SM values, respectively. From Eq. (20), we see that if the new physics contributions vanish, one obtains the conventional interpretation of the experimental results:

$$|V_{ud}|_{EX}^2/|V_{ud}|_{SM}^2 < 1 \qquad (24)$$
$$Q_W^{EX}/Q_W^{SM} < 1. \qquad (25)$$

However, an equally acceptable explanation is to assume $|V_{ud}|^2$ and Q_W assume their SM values and that

$$\Delta_\beta^{NEW} - \Delta_\mu^{NEW} < 1 \qquad (26)$$
$$\Delta_{PV}^{NEW} - \Delta_\mu^{NEW} < 1. \qquad (27)$$

In particular, if both Δ_β^{NEW} and Δ_{PV}^{NEW} vanish and if $\Delta_\mu^{NEW} > 0$, the measured effective Fermi constants in β-decay and cesium APV would be smaller in magnitude than the SM predictions.

The R parity-violating interaction of Eq. (15) generates just such a positive value for Δ_μ^{NEW}:

$$\Delta_\mu^{NEW} = \frac{\lambda_{12k}^2}{4\sqrt{2} G_\mu M_{\tilde{\phi}_{kR}^e}^2}. \tag{28}$$

Using the present experimental results and Eq. (20) one obtains

$$\lambda_{12k} = (0.027 \pm 0.007)(M_{\tilde{e}_k}/100 \text{ GeV}) \tag{29}$$

from superallowed decays and

$$\lambda_{12k} = (0.13 \pm 0.05)(M_{\tilde{e}_k}/100 \text{ GeV}) \tag{30}$$

from cesium APV. Although these results differ by more than one σ, one should keep in mind that the cesium result is the first PV result to differ from the SM, whereas the superallowed results depend on an average of ft values for nine different decays, several of which have been measured more than once. In short, the precise magnitude of the deviation leading to Eq. (30) may not be as robust as that observed in β-decay. The primary point here is that the magnitudes of the results in Eqs. (29–30) are not too distinct, and the signs of the observed deviations are both consistent with the R parity-violating effects in Eqs. (15) and (28). It will be interesting to see whether future electron PV experiments also produce deviations from the SM predictions consistent with this SUSY scenario[1].

V INTERPRETATION ISSUES AND NEUTRON DISTRIBUTIONS

In general, the interpretation of precision, low-energy measurements raises thorny issues not relevant to high-energy measurements. The PV processes discussed here are no exception. To illustrate, I consider the interpretation of atomic PV. As noted above, the dominant error in the cesium weak charge comes from atomic theory. Although this theory error appears to have been reduced in light of new measurements of parity-conserving atomic transitions, it is questionable whether further reductions can be achieved. A clever strategy for evading this atomic structure uncertainty is to measure ratios of APV observables along an isotope chain. A representative ratio is

$$\mathcal{R} = \frac{A_{PV}^{NSID}(N') - A_{PV}^{NSID}(N)}{A_{PV}^{NSID}(N') + A_{PV}^{NSID}(N)} \tag{31}$$

where $A_{PV}^{NSID}(N)$ is an APV nuclear spin-independent observable for an atom with neutron number N. Since the atomic electronic structure contributions $A_{PV}^{NSID}(N)$ and $A_{PV}^{NSID}(N')$

---

[1] Another constraint on R parity-violating SUSY may be obtained from relations among electroweak parameters. The constraints imposed by these relations on some types of new physics have been analyzed in Ref. [23]. The corresponding SUSY constraints will be discussed in a forthcoming publication

are relatively constant (for a given Z), the atomic structure-dependence drops out of the ratio $\mathcal{R}$ and one has

$$\mathcal{R} \approx \frac{Q_w(N') - Q_w(N)}{Q_w(N) + Q_w(N')} \equiv \mathcal{R}_{SM}(1 + \delta_\mathcal{R}) \tag{32}$$

where $\mathcal{R}_{SM}$ is the value of the ratio in the SM.

The correction $\delta_\mathcal{R}$ contains contributions from possible new physics. As first pointed out by Fortson, Wilets, and Pang, however, there is also a second effect due to the variation of the neutron density $\rho_n(r)$ along the isotope chain [24]. To get an idea of the relative importance of these two contributions, one can model the nucleus as a sphere of constant neutron and proton density out to radii R_N and R_P, respectively. In this case, one has

$$\delta_\mathcal{R} \approx \left(\frac{2Z}{N+N'}\right) \Delta Q_w^P - \left(\frac{N'}{\Delta N}\right)(Z\alpha)^2(3/7)\delta(\Delta X_N) \tag{33}$$

where ΔQ_w^P is the shift in the proton's weak charge due to new physics,

$$\Delta X_N = \frac{R_{N'} - R_N}{R_P} \tag{34}$$

is the shift in the mean square neutron radius (relative to the proton radius) along the isotope chain, and $\delta(\Delta X_N)$ is the uncertainty in this shift.

Several features of Eq. (33) are worth noting. First, the shift in the ratio $\mathcal{R}$ due to new physics depends primarily on the shift in the weak charge of the proton. The shift in the weak charge of the neutron largely cancels out of the ratio, to first order in small shifts. Whereas the weak charge of a single isotope is slightly more sensitive ΔQ_w^N than to Q_w^P, the sensitivity of $\mathcal{R}$ to new physics is dominated by ΔQ_w^P. Second, the dependence of $\mathcal{R}$ on variations in neutron radii along the isotope shift is enhanced by a factor of $N'/\Delta N$. For a heavy atom like cesium or barium, for example, this enhancement factor can be on the order of 5. Thus, if one is going to use APV isotope ratio measurements to learn about ΔQ_w^P, one must have extremely precise knowledge of the shift in neutron radii.

At present, there exist no high-precision experimental determinations of the neutron radii of heavy nuclei. Consequently, nuclear theory must be used to determine the second term on the RHS of Eq. (33). To set the scale of the level of accuracy nuclear theory must achieve to make the isotope ratio measurements useful, supposed we require the uncertainty in the neutron radius term to be as small as the prospective experimental uncertainty in the value of $\mathcal{R}$, namely, 0.1 %. Pollock [25] and Chen and Vogel [26,27] have analyzed the nuclear model spread in ΔX_N; from their analyses, we learn that nuclear theory is at least a factor of two away from achieving the requisite precision (for a summary of the theoretical situation, see Ref. [13]). In principle, this presents a stumbling block for the isotope ratio program.

There exist two strategies for overcoming this difficulty. One is to perform a direct measurement of ΔQ_w^P using PVES from a proton target. From Eq. (4), we may write the proton asymmetry as

$$A_{\text{LR}}(^1\text{H}) = a_0 Q^2 \left[Q_{\text{w}}^P + F^P(q) \right] \tag{35}$$

where Q_{w}^P is the proton weak charge. The form factor term $F^P(q)$ vanishes in the forward angle limit. Thus, by going to forward angle kinematics, the Q_{w}^P can be separated from $F^P(q)$. The form factor term is presently under study in the strange quark experiments. Upon completion of the strange quark program, this term should be known with sufficient precision over a large enough kinematic range to afford a precise separation of Q_{w}^P in a future, forward angle measurement. A letter of intent for such a measurement has recently been issued [11]. The proposed measurement would employ a re-configured G0 apparatus in order to reach sufficient forward angle kinematics. It is hoped that this measurement will yield a 3–5% determination of Q_{w}^P. This level of precision would be comparable to a 0.1–0.2% determination of $\mathcal{R}$, if the interpretation of the latter were not clouded by $\rho_n(r)$ uncertainties.

A second for getting around the $\rho_n(r)$ problem in Eq. (33) involves measuring the neutron distribution of a heavy nucleus using PVES. It is possible that a sufficiently precise determination of $\rho_n(r)$ on a single isotope would sufficiently constrain nuclear theory that the nuclear model-dependence in the isotope shifts, $\delta(\Delta X_N)$ would be reduced to an acceptable level. The idea for using PVES to determine $\rho_n(r)$ was first suggested by Donnelly, Duback, and Sick [28]. These authors noted that the Z^0 preferentially sees neutrons over protons, since at tree-level in the SM, $Q_{\text{w}}^P = 1 - 4\sin^2\theta_{\text{w}} \sim 0.1$ whereas $Q_{\text{w}}^N = -1$. Thus, the PV asymmetry for scattering from a heavy nucleus should be quite sensitive to the neutron distribution. To illustrate this idea, consider PVES from a $(J^\pi, T) = (0^+, 0)$ nucleus. The asymmetry has the form [28,14]

$$-\left[\frac{4\sqrt{2}\pi\alpha}{G_F |Q^2|}\right] A_{\text{LR}} = Q_{\text{w}}^P + Q_{\text{w}}^N \frac{\int d^3x \, j_0(qx) \rho_n(\vec{x})}{\int d^3x \, j_0(qx) \rho_p(\vec{x})}. \tag{36}$$

Since $\rho_p(\vec{x})$ is typically known with very high accuracy, the PV asymmetry essentially becomes a "meter" of $\rho_n(q)$. This idea is being exploited in a proposal before the Jefferson Lab PAC [12].

It goes without saying that a precise determination of $\rho_n(q)$ for any heavy nucleus is of fundamental interest for nuclear structure physics. From this standpoint alone, the investment of effort in making the measurement is well-justified. It remains to be seen, however, whether the information gleaned from a precise determination of $\rho_n(q)$ for ^{208}Pb at one or two kinematic points will suffice to reduce the nuclear structure uncertainty in Eq. (33). For example, it is unlikely that lead atoms will be used in the APV isotope ratios. The isotopes of Ba and Yb are currently under study in Seattle and Berkeley. Moreover, the interpretation of $\mathcal{R}$ requires knowledge of $\rho_n(r)$ in more detail than implied by the simplified expression in Eq. (33). Whether knowledge of the momentum-space distribution at a few points will supply the necessary details about $\rho_n(r)$ is an open question. Finally, the constraints which knowledge of $\rho_n(r)$ for a single isotope would place on calculations of isotope shifts has yet to be quantified. In short, there exist several challenges for nuclear theory in making a PVES determination of $\rho_n(q)$ useful for the APV isotope ratios (for a recent discussion of these issues, see Ref. [29]). From this standpoint, a measurement of the PV $\vec{e}p$ asymmetry provides a cleaner and more direct window on ΔQ_W^P.

VI CONCLUSIONS

The field of parity-violation with electrons has made tremendous strides in 25 years. I hope this discussion has convinced the reader that its future prospects are just as exciting as its history. For the next decade at least, it is likely that PV with electrons will provide one of the most powerful probes of new physics at the TeV scale, complementing information to be gained from high-energy collider experiments. At the same time, it will remain a focal point for interdisciplinary activity, bringing together insights from particle, nuclear, and atomic physics. One may only speculate as to the *new* insights PV with electrons will provide for each field by the time a Bates-35 celebration is planned.

ACKNOWLEDGMENTS

It is a pleasure to thank W.J. Marciano, D. Budker, R. Carlini, J.M. Finn, E.N. Fortson, S.J. Pollock, and P. Souder for useful discussions and S.J. Puglia for assistance in preparing the manuscript. This work was supported in part under U.S. Department of Energy contract #DE-AC05-84ER40150 and a National Science Foundation Young Investigator Award.

REFERENCES

1. M.A. Bouchiat and C. Bouchiat, Phys. Lett. **B48**, 111 (1974); J. Phys. (Paris) **35**, 899 (1974).
2. D. Budker, "Parity Nonconservation in Atoms", to appear in proceedings of WEIN-98, C. Hoffman and D. Herzceg, Eds., World Scientific, Singapore, 1998.
3. C.S. Wood *et al.*, Science **275** (1997) 1759; S.C. Bennett and C.E. Wieman, Phys. Rev. Lett. **82**, 2484 (1999).
4. C.Y. Prescott *et al.*, Phys. Lett. **B77** (1978) 347; Phys. Lett. **B84** (1979) 524.
5. W. Heil *et al.*, Nucl. Phys. **B327** (1989) 1.
6. P.A. Souder *et al.*, Phys. Rev. Lett. **65** (1990) 694.
7. MIT-Bates experiment 89-06 (1989), R.D. McKeown and D.H. Beck spokespersons; MIT-Bates experiment 94-11 (1994), M. Pitt and E.J. Beise, spokespersons; Jefferson Lab experiment E-91-017 (1991), D.H. Beck, spokesperson; Jefferson Lab experiment E-91-004 (1991), E.J. Beise, spokesperson; Jefferson Lab experiment E-91-010 (1991), M. Finn and P.A. Souder, spokespersons; Mainz experiment A4/1-93 (1993), D. von Harrach, spokesperson.
8. B. Mueller et al., SAMPLE Collaboration, Phys. Rev. Lett. **78** (1997) 3824; D.T. Spayde et al., SAMPLE Collaboration, [nucl-ex/9909010].
9. K.A. Aniol et al., HAPPEX Collaboration, Phys. Rev. Lett. **82** (1999) 1096, [nucl-ex/9810012].
10. SLAC Proposal E158 (1997), K. Kumar, spokesperson.

11. R. Carlini, J.M. Finn, and M.J. Ramsey-Musolf, Letter of Intent to the Jefferson Laboratory PAC, unpublished (1999).
12. Proposal to the Jefferson Laboratory PAC, R. Michels and P.A. Sounder, spokespersons (1998).
13. M.J. Ramsey-Musolf, Phys. Rev. **C60**, 015501 (1999).
14. M.J. Musolf et al., Phys. Rep. **239** (1994) 1.
15. R.N. Mohapatra, *Unification and Supersymmetry*, Springer-Verlag, New York, 1992.
16. M.J. Musolf and B.R. Holstein, Phys. Rev. **D 43**, 2956 (1991).
17. A. Czarnecki and W.J. Marciano, Phys. Rev. **D53** (1996) 1066.
18. I.S. Towner and J.C. Hardy, "Currents and their Couplings in the Weak Sector of the Standard Model", in *Symmetries and Fundamental Interactions in Nuclei*, W.C. Haxton and E.M. Henley, Eds., World Scientific, Singapore, 1995, p. 183.
19. E. Hagberg et al., in *Non-Nucleonic Degrees of Freedom Detectedin the Nucleus*, T. Minamisono et al., Eds., World Scientific, 1996, Singapore.
20. P. Langacker, "Tests of the Standard Model and Searches for New Physics", in *Precision Tests of the Standard Electroweak Model*, P. Langacker, Ed., World Scientific, Singapore, 1995, p.883.
21. D. London and J.L. Rosner, Phys. Rev. **D34** (1986) 1530.
22. V. Barger, G.F. Guidice, T. Han, Phys. Rev. **D40** (1989) 2987.
23. W.J. Marciano, Phys. Rev. **D60**, 093006 (1999).
24. E.N. Fortson, Y. Pang, L. Wilets, Phys. Rev. Lett. **65** (1990) 2857.
25. S.J. Pollock, E.N. Fortson, L. Wilets, Phys. Rev. **C46** (1992) 2587.
26. B.Q. Chen and P. Vogel, Phys. Rev. **C48** (1993) 1392.
27. P. Vogel, "Atomic Parity Non-conservation and Nuclear Structure" in *Nuclear Shapes and Nuclear Structure*, M. Vergnes, D. Goute, P.H. Heenen, and J. Sauvage, Eds., Edition Frontieres, 1994, Gif-sur-Yvette.
28. T.W. Donnelly, J. Duback, I. Sick, Nucl. Phys. **A503** (1989) 589.
29. C.J. Horowitz, S.J. Pollock, P.A. Souder, and R. Michaels [nucl-th/9912038] (1999).

A Look to the Future

Richard G. Milner

MIT-Bates Linear Accelerator Center, P.O. Box 846, Middleton, MA 01949, USA

Abstract. The MIT-Bates Linear Accelerator Center carries out research into the structure of hadrons using electron beams at energies up to 1 GeV. The major research thrust for the future utilizes the South Hall Ring (SHR). Experiments using extracted beam and the out-of-plane spectrometer system (OOPS) will probe the shape and structure of the proton as well as the electromagnetic interaction with the deuteron. Experiments will get underway in 2000. The unique 1 GeV stored beam capability of the SHR will be used to carry out a program of measurement of spin-dependent electron scattering from light nuclei. This program will utilize the Bates Large Acceptance Spectrometer Toroid (BLAST), which is at present under construction. Data taking with BLAST is scheduled to commence in Fall 2001.

INTRODUCTION

The study of strongly interacting matter with lepton probes has been one of the most successful areas of physics in the last fifty years. Determination of the basic distribution of magnetism and charge in atomic nuclei was an early success [1]. Later, the experimental discovery of the fundamental pointlike constituents of matter via deep inelastic scattering [2] was of profound significance. It is one of the underpinnings of the theory of the strong interaction, quantum chromodynamics (QCD). In the 1980's, electron scattering at medium energies greatly increased our understanding of the distribution of momentum among the nucleons in the nucleus [3].

At present a primary focus of electromagnetic hadronic physics is the detailed structure of the nucleon. Experiments are in progress or planned to study its shape, charge, magnetism, flavor and spin structure in terms of the fundamental constituents. QCD tells us that the observed colorless hadrons are effective degrees of freedom composed of the colored quarks and gluons. Decades of study of hadrons, including in particular the nucleon-nucleon interaction, allows the structure of light nuclei (e.g. deuterium and ^{3}He) to be calculated to high precision using new theoretical techniques. Further, in heavier nuclei fundamental issues remain to be explored such as the origin of nuclear binding and the role of non-nucleonic components.

The study of strongly interacting matter using lepton scattering is now a very active field worldwide with experiments being carried out at a number of complementary facilities, e.g. Bates, Mainz, Bonn, Jefferson Laboratory, SLAC, DESY, and CERN. The present experiments utilize new technology which has only become available in the last decade: e.g. high duty factor beams, polarized beams and targets, and storage rings with internal targets. The insight obtained at these facilities is often directly related, even though the experiments are very different. For example, the SAMPLE experiment probes the strangeness content of the proton at Bates via parity violating electron elastic scattering at 200 MeV. This is closely related to the strange quark contribution to the spin of the proton as measured with 200 GeV muons in deep inelastic scattering at CERN.

The present vitality in the field of lepton scattering is in part due to significant theoretical progress. Over the last decade there have been important advances in effective field theory and chiral perturbation theory [4], exact calculation of light nuclei [5] and in the use of lattice techniques to study QCD [6]. This last field holds the promise of producing exact calculations using the fundamental theory.

The MIT-Bates Linear Accelerator Center studies QCD at long distances using electron scattering up to 1 GeV energy. At Bates over the last 25 years, many of the techniques now in general use in electromagnetic nuclear physics have been pioneered. Examples include: high resolution spectrometry, recoil polarization, polarized beams, polarized targets, out-of-plane spectrometery, and polarized internal gas targets. The cornerstone of the Laboratory's future is the South Hall Ring (SHR) with its high duty factor and internal target capability.

THE MIT-BATES ACCELERATOR COMPLEX

The MIT-Bates linear accelerator is 160 m in length and accelerates pulsed beam using 2856 MHz radiofrequency electromagnetic waves to a maximum energy of 500 MeV. Figure 1 shows a schematic layout of the facility. The recirculator system can double the beam energy to 1 GeV.

The South Hall Ring has a circumference of 190 m and can be operated both in stretcher and storage modes. When configured to stretch the pulsed beam, beam with high duty factor ($>50\%$) is extracted and delivered to the South Experimental Hall. Pulsed beam currents can be as high as 100 μA on target, while extracted beam currents are typically of order 10 μA.

When the South Hall Ring is operated in storage mode, the internal beam intensity is obtained by repeated injection, called stacking. Over 200 mA of beam has been stored in this way (see Fig. 2). The stored beam has been passed through an internal gas target, as will be used for BLAST. The beam characteristics in terms of position, stability, lifetime and halo have been found to be excellent for internal target experiments.

A vital system at Bates is the polarized electron source [7], which is necessary for essentially all of the experiments underway. The Bates source uses optical pumping

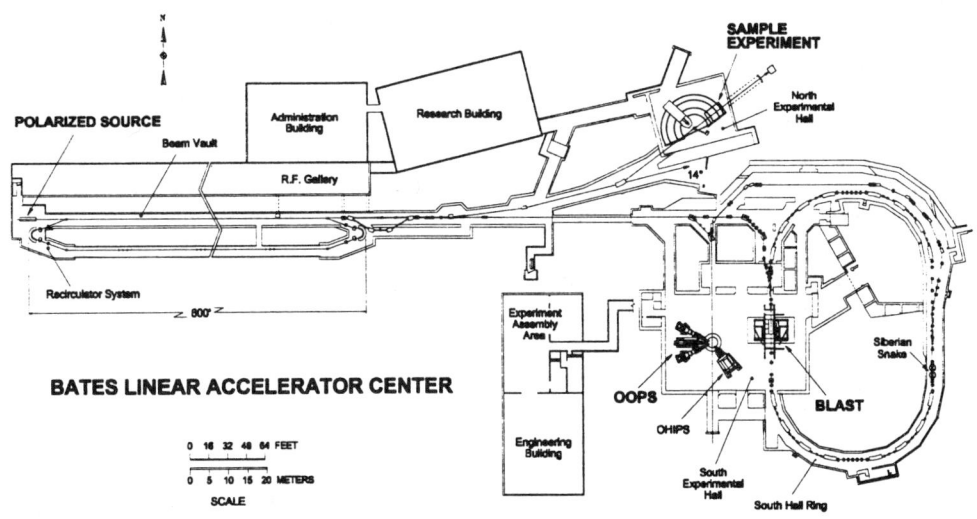

FIGURE 1. A schematic layout of the accelerator complex at the MIT-Bates Linear Accelerator Center.

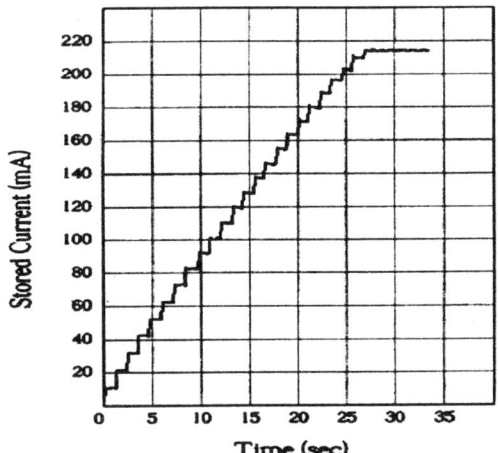

FIGURE 2. Electron beam current stored in the South Hall Ring as a function of time. The stepped structure shows repeated injection pulses stacking the beam to higher current. The flat top at 215 mA is a result of the current exceeding the range of the measuring device.

of Gallium Arsenide to deliver beam polarized to about 40%. A new polarized source test bench is nearing completion and will be used to carry out an R&D program on lasers and crystals. A major goal will be to provide increased beam polarizations of order 70%. In addition, an effort is getting underway to design and install a magnetic system to rapidly reverse the stored electron beam spin in the SHR [8]. The goal is to reverse the beam spin several times with high efficiency within one fill. This would allow the formation of scattering spin asymmetries within each fill and thus significantly reduce systematic uncertainties.

Over the last seven years, a systematic program of refurbishing the accelerator complex has been in progress. This was undertaken by my predecessor, Stanley Kowalski. The recirculator system has been completely rebuilt and the vacuum system greatly improved. Additional beam instrumentation has been added and the entire linac realigned. This has resulted in an accelerator with significantly improved reliability and stability, which can be tuned in a deterministic and reproducible manner. The RF transmitters are being upgraded. A unique solid-state switch tube hybrid modulator was successfully operated in the last year and it is planned that all transmitters will be upgraded by the end of 2001. This may allow an increase in the average current of the polarized extracted beam from the SHR. In addition, a new high peak power and high efficiency klystron was developed [9]. These new klystrons should allow comfortable accelerator operation at an energy of 1 GeV. Further, the conversion of the control system to EPICS is scheduled to be completed in 2000.

Thus, although the Bates linear accelerator has delivered beam for 25 years, it is largely composed of new equipment and is now instrumented to operate reliably and efficiently for many years to come.

SCIENTIFIC PROGRAM FOR NEXT FIVE YEARS

At this time, there are three principal thrusts to the research program at Bates:

- study of the proton's magnetism (SAMPLE). This experiment has measured the strange quark contribution to the instrinsic magnetism of the proton.

- study of the structure and shape of the proton and deuteron [15]. This program uses a unique out-of-plane spectrometer system (OOPS) to determine fundamental properties of the proton and deuteron.

- study of the spin structure of the few-body systems. This program will utilize the stored beam of the Bates South Hall Ring, polarized internal gas targets, and a new large acceptance detector (BLAST) [19] to probe spin-dependent electron scattering from light nuclei.

A The SAMPLE Experiments

In September 1999, the SAMPLE experiments on hydrogen and deuterium completed data taking. The final result on hydrogen has been reported [10] and indicates a surprisingly large and positive value for the strange magnetic moment of the proton, assuming a calculated value for an axial radiative correction. This last assumption is tested with the deuterium data taken in 1999. As the SAMPLE experiment is described in detail elsewhere in this proceedings [11] and as the experiments are completed, I will only add that a future measurement on deuterium at lower incident energy is under consideration.

B The OOPS Program

The OOPS system is a unique facility of multiple magnetic spectrometers specifically designed for accurate extraction of the five response functions accessible through the (e,e'p) reaction with polarized beam [12]. Four spectrometers have been constructed and located in the South Experimental Hall [13]. Further, an elaborate support system allows precise and reproducible positioning of the 16 ton spectrometers to better than ± 1 mm and ± 1 mrad. The scattered electron is detected by a high resolution spectrometer and the OOPS proton spectrometers are arranged symmetrically around the direction of the momentum transfer vector, $\vec{q}$. Schematically this is shown in Fig. 3.

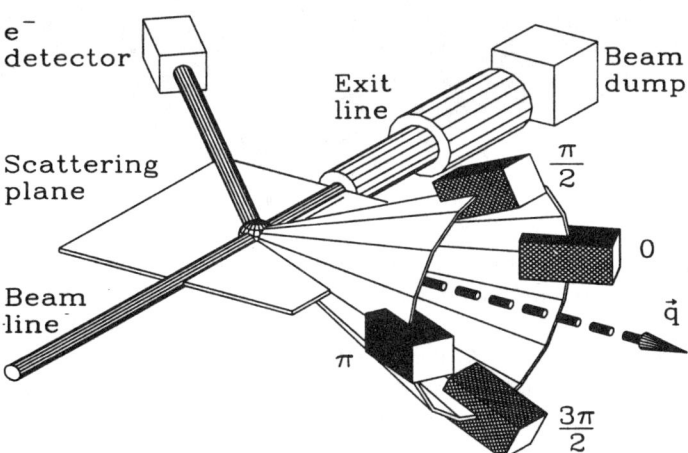

FIGURE 3. *A schematic representation of the experimental geometry in one of the possible configurations.*

Assembly of the full spectrometer system in the South Experimental Hall is almost complete and first experiments are planned for 2000. The OOPS program is being carried out by physicists from Arizona State Univ., Univ. of Athens (Greece), Cal. State Univ. Los Angeles, Florida State Univ., Univ. of Illinois, MIT, Univ. of Massachusetts, Univ. of New Hampshire, Old Dominion Univ., Tohoku Univ. (Japan), and Shizuoka Univ. (Japan). A computer generated schematic layout of the OOPS facility is shown in Fig. 4.

FIGURE 4. *A computer generated schematic layout of the OOPS facility showing the OHIPS electron spectrometer (left) and the array of four out-of-plane spectrometers (right).*

Although the full OOPS instrument is now nearing completion, in recent years several experiments using a subset of the full system have been carried out on the proton, deuteron and ^{12}C. There are three major directions to the OOPS scientific program:

- *the shape of the proton.* This is studied by measurement of the electroexcitation transition of the proton to its first excited state, the $\Delta(1232)$. In 1995, the first data at Bates to probe this question were taken with conventional in-plane spectrometers, at incident energies between 700 and 800 MeV to search for quadrupole strength in electroexcitation of the first resonance.

The coincident cross section was measured [15] for the process H(e,e'p)π^0 in the vicinity of $W = 1232$ MeV, the location of the Δ resonance. The OOPS data highlight the significance of non-resonant background amplitudes. The longitudinal-transverse interference response R_{LT} and the related asymmetry A_{LT} were also extracted. These quantities are sensitive to the quadrupole amplitudes. The theoretical prediction which best described the cross section data is now in poor agreement. Moreover, the experimental results cannot be reproduced by any of the available theoretical predictions. In 1998, the first data on the question of proton deformation with OOPS were taken using polarized electron beam and two out-of-plane spectrometers. New structure functions were measured for the first time in the H(e,e'p)π^0 and H(e,e' π^+)n reactions. These polarization observables should vanish in the absence of non-resonant amplitudes, but preliminary results show that they are several sigma away from zero. This result, along with the 1995 data, points to the presence of a strong non-resonant background in the electroexcitation of the Δ resonance. Thus, they demonstrate the inadequacy of the previously practiced analysis scheme [16], and the need to investigate the contributions of all coherent small amplitudes before conclusions concerning the issue of proton deformation can be reached. In 2000, it is planned to take data on the R_{TT} response.

- *measurement of virtual Compton scattering on the proton.* From the initial proposal, the OOPS system was designed to measure the reaction (e,e'γ). By detecting particles scattered out of the electron scattering plane, a significant enhancement of the virtual Compton scattering (VCS) mechanism over the dominant Bethe-Heitler process can be accomplished. An experiment to measure VCS on the proton with the OOPS system is planned for 2000 [17]. The object is to extract the electric and magnetic polarizabilities of the proton at a momentum transfer of 240 MeV/c. The projected results are expected to yield values for the polarizabilities which will have significantly improved accuracy over the values at zero momentum transfer. The latter are known from the world's data from real Compton scattering [18].

- *the electromagnetic structure of deuterium.* The ability of OOPS to isolate small (e,e'p) interference responses and asymmetries allows for precision studies of the transition currents in the deuteron. Each response exhibits sensitivity to various components of the currents (e.g. meson exchange currents, isobar components, relativistic effects and final state interactions) within this fundamental two-nucleon system. In 1997 (e,e'p) data on deuterium were taken in the quasielastic and dip kinematics. This provided new information on the longitudinal-tranverse asymmetry and the OOPS data are in good agreement with the predictions of Arenhoevel et al [14]. Through an ongoing systematic program of interference response measurements over a broad kinematic range, the OOPS facility will provide sensitive tests of these currents, including the role of the $\Delta(1232)$ within the deuteron.

OOPS is on the threshold of a new era with the realization of the full instrument as well as the upgrade of OHIPS. Starting in 2000 the availability of high duty factor beam from the SHR will provide important new data on the structure of the proton via VCS, the shape of the proton via the $N \to \Delta$ transition, and on the structure of deuterium.

C The BLAST Program

The Bates Large Acceptance Spectrometer Toroid (BLAST) is a detector designed to study in a comprehensive and precise way the spin dependent electromagnetic response of few-body nuclei [19]. These systems (A = 1,2,3) consist of the free proton, the weakly bound deuteron system and the three body ^{3}He system. The BLAST scientific program [20,21] is focussed on the study of these systems in terms of nucleon structure, the ground state few body structure built from the nucleon-nucleon interaction and the nature of the interaction of the virtual photon for $Q^2 \leq 1$ (GeV/c)^2. A major consideration in the design of BLAST has been the realization that these aspects of the study of the electromagnetic response are interrelated in a complicated way and can only be unambiguously separated by a broad study of the few body systems. In addition, both the choice of few body systems and the relatively low momentum transfers should allow accurate comparison with theoretical calculations.

To accomplish its ambitious scientific goals, BLAST will utilize the latest technology available in the form of polarized electron scattering from pure, polarized internal gas targets. The internal target technique has now been successfully implemented at several laboratories worldwide (Novosibirsk, IUCF, NIKHEF, DESY) and it has been widely recognized that this arrangement can provide the lowest combined statistical and systematic error. The newly commissioned Bates South Hall Ring will deliver polarized longitudinally polarized electrons at the location of the BLAST detector. The polarized internal targets are highly polarized > 50 %, do not have any dilution from non-polarized species, are rapidly reversible, can be oriented with low magnetic fields and the resulting luminosity is well matched to that of a large acceptance detector. Further, thin walled target cells allow straightforward detection of low energy recoil particles which will allow complete reconstruction of the final state in the electrodisintegration of few body systems. The ability with BLAST to carry out multiparticle detection over a large solid angle from polarized internal targets will provide an unprecedented and unique oportunity to study simultaneously the spin structure of the few-body nuclear ground states, the reaction mechanism, and nucleon form factors.

The BLAST detector consists of an eight-sector copper coil array producing a toroidal magnetic field, instrumented with two opposing wedge-shaped sectors of wire chambers, scintillation detectors, Čerenkov counters, neutron detectors, a lead glass calorimeter and recoil detectors [19]. The open geometry maximizes acceptance while allowing good momentum and angular resolution, and with a luminosity capability that is matched to the densities of the polarized internal targets. The BLAST design emphasizes proven technology, commercial electronics, and existing data acquisition system software to achieve low cost and a short implementation time. Clear upgrade possibilities exist so that the detector can evolve to match developing physics priorities.

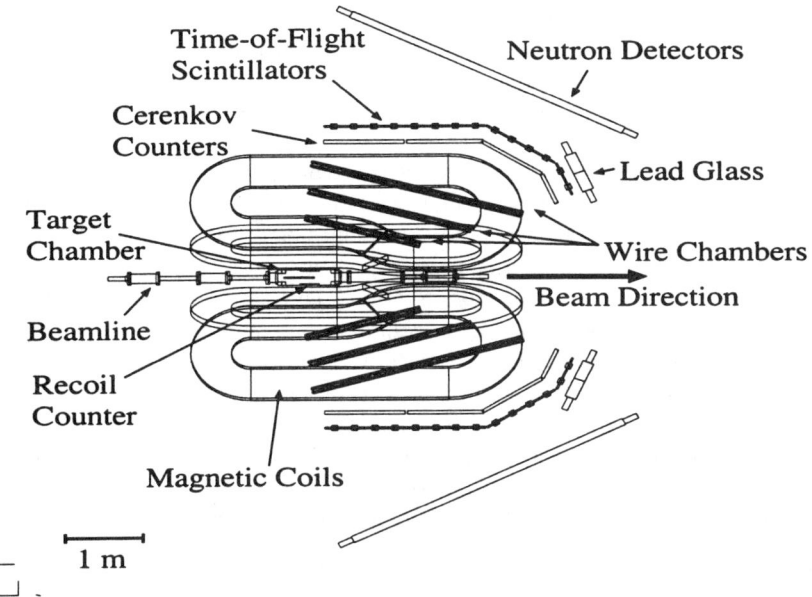

FIGURE 5. *A view of the BLAST spectrometer showing the magnetic coils, the two sectors to be initially instrumented and the different detector elements.*

Fig. 5 is a top view of the BLAST spectrometer showing the magnetic coils, the two sectors to be initially instrumented and the different detector elements. The tracking drift chambers are located between the coils followed by the Čerenkov detectors, the timing scintillators and the neutron detector array.

BLAST is under construction by an international collaboration of physicists from 11 institutions: Arizona State Univ., Boston Univ., ETH-Zürich (Switzerland), MIT, Univ. of New Hampshire, North Carolina AT&T Univ., Ohio Univ., Univ. of Virginia, Free Univ. Amsterdam (The Netherlands), Univ. of Wisconsin, and Yerevan Physics Institute (Armenia). The coils have been constructed and, with the appropriate mechanical supports, are scheduled to be installed in the South Hall Ring of Bates during the fall of 2000. A polarized proton and deuteron target [30] has been transferred from NIKHEF and installed in the South Hall Ring. Initial operation has been established and full tests within BLAST will be carried out during the latter half of 2000. Construction of the detectors is well underway and complete prototypes will be tested in the South Hall Ring in early 2000. A significant fraction of the detectors is expected to be available for installation in the fall of 2000.

At present the South Hall Ring delivers high currents (100-200 mA) of unpolarized electron beams through gas targets of the required density, with lifetimes of several minutes. In the year 2000 polarized internal beams will be developed. These will require the operation of the Siberian snake magnets to ensure longitudinal polarization at the position of the BLAST targets. A laser Compton back-scattering electron polarimeter is expected to be installed early in 2000. Commissioning of the full detector is scheduled to begin in summer 2001.

BLAST will carry out a number of important measurements:

- *precise information on nucleon form factors for momentum transfers up to about 1 (GeV/c)²*. In particular, BLAST will provide data on the neutron magnetic and electric form factors with both deuteron [31] and ^{3}He [32] targets. These are fundamental quantities which are essential to any description of electromagnetic scattering from nuclei. A compilation of recent data is shown in Fig. 6. Because of the significantly larger binding energy of the neutron in ^{3}He compared to that in deuterium, it is possible that the neutron charge distribution in ^{3}He may be modified from its free value [29]. BLAST has the necessary precision to probe for such an effect. Fig. 7 shows the projected statistical errors for the extraction of G_E^n from BLAST data as a function of Q^2. BLAST will measure G_E^n from 0.1 to 0.8 (GeV/c)² with high precision both on deuterium and ^{3}He within the same apparatus.

- *a precise measurement of the spin-dependent momentum distribution in few-body nuclei*. These data will test our understanding of the spin structure of few-body systems in terms of the successful theoretical framework which has been developed primarily for unpolarized scattering [19,44]. Effects such as final-state interactions, meson exchange currents, and the off-shell nature of the bound nucleon can be studied over the broad kinematic range provided by the BLAST detector. In addition, the spin-dependent momentum distributions are used as input for calculations of spin-dependent scattering in the deep-inelastic region where polarized deuterium and ^{3}He are used to determine the neutron spin structure at the quark level.

- *precise data from elastic electron scattering from tensor polarized deuterium (T_{20})*, particularly in the region of the first minimum of the charge form factor of the deuteron [19]. Fig. 8 shows the projected BLAST data compared to the existing world data. In addition, data will be obtained in the same experiment for the exclusive scattering channels.

- *measurement of spin-dependent charged pion electroproduction on few-body systems from threshold to beyond the Δ-resonance*. Such studies are important for understanding the role of the nucleon resonance in few-body systems. BLAST will allow reconstruction of the resonance from its π-nucleon decay channel. In this way, for example, the presence of pre-existing Δ-components in the ^{3}He ground state can be studied [45].

- *study of the N→Δ transition to isolate components beyond the dominant M1 transition* can be carried out using a polarized proton target and polarized beam.

- *a precise determination of the proton charge radius* using lower beam energies is planned [46].

- *a program of measurements using unpolarized targets*. Examples include the study of multinucleon processes [47,48] which are well suited for a large acceptance detector like BLAST, measurement of the neutron momentum distributions in nuclei, and the detection of heavy recoils in processes of astrophysical significance [49].

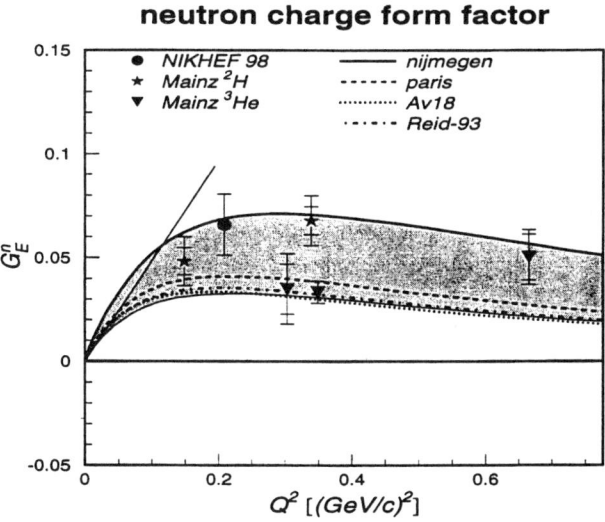

FIGURE 6. *A selection of recent data [22-27] for the charge form factor of the neutron from spin-dependent electron scattering. The curves correspond to the parametrization of the data from Platchkov et al. [28].*

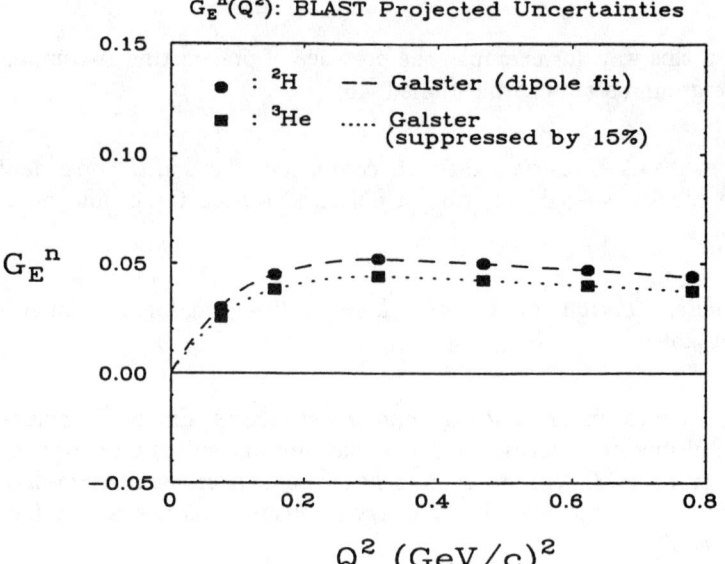

FIGURE 7. Projected statistical accuracies for G_E^n as function of Q^2. Luminosities of 10^{33} and 2×10^{32} (atoms/cm^2/s) are assumed for the polarized ^{3}He and deuterium measurements, respectively. 1000 hours of running with each target is also assumed.

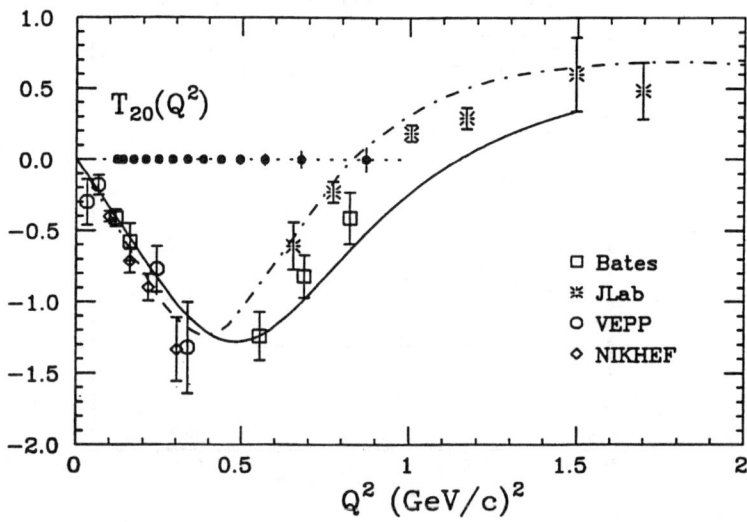

FIGURE 8. Data and theoretical predictions for T_{20} as a function of momentum transfer. The data are from [33,34,36,38–41]. The curves represent two theoretical models [42,43]. Also indicated are the expected data (solid circles) with BLAST for 1000 hours of data taking at a luminosity of 10^{32} atoms cm^{-2}s^{-1}.

SOME THOUGHTS ON THE FUTURE OF ELECTROMAGNETIC NUCLEAR PHYSICS

On the auspicious occasion of the 25th anniversary of first beam to experiment at Bates, it is interesting to consider the general status and future of electromagnetic hadronic physics. The field is very active with new facilities and technical capabilities just now becoming available. In Fig. 9 is shown a plot of the maximum luminosity vs. center-of-mass (CM) energy for an illustrative selection of facilities (both in operation and planned) worldwide which study hadron structure using lepton scattering. There is a significant concentration of capability at low CM energy, the regime of traditional nuclear physics. Essentially, all hadronic matter is composed of *up* and *down* quarks so experiments at Bates, Mainz, Bonn, and JLab are in this energy range. Another major focus is deep inelastic scattering, which is studied at CERN, DESY, and SLAC. At these laboratories the fundamental quark structure of hadrons is directly explored.

For the future, a staged upgrade of CEBAF and a similar machine in Europe (ELFE) are under discussion. These would extend high-duty factor, high luminosity, fixed target capability into the low Q^2 DIS regime. While these machines are of significant interest and can certainly do first rate physics, it seems to me that the CM energy is rather low for a comprehensive study of QCD. I would like to motivate some basic design issues with respect to a machine which would directly study QCD. These include:

- the machine should allow direct study of the fundamental constituents of QCD, namely the quarks and gluons. Quarks are accessed via deep inelastic scattering and gluons via the photon-gluon fusion diagram. A minimum CM energy of about 10 GeV is required for this. An energy of about 20 GeV is more desirable, particularly to study the Q^2 evolution of processes.

- the machine should have sufficient luminosity. My estimate is that a minimum would be about 10^{33} cm^{-2} s^{-1} to carry out a precise set of measurements on the fundamental quark and gluon constitutents of the nucleon.

- it is desirable to have the capability to probe as many of the quark flavors as possible. On the figure are indicated the CM energy thresholds corresponding to meson production thresholds for the different flavors of quarks. It is seen that a minimum CM energy of about 10 GeV is required to probe the five lowest mass quark flavors.

- it is essential that both beam and nucleon target be polarized. With present technology such experiments are limited in luminosity to about 10^{35} cm^{-2}s^{-1}.

- it is desirable to have the possibility to scatter from nuclear targets as well as the proton.

A possible scheme which has the potential to fulfill the above constraints is an electron-hadron collider. This has been previously studied at GSI [50] and

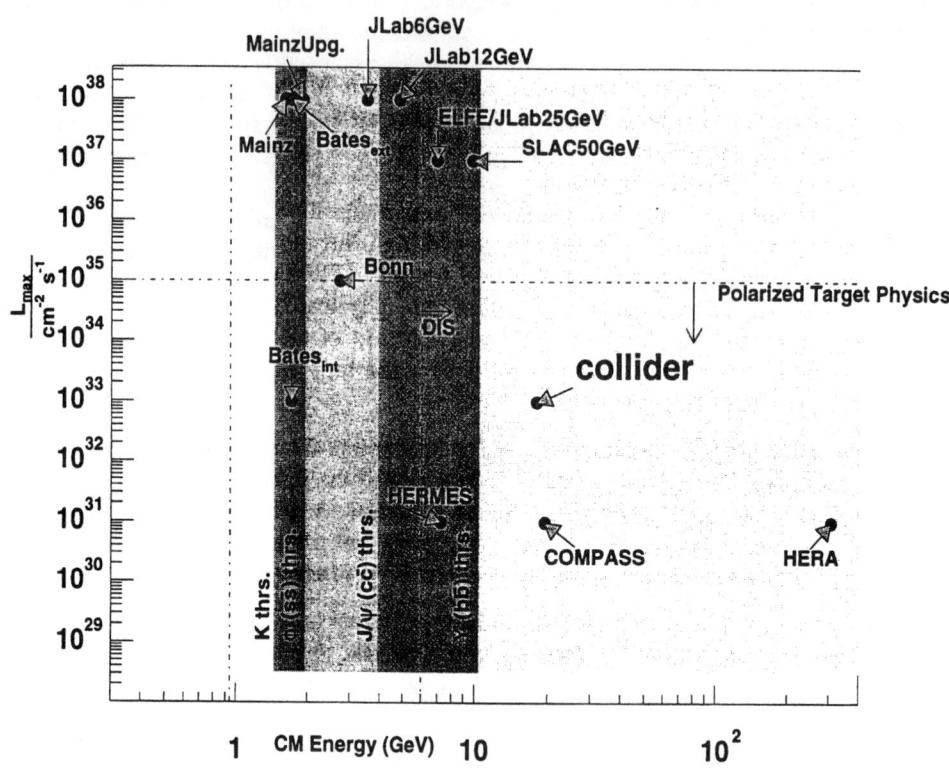

FIGURE 9. *A plot of maximum luminosity vs. center-of-mass energy for a selection of facilities worldwide which study hadron structure using lepton scattering.*

IUCF [51]. For example, a 4 GeV electron beam colliding with a 15 GeV proton beam has a CM energy of 15 GeV. The location of such a machine is marked in Fig. 9. This is equivalent to a fixed target experiment with 100 GeV incident lepton beam. Further, the collider has a significant advantage over the fixed target configuration in that scattered particles at high momentum transfer are detected at large angles with respect to the beam direction.

What kinds of experiments would be carried out at such a machine? One possibility is the measurement of the total angular momentum of the quarks in the proton via deeply virtual Compton scattering [52]. A complete and precise determination of this quantity would have profound impact on our understamding of the spin structure of the nucleon. A second area is the measurement of the gluon momentum distribution in the nucleus. Is this modified from the gluon momentum distribution in the nucleon? Does the gluon have a role in nuclear binding?

It is clear that the timescale to realize such a machine is of order a decade. A significant effort on both the science and technical design is required to develop the optimal machine. One crucial issue is that the machine must have sufficient luminosity to carry out the important science. Bates could play a leadership role in such an endeavor, contributing unique expertise on the electron beam. It is not excluded that an appropriately modified configuration of the existing SHR could serve as the electron arm of such a collider.

SUMMARY

For the foreseeable future, the main focus of the research program at the MIT-Bates facility will be centered around the South Hall Ring. As over the past 25 years, the future Bates program will comprise experiments using electron scattering to address fundamental problems in the structure of matter using pioneering instrumentation.

In the next several years, the OOPS program will produce high quality data on the structure of the proton and the deuteron: comprehensive studies of the N→Δ transition, virtual Compton scattering, and the nature of the electromagnetic deuteron current.

After 2001 the main focus of the laboratory will be directed at the program of measurements of the spin dependent electromagnetic response of the few body systems to be carried out at Bates with the novel combination of CW polarized electron beams, polarized internal gas targets and the large acceptance BLAST detector. Construction of BLAST is scheduled to be completed in summer 2001. Precise data from BLAST on the electric form factor of the neutron, the tensor analyzing power T_{20} in elastic electron-deuteron scattering, the spin structure of light nuclei and other fundamental observables are eagerly awaited.

the BLAST detector, Cambridge, Massachusetts, May, 1998, edited by R. Alarcon and R. Milner (World Scientific, 1999).
22. M. Meyerhoff et al., Phys. Lett. B **327**, 201 (1994).
23. J. Becker et al., Ph.D. thesis, Univ. Mainz, 1998 (to be published).
24. C. Herberg et al., Eur. Phys. J. A **5**, 131 (1999).
25. M. Ostrick et al., Phys. Rev. Lett. **83**, 276 (1999).
26. I. Passchier et al., Phys. Rev. Lett. **82**, 4988 (1999).
27. D. Rohe et al., Phys. Rev. Lett. **83**, 4257 (1999).
28. S. Platchkov et al., Nucl. Phys. A **510**, 740 (1990).
29. D.H. Lu et al, Phys. Lett. B **441**, 27 (1998).
30. Z.-L. Zhou et al., Nucl. Instr. Meth. A **378**, 40 (1996).
31. Bates proposal 91-10, spokespersons R. Alarcon and J. van den Brand.
32. Bates proposal 89-12, spokespersons J. van den Brand and R.G. Milner.
33. I. The et al., Phys. Rev. Lett. **67** (1991) 173; M. Garçon et al., Phys. Rev. C **49** (1994) 2516.
34. M. Ferro-Luzzi et al., Phys. Rev. Lett. **77** (1996) 2630.
35. M. Bouwhuis et al., Phys. Rev. Lett. **82** (1999) 3755.
36. Experiment E94-018 at TJNAF, Newport News (VA), USA, spokespersons: S. Kox and E. J. Beise.
37. H. Henning, J. Adam, Jr., P.U. Sauer, and A. Stadler, Phys. Rev. C 52 (1995) R471.
38. V.F. Dimitrev et al., Phys. Lett. B **157**, 143 (1985).
39. R. Gilman et al., Phys. Rev. Lett. **65**, 1733 (1990).
40. S.G. Popov et al., in *Proceedings of the 8th International Symposium on Polarization Phenomena in Nuclear Physics, Bloomington, Indiana 1994*, AIP Conference Proceedings 339.
41. M.E. Schulze et al., Phys. Rev. Lett. **52**, 597 (1984).
42. R. Schiavilla and D.O. Riska, Phys. Rev. C **43**, 437 (1990).
43. E. Hummel and J.A. Tjon, Phys. Rev. C **42**, 423 (1990).
44. R.G. Milner et al., Phys. Lett. B **379**, 67 (1996).
45. R.G. Milner and T.W. Donnelly, Phys. Rev. C **37**, 870 (1988).
46. H. Gao, spokesperson, Bates Letter of Intent 99-01.
47. R.P. Redwine, spokesperson, Bates proposal 89-21.
48. W. Hersman, spokesperson, Bates proposal 91-10.
49. G. Tsentalovich, spokesperson, Bates Letter of Intent 99-02.
50. A. N. Skrinsky, Contribution to the EPIC Workshop, Bloomington, Indiana, USA, March 1999.
51. J.N. Cameron, Contribution to the PANIC Conference, Uppsala, Sweden, June 1999.
52. X. Ji, Phys. Rev. D **54**, 6897 (1996).

List of Participants

Ricardo Alarcon
Arizona State University
c/o MIT-Bates Laboratory
P. O. Box 846
Middleton, MA 01949

Peter D. Barnes
Los Alamos National Laboratory
Physics Division, MS D434
Los Alamos, NM 87545

Captain H. Raymond Bates
51 Trinity Road
Marblehead, MA 01945

Douglas Beck
University of Illinois
Loomis Laboratory of Physics
1110 West Green Street
Urbana, IL 61801

Aron Bernstein
MIT
Bldg. 26-419
77 Massachusetts Avenue
Cambridge, MA 02139

Ingvar Blomqvist
Danfysik A/S
Mollehaven 31
DK-4040 Jylinge
Denmark

Tancredi Botto
MIT
Bldg. 26-441
77 Massachusetts Avenue
Cambridge, MA 02139

Maria Barbaro
Universita' di Torino
via P. Giuria 1
Departimento di Fisica Teorica
I-10125 Torino
Italy

Larry Bartoszek
818 W. Downer Place
Aurora, IL 60506

Gunter Baum
University Bielefeld
Physics Department
Universitaetsstr. 25
D-33615 Bielefeld
Germany

Elizabeth Beise
University of Maryland
Physics Department
College Park, MD 20742

William Bertozzi
MIT
Bldg. 26-437
77 Massachusetts Avenue
Cambridge, MA 02139

Edward Booth
Boston University
c/o MIT - Bates Laboratory
P. O. Box 846
Middleton, MA 01949

Rick Bradanick
VAT, Inc.
500 West Cummings Park
Woburn, MA 01801

Larry Cardman
Jefferson Laboratory
MS 12H
12000 Jefferson Avenue
Newport News, VA 23606

E. Moya de Guerra
IEM-CSIC
Serrano 123
28006 Madrid
Spain

Hans de Vries
NIKHEF
P. O. Box 41882
1009 DB Amsterdam
The Netherlands

T. W. Donnelly
MIT
Bldg. 6-300
77 Massachusetts Avenue
Cambridge, MA 02139

Manouch Farkhondeh
MIT-Bates Laboratory
P. O. Box 846
Middleton, MA 01949

Robert Frankland
Del Electronics, Power Conversion Products
One Commerce Park
Valhalla, NY 10595

James Friar
Los Alamos National Laboratory
Physics Division, MS D434
Los Alamos, NM 87545

Fabio Casagrande
MIT-Bates Laboratory
P. O. Box 846
Middleton, MA 01949

Kees de Jager
Jefferson Laboratory
12000 Jefferson Avenue
Newport News, VA 23606

George Dodson
MIT-Bates Laboratory
P. O. Box 846
Middleton, MA 01949

Karen Dow
MIT-Bates Laboratory
P. O. Box 846
Middleton, MA 01949

Herman Feshbach
MIT
Bldg. 6-307
77 Massachusetts Avenue
Cambridge, MA 02139

Wilbur Franklin
MIT
Bldg. 26-402
77 Massachusetts Avenue
Cambridge, MA 02139

Haiyan Gao
MIT
Bldg. 26-413
77 Massachusetts Avenue
Cambridge, MA 02139

Michel Garcon
CEA/Saclay
DAPNIA/SPHN, CE SACLAY
F-91191 Gif-sur-Yvette
Cedex
France

Marcel P. J. Gaudreau
Diversified Technologies, Inc.
35 Wiggins Avenue
Bedford, MA 01730

Donald F. Geesaman
Argonne National Laboratory
Physics Division, Bldg. 203
9700 S. Cass Avenue
Argonne, IL 60439

Claus-Konrad Gelbke
Michigan State University
NSCL
East Lansing, MI 48824

Shalev Gilad
MIT
Bldg. 26-449
77 Massachusetts Avenue
Cambridge, MA 02139

Jake Haimson
Haimson Research Corporation
3350 Scott Boulevard
Building 60
Santa Clara, CA 95054

Stephen Harvell
MDC Vacuum
23842 Cabot Boulevard
Hayward, CA 94545

Douglas Hasell
MIT
Bldg. 26-415
77 Massachusetts Avenue
Cambridge, MA 02139

Jochen Heisenberg
University of New Hampshire
Physics Department
Durham, NH 03824

Barry Holstein
University of Massachusetts
Department of Physics and Astronomy
Amherst, MA 01003

Akio Hotta
University of Massachusetts
Department of Physics and Astronomy
Amherst, MA 01003

Kenneth Jacobs
MIT-Bates Laboratory
P. O. Box 846
Middleton, MA 01949

Marc Kastner
MIT
Bldg. 6-113
77 Massachusetts Avenue
Cambridge, MA 02139

Brad Keister
National Science Foundation
Physics Division 1015N
4201 Wilson Boulevard
Arlington, VA 22230

Michael Kempkes
Diversified Technologies
35 Wiggins Avenue
Bedford, MA 01730

Stanley Kowalski
MIT
Bldg. 26-427
77 Massachusetts Avenue
Cambridge, MA 02139

Robert Lourie
Renaissance Technologies
600 Route 25A
Setauket, NY 11733

June Matthews
MIT
Bldg. 26-433
77 Massachusetts Avenue
Cambridge, MA 02139

Robert McKeown
California Institute of Technology
106-38 Kellogg Radiation Laboratory
Pasadena, CA 91125

J. M. Udias Moinelo
University of Madrid
Dpt. Fisica Atomica, Fac. CC. Fisicas
Universidad Comlutense
E 28040 Madrid
Spain

Bruno Mosconi
Largo E. Fermi, 2
Department of Physics
I-50125 Florence
Italy

Dennis Kovar
U. S. Department of Energy
ER23
19901 Germantown Road
Germantown, MD 20874

Tong-Uk Lee
MIT
Bldg. 26-402
77 Massachusetts Avenue
Cambridge, MA 02139

Chiara Maieron
MIT
Bldg. 6-408A
77 Massachusetts Avenue
Cambridge, MA 02139

Kevin McIlhany
MIT
Bldg. 26-551
77 Massachusetts Avenue
Cambridge, MA 02139

Richard Milner
MIT-Bates Laboratory
P. O. Box 846
Middleton, MA 01949

Alfredo Molinari
Universita' di Torino
via P. Giuria 1
Departimento di Fisica Teorica
I-10125 Torino
Italy

John Negele
MIT
Bldg. 6-308
77 Massachusetts Avenue
Cambridge, MA 02139

Lou Nielson
CML Engineering
2020 Ridgecrest Place
Escondido, CA 92029

V. R. Pandharipande
University of Illinois
Loomis Laboratory of Physics
1110 West Green Street
Urbana, IL 61801

Peter Paul
Brookhaven National Laboratory
Science and Technology
P. O. Box 5000, Bldg. 460
Upton, NY 11973

Alex Pozamantir
Purdue University
Lafayette, IN 47908

Michael J. Ramsey-Musolf
University of Connecticut
Department of Physics, U46
2152 Hillside Road
Storrs, CT 06269

Paul J. Reardon
Proton Therapy Corp. of America
32 Lochatong Road
West Trenton, NJ 08628

Don Rosenbaum
Litton Electron Devices
960 Industrial Road
San Carlos, CA 94070

Bill North
MIT-Bates Laboratory
P. O. Box 846
Middleton, MA 01949

Costas Papanicolas
IASA-University of Athens
P. O. Box 17214
Athens, 10024
Greece

Gerry Peterson
University of Massachusetts
Lederle GRC 417
Amherst, MA 01003

Marco Radici
Universita di Pavia
Dept. Nucl. Teor. Fis.
Via Bassi 6
I-27100 Pavia
Italy

George Rawitscher
University of Connecticut
Department of Physics, U46
2152 Hillside Road
Storrs, CT 06269

Robert Redwine
MIT
Bldg. 26-505
77 Massachusetts Avenue
Cambridge, MA 02139

Rob Russell
Del Electronics, Power Conversion Products
One Commerce Park
Valhalla, NY 10595

Ingo Sick
University of Basel
Department of Physics
Klingelbergstrasse 82
CH-4056 Basel
Switzerland

Timothy P. Smith
MIT-Bates Laboratory
P. O. Box 846
Middleton, MA 01949

Paul Souder
Syracuse University
201 Physics Building
Syracuse, NY 13244

Daniel Tieger
MIT-Bates Laboratory
P. O. Box 846
Middleton, MA 01949

Christoph Tschalaer
MIT-Bates Laboratory
P. O. Box 846
Middleton, MA 01949

William Turchinetz
MIT-Bates Laboratory
P. O. Box 846
Middleton, MA 01949

J. W. Van Orden
Jefferson Laboratory
12000 Jefferson Avenue
Newport News, VA 23606

Thomas Walcher
Institut fur Kernphysik
Joh.-Joachim-Becher-Weg 45
55099 Mainz
Germany

Ed Six
Arizona State University
Physics Department
Box 871504
Tempe, AZ 85287

Steve Soltis
Del Electronics, Power Conversion Produc
One Commerce Park
Valhalla, NY 10595

Stephen Steadman
U. S. Department of Energy
ER23
19901 Germantown Road
Germantown, MD 20874

Richard True
Litton Electron Devices
960 Industrial Road
San Carlos, CA 94070

Evgeni Tsentalovich
MIT-Bates Laboratory
P. O. Box 846
Middleton, MA 01949

Johannes van den Brand
NIKHEF
P. O. Box 41882
1009 DB Amsterdam
The Netherlands

Erich Vogt
TRIUMF
404 Wesbrook Mall
Vancouver, BC V6T 2A3
Canada

Defa Wang
MIT-Bates Laboratory
P. O. Box 846
Middleton, MA 01949

Fuhua Wang
MIT-Bates Laboratory
P. O. Box 846
Middleton, MA 01949

Claude Williamson
MIT
Bldg. 26-431
77 Massachusetts Avenue
Cambridge, MA 02139

Zilu Zhou
MIT
Bldg. 26-452
77 Massachusetts Avenue
Cambridge, MA 02139

Vitaliy Ziskin
MIT
Bldg. 26-457
77 Massachusetts Avenue
Cambridge, MA 02139

Abbasali Zolfaghari
MIT-Bates Laboratory
P. O. Box 846
Middleton, MA 01949

Townsend Zwart
MIT-Bates Laboratory
P. O. Box 846
Middleton, MA 01949

AUTHOR INDEX

B

Beck, D. H., 144
Beise, E. J., 305
Bernstein, A. M., 254
Bertozzi, W., 48

D

de Jager, K., 225

F

Friar, J. L., 168

G

Gao, H., 181
Garçon, M., 115
Geesaman, D. F., 5

H

Heisenberg, J. H., 20
Holstein, B. R., 271

M

Milner, R. G., 329

N

Negele, J. W., 209

P

Pandharipande, V. R., 101
Papanicolas, C. N., 237
Peterson, G. A., 87

R

Ramsey-Musolf, M. J., 313

S

Schmieden, H., 196
Sick, I., 33
Souder, P. A., 291

V

van den Brand, J. F. J., 157
Van Orden, J. W., 130

W

Walcher, T., 66